Some Key Topics in Chemistry and Biochemistry for Biotechnologists

Editor

Munishwar Nath Gupta
Former Emeritus Professor
Department of Biochemical Engineering and Biotechnology
Indian Institute of Technology Delhi
New Delhi, India

CRC Press
Taylor & Francis Group
Boca Raton London New York

CRC Press is an imprint of the
Taylor & Francis Group, an **informa** business

A SCIENCE PUBLISHERS BOOK

Cover credit: Image taken from Chapter 4 of this book. Reproduced by kind courtesy of authors of the chapter.

First edition published 2023
by CRC Press
6000 Broken Sound Parkway NW, Suite 300, Boca Raton, FL 33487-2742

and by CRC Press
4 Park Square, Milton Park, Abingdon, Oxon, OX14 4RN

© 2023 Munishwar Nath Gupta

CRC Press is an imprint of Taylor & Francis Group, LLC

Library of Congress Cataloging-in-Publication Data (applied for)

ISBN: 978-1-032-26301-4 (hbk)
ISBN: 978-1-032-26303-8 (pbk)
ISBN: 978-1-003-28759-9 (ebk)

DOI: 10.1201/9781003287599

Typeset in Times New Roman
by Radiant Productions

Preface

Covid-19 brought out some fault lines in our preparation for a crisis which needed coordinated efforts from scientists and technologists belonging to different disciplines. Lately, there has been a debate about the relative values of a specialist versus a generalist. The experience with Covid-19 perhaps teaches us that it is best to have specialists who are also reasonably familiar with basic concepts of other disciplines which have interfaces with their own special subject. It is in line with this thinking that engineers and doctors are taught chemical sciences at an undergraduate level. As science and technology grows, it is becoming increasingly difficult to optimise this exposure. Not everything is covered in standard text books. The present volume is an effort to bring together some topics which are generally not covered adequately in the books/syllabi of undergraduate or even post graduate courses. The idea is to encourage scientists/technologists to read more widely. That would enable these persons to deal more effectively with such crisis by rounding off their knowledge in some basic areas in chemical and biochemical sciences.

Chapter 1 covers flow-based systems which are increasingly becoming important in both chemical and biological sciences. Biologists would profit from learning how rates of biocatalytic reactions are monitored in such systems. Chapters 2 and 3 describe chemiproteomics and high throughput screening methods. Both topics are valuable in the context of drug discovery/design.

Enzymes function in many different milieu. Among such non-conventional media, solid-gas biphasic medium is perhaps most under appreciated. Gases like hydrogen sulphide and nitric oxide are now known to be metabolic intermediates of high biological significance. Hence the choice of the topic of Chapter 4 as solid-gas biocatalysis. Biocatalysis in non-conventional media and at high temperatures needs enzymes with high stability. More robust enzymes derived from salty environments (marine resources being the most important one) are described in Chapters 5 and 6.

Nanosciences/nanotechnology is now relevant to many area of biotechnology and medicine. Quantum dots, extremely small nanoparticles have some fascinating properties. Among these is their value as fluorescent materials. Chapter 7 discusses these nanoparticles. While most reviews focus on their applications; this chapter highlights synthesis. This may enable one to design and tailor QDs for specific applications. While chirality of organic molecules is generally a familiar topic among chemists; chirality of nanomaterials is not so well known. Chapter 8 deals with this somewhat obscure topic.

Chapter 9 covers basics of colloids and immunochemistry in the context of Covid-19. This last chapter points out how being clear about some basic chemistry and biochemistry would have prevented some mis-steps in dealing with Covid-19 at various levels.

I hope this volume would be useful as a reference as well as a supplementary reading material by chemists, biotechnologists and those in the medical profession.

I thank various authors. I thank Mr. Raju Primlani (from Science publishers-an imprint of CRC which is publishing this book) to encourage me to put together this volume. I also thank others from CRC who have worked in producing this volume.

7th September, 2022   Munishwar Nath Gupta
Delhi

Contents

Chapter 1

Uses of Flow-based Systems for Enzyme Kinetics

Supaporn Kradtap Hartwell

Introduction

Enzymes are proteins that accelerate chemical reactions upon binding to a specific reagent called a substrate. Enzymes have been incorporated in numerous chemical and biochemical analyses and biotechnological processes to catalyze the target reactions. The shape and properties of the binding sites on the enzyme, arisen from the unique amino acids sequence, is complimentarily specific for a certain substrate and is conventionally known as a Lock and Key fit. The induced fit model further explains that a binding site on an enzyme is not rigid and can adjust its shape to better accommodate a substrate molecule (Koshland 1958, Boyer 2022, Vasella et al. 2022). Some enzymes facilitate bond breaking and transition state formation by physically altering the substrate molecules (Garcia-Viloca et al. 2004). Even though some enzymes may form chemical bonds with the substrate during the reaction, these bonds are temporary and active sites will return to normal after releasing the products. In other words, an enzyme catalyzes a chemical reaction without itself being altered, and therefore, can resume its function after removal of products. Figure 1 shows the reversible binding of enzyme-substrate, followed by the releasing of products and freeing of enzyme. Thus, apart from having high selectivity, enzymes have signal amplification ability by continuously reacting the released enzyme with more incoming substrate molecules. This helps to increase the number of products and promote sensitivity, which is beneficial when using enzymatic reaction as part of a chemical detection process such as in enzyme-based immunoassay techniques (Calabria et al. 2021). In addition, the fact that enzyme molecules can resume their

Department of Chemistry, 3800 Victory Parkway, Xavier University, Cincinnati, Ohio, USA 45207.
Email: kradtaps@xavier.edu

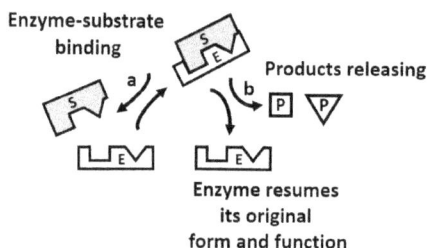

Figure 1. Enzyme-substrate reaction a) Reversible binding of enzyme-substrate, and b) irreversible releasing of products. E, S, and P represent enzyme, substrate, and product, respectively.

function makes it feasible to study enzyme kinetics in the heterogeneous format because the immobilized enzyme molecules can be reused without having to flush them out of the system.

Knowledge about enzyme kinetics is essential in evaluating characteristics and possible applications of a particular enzyme and in identifying suitable substrate for that enzyme. Kinetic parameters of an enzyme aid in understanding of the mechanism of enzyme-catalyzed reactions which allow better manipulation and optimization of the conditions for effective enzymatic reaction. Enzymes serve as catalysts, which are not consumed in the reaction, and so their concentrations are unchanged. Therefore, enzyme kinetics are usually evaluated from the rate of change in concentration of substrate or products either in transient state (before reaching steady state) or at steady state when concentration of enzyme-substrate product no longer changes. The steady-state approach often cannot reveal detailed understanding of enzyme mechanisms, whereas the pre-steady state condition, referred to when enzyme and substrate are mixed at $t = 0$ until either a steady state or equilibrium is established, provides more details about mechanism of the catalytic process (Fisher 2005). In steady state approach, the concentration of the intermediate E-S does not change. In the equilibrium approach, the rate of formation of E-S is equal to its rate of transformation to either back to the substrate or product. Any standard book dealing with enzyme kinetics can be consulted to learn how both approaches/assumptions lead to a similar equation called Michaelis-Menten equation (please see below for a brief introduction to Michaelis-Menten model). For the system with multiple substrates, enzyme kinetics can also predict the sequence of substrates binding as well as the sequence of products releasing.

A simple model developed by Michaelis and Menten to explain enzyme kinetics assumes that enzyme (E), substrate (S), and enzyme-substrate complex (ES) are in rapid equilibrium. Reactivity of an enzyme towards a particular substrate can be expressed with kinetic parameters namely K_{cat}, V_{max}, and K_m. The catalytic constant K_{cat} (or turnover number) represents the number of substrate molecules that can be converted to product per enzyme molecule per unit time. Michaelis constant K_m and maximum velocity V_{max} indicate how easily the enzyme becomes saturated with a particular substrate. V_{max} is basically the maximum rate that the enzyme reaches at the point of being saturated by the substrate, where K_m is the substrate concentration at which the reaction velocity is half of the maximum velocity (or $V = V_{max}/2$).

Enzyme kinetic assays are generally carried out batch-wise by mixing enzyme at a constant concentration with a substrate at various concentrations in several cuvettes. The initial rates (velocity V) of the individual reactions are observed for a period of time, e.g., measurement of products in a spectrophotometer for a set period of time such as a few seconds or a few minutes; $V = d[P]/dt$. Initial rate is preferred because it is easiest to observe the rate of reaction when the reaction is just started without the effect of accumulated products that slow down the reaction. At low substrate concentration ($[S] \ll K_m$), the reaction is first order with respect to substrate as rate of the reaction is directly proportional to the concentration of substrate. At high substrate concentration ($[S] \gg K_m$) where it is assumed to be a constant parameter in the rate law, the reaction is assumed zero order with respect to substrate as rate of the reaction is at maximum and independent of substrate concentration (Atkins and de Paula 2017). Figure 2 shows three common plots used for determination of the important kinetic parameters K_m and V_{max} of an enzyme. Plot of V *vs* [S] is non-linear. Transformation of the curve to linear graph, either by plotting 1/V *vs* 1/[S] as proposed by Lineweaver and Burk, or V *vs* V/[S] as proposed by Eddie and Hofstee, makes it easier to estimate K_m and V_{max} from slope and XY intercepts.

Apart from physical conditions such as temperature, pH, and ionic strength, the presence of enzyme inhibitors can alter V_{max} or K_m of the enzyme. Inhibitors may be in the form of small organic molecules or small peptide analog of the natural substrate. Table 1 summarizes common types of inhibitors and their effects on kinetic parameters of an enzyme (Kuddas 2019, Strelow et al. 2012). Nevertheless, inhibitors can be used as tools to validate purity of an enzyme. A detailed discussion on this is beyond the scope of this chapter but is available elsewhere (Scott and Williams 2012).

As mentioned, enzyme kinetic assays usually involve mixing enzyme at a constant concentration with various concentrations of substrate. When performed manually, the process can be tedious and slow. The commonly used 96 well-plate does not always answer the demand for high throughput analysis (Pereira et al. 2020), for example to evaluate sources of enzyme, and drug discovery. Limitations of the common 96 well-plate include insufficient space to accommodate the experiments that involve many different conditions to be studied. It also requires relatively large volumes of reagents, is not portable nor fully automated, and still requires well trained

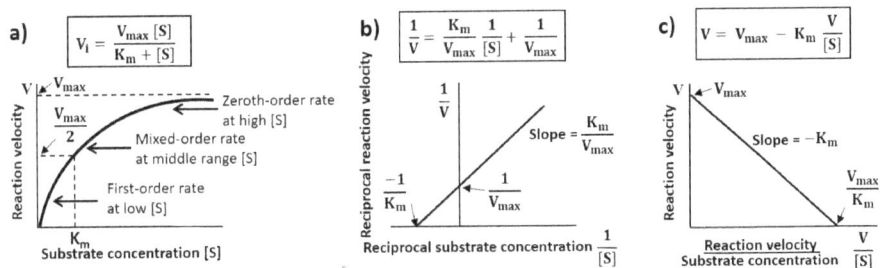

Figure 2. Graphs representing relationship of reaction velocity and concentration of substrate. a) Michaelis-Menten plot, b) Lineweaver-Burk (double reciprocal) plot, c) Eddie-Hofstee plot.

Table 1. Common types of inhibitors and their effects on V_{max} and K_m.

Type of inhibitors	Characteristics of inhibition	Effect on K_m and V_{max}
Competitive inhibitors	• Inhibitors bind only to free enzyme. • Inhibitors compete for the active sites on the enzyme, preventing the real substrate from binding. • Possible ways of inhibition include steric hindrance blocking accessibility of substrate to the active site of the enzyme, overlapping of binding site, binding to the same active site or sharing some binding pockets, or causing conformational change of the active site.	V_{max} is unchanged, but K_m is higher. The level of inhibition depends on the relative concentration of substrate and inhibitor. To achieve V_{max}, higher substrate concentration is required to compete and beat inhibitor to the enzyme.
Noncompetitive inhibitors	• Inhibitors bind equally well to both free enzyme and the enzyme-substrate complex. • Inhibitors bind at a separate site from active site for substrate. • Possible ways of inhibition include deformation of the structure of the enzyme, conformational change that inhibits catalytic capability of the enzyme, or hindering the binding and releasing of substrate.	V_{max} is lower, but K_m is unchanged. Noncompetitive inhibitors are not affected by substrate concentration.
Uncompetitive inhibitors	• Inhibitors bind only to the enzyme-substrate complex at a location outside the active site. • The resulting enzyme-substrate-inhibitor complex is enzymatically inactive.	Both V_{max} and K_m are lower.
Allosteric inhibitors	• Inhibitors bind to allosteric site other than or in addition to the active site, causing conformational change of the enzyme. • Possible ways of inhibition include affecting the formation of the enzyme-substrate complex, stabilization of the transition state, or reducing the ability to lower the activation energy of catalysis.	Depend on type of inhibition, an allosteric inhibitor may display a competitive, noncompetitive, or uncompetitive inhibition.

personnel to operate devices, such as a pipette, precisely. These limitations drive the development for automatic and high throughput systems for enzyme kinetics. The following sections present flow based techniques including flow injection and sequential injection analysis techniques, microfluidic systems, as well as paper based microfluidic devices for studies related to enzyme kinetics.

Flow based analysis formats

Flow injection and sequential injection systems

Introduced in the 1970s (Ružička and Hansen 1975), flow injection has been known as a low cost automatic chemical analysis technique that can reduce analysis time and, in many cases, amount of chemicals. The flow injection technique involves introducing a plug of a chemical reagent, usually via an injection valve, into a flowing stream of another reagent that is forced into a small tubing by pumping. Chemical reaction takes place while the reagents are merging and flowing downstream. Product is detected when the reaction plug flows through the detector. For studying enzyme kinetics, volumes of enzyme/substrate and degree of mixing can be optimized by controlling flow rates. Detection of product or substrate concentration, especially at non-equilibrium state, can be done rapidly and at a more precise time for each replicate, as compared to the batch method. The process becomes more automatic and helps to minimize inconsistency owing to manual operations. This is especially useful in measuring initial rate that is often hard to do precisely in batch mode due to the difficulty in conducting prompt and precise solution transferring, mixing, and maneuvering cuvettes. The ability to measure initial rate is attractive because concentration of the main substrate of interest is still abundant at the beginning of enzymatic reaction. Therefore, it can suppress interferences from other lower concentration substrates that may become pronounced as the reaction progresses and the concentration of the main substrate is depleted. Also, as the product concentration[s] are low, product inhibition does not complicate the situation.

Conventional flow injection analysis (FIA) systems for enzyme kinetics have been reviewed (Hartwell and Grudpan 2012). A typical flow injection system, composed of peristaltic pump, a 6-port injection valve, tubing, and detector, can easily be operated at milliliter level. Volume of sample/reagent is dictated by the size of the injection loop connected to the ports on the valve, see Fig. 3a. In loading position, the injected solution fills the loop of a selected size/volume and excess amount goes to waste. In injection position, the valve is switched to let the carrier solution push out the injected solution inside the loop and mix with it while travelling through the detector. In enzyme kinetics, enzyme may be injected into the flowing substrate carrier solution or vice versa, depending on cost of reagents and experimental design. Due to the continuous flow in one direction, it is more convenient to use enzyme solution at constant concentration as a carrier, and inject a substrate of various concentrations. However, high consumption of enzyme solution can increase analysis cost.

A more modern sequential injection analysis (SIA) system with a computer controlled syringe pump can perform chemical analysis at lower volume at microliter level. The flow control is not continuous, instead direction and movement of liquid is programmed based on volume and flow rate selection. Another main component of the SIA system is a selection valve for aspiration and dispensing of all the solutions. Flow reversal of the syringe pump controlled by a computer software which allows multiple solutions as well as air segments to be introduced as stacked zones into a long holding coil located between the syringe pump and the selection valve, see Fig. 3b. In enzyme kinetics, aliquots of enzyme, substrate, and co-factor solutions are aspirated one at a time through different ports on the selection valve into a holding coil. The sequence of aspiration should be optimized to allow

a)

injection port

waste

Injection
loop

carrier →

detector
Sample loading
position

injection port

waste

Injection
loop

carrier →

detector
Sample injection
position

Aspiration
(flow toward carrier container)

3   R1   R2

      R1   R3

2   R2

1   carrier   R1

Dispensing (flow toward detector)

carrier   R1   R2   carrier   Mixture R1 R2 R3

                  R3

Multi-port
selection valve

b)

Holding coil   Detector

Waste
Multi-port
selection valve

Carrier solution

Syringe pump   R1   R3
                R2

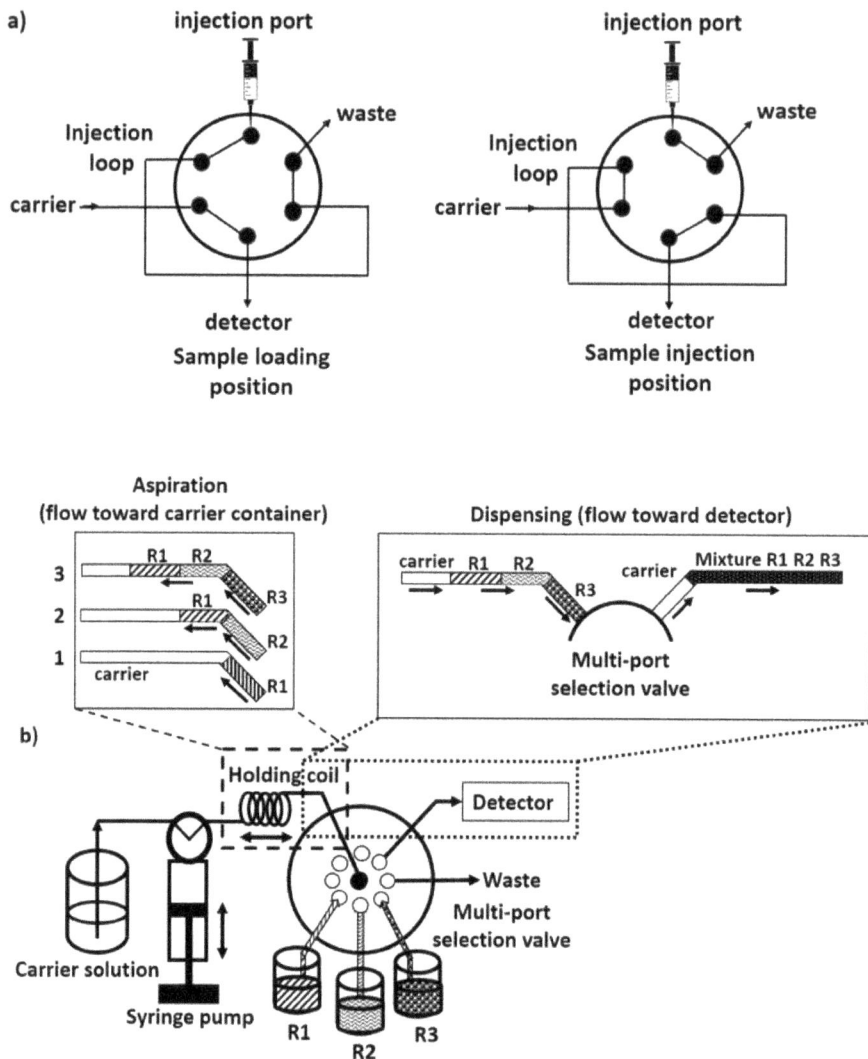

Figure 3. Diagrams and operation of a) 6-port injection valve used in FIA, and b) multi-port selection valve used in SIA. R1, R2, R3 are reagents such as enzyme, substrate, and co-factor.

efficient reaction between enzyme and substrate while minimizing dispersion of the injected enzyme-substrate reaction plug through the interface with the carrier solution. The stacked zones are mixed by the flow reversal when those zones are being sent through one of the ports on the selection valve into the detector. The multi-port selection valve (usually 8-port valve) used in SIA system can accommodate aspiration of multiple reagents/samples, whereas a 6-port injection valve used in FIA system can only be used to inject reagents/samples into the system one at a time through the injection port. Because these injection/selection valves are the heart of the systems, understanding the operation principles of the sample/reagent introduction is important for designing the analysis. Consider Fig. 3 for comparison of operational principles of the FIA 6-port injection valve and the SIA 8-port selection valve, as described above, for better understanding of flow direction manipulation and accommodation.

A review on applications of SIA for enzyme kinetics include various examples of research (van Staden and Raluca 2002, Silvestre et al. 2011), mainly in homogenous phase where enzyme is in solution (non-immobilized enzyme). Incorporation of the mixing chamber to one of the ports on the selection valve is most common in order to promote mixing by increasing radial movement that may be insufficient in the holding coil alone. Detection, then, can be done directly on the mixing chamber (e.g., using fiber optics), or by sending an aliquot of product solution from the chamber to the detector unit. Lab-on-Valve (LOV) and Lab-at-Valve (LAV) commonly used with the SIA system are ways to incorporate the detection unit within or closely adjacent to a multi-position selection valve, in order to enable higher throughput analysis and lower volume consumption of reagents (from hundreds to tenths of microliter). The low dead volume in the SIA system allows for faster analysis because the new portion of solution can be sent to the detection at the same time that the previous portion in the detector is pushed out. While the most common detector used for the LOV system is fiber optics which can fit within the channel specially made for it located as part of the selection valve, other devices such as mixing chamber, filtration unit, and dialysis unit can also be incorporated around the valve which are usually referred to as LAV. Schematic representations of basic FIA, SIA, SIA-LOV, and SIA-LAV systems are depicted in Fig. 4. Please note that figures shown are common conventional flow based systems. Implementation of other types of valves such as a solenoid valve to perform hydrodynamic injection is also possible (Khongpet et al. 2018).

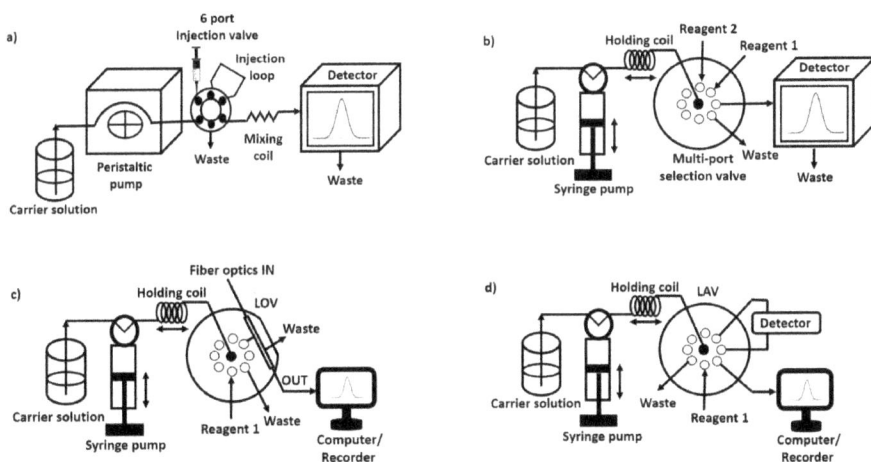

Figure 4. Common flow based systems a) FIA using a peristaltic pump and 6-port injection valve, b) SIA using a syringe pump and 8-port injection valve, c) SIA-LOV where the detector is integrated onto the selection valve, d) SIA-LAV where the detector is adjacent to the selection valve.

In general the flow of solution in the FIA system is continuous, while the flow of solution in the SIA system is stop and go due to reversed flow operation of the syringe pump. However, both FIA and SIA systems can be designed and adapted to manipulate various other solution flow formats (such as continuous flow, stopped flow, segmented flow, and quenched flow) to serve particular applications. Briefly, continuous flow systems are based on mixing of the merged sample and reagent solutions injected into the continuous flow of the carrier solution. In order to allow more reaction time between enzyme and substrate, and hence increased sensitivity, stopped flow format can be used in which the reaction zone is physically stopped for a period of time. This is done by stopping the pump, before online detection. This is similar but is not the same as quenched flow where the reaction zone is stopped both physically and chemically by stopping the pump and introducing a quenching solution. In this case, a typical FIA system with a single 6-port injection valve and one direction flow has limitation due to number of solutions that can be injected being limited to one. An additional valve or a pump is needed to introduce a quencher. The 8-port selection valve with capability of reverse flow enables introduction of more solutions using a single valve. The stopped flow technique can be used to detect intermediates, while the quenched flow technique is mostly limited for detection of product. The two techniques can be used complementarily (Fisher 2005). In the FIA and SIA systems, band broadening and dispersion of the sample/ reagent zone that occur at the interfaces with the carrier stream may cause dilution and low sensitivity of detection. However, the systems can introduce air segments before and after each reaction zone to prevent mixing/minimize dispersion, which in turn improves sensitivity and reduces consumption of reagents (Patton and Crouch 1986).

Microfluidics

Due to the great demand for high throughput analysis along with the need for low volume consumption of chemical/biological reagents, downscaling the analysis system has been a major focus. Study of enzyme, in particular, requires numerous observations of the catalytic capability of the enzyme under various conditions of enzyme-substrate reactions which is time consuming, labor intensive, and can be expensive due to high total volume of reagents. The progress in downscaling of the chemical analysis system has been gearing toward a small portable unit known as lab-on-chip or microfluidic systems.

Microfluidic devices are usually composed of a network of microchannels of the size in the range of sub-millimeters engraved on a solid material such as glass or plastic that can accommodate the chemical reaction in the volume of microliter or lower. Although in some applications, top-open channels can be used (Casavant 2013), the microchannels are usually closed with another solid plate that is secured to the plate with engraved microchannels using such physical and chemical methods as double sided tape, magnetic clamps, or plasma treatment (Tkachenko et al. 2009, Tsao and DeVoe 2009, Khongpet et al. 2020).

Pumping systems for microfluidics

Liquid pumping systems used for introduction of liquid into microchannels are commonly based on either pressure or electroosmotic force (Schilling 2001). Due to the small diameter of the microchannel, liquid inside the microfluidic system is mainly governed by a laminar flow. Without the turbulent flow that can cause molecules to flow in random directions, laminar flow allows the benefit of predictable movement and, therefore, physical/chemical conditions of liquid when flowing through the microchannel. For the pressure driven pump that produces the force from one end of the channel, velocity of liquid adjacent to the inner wall of a microchannel is lower than velocity of liquid in the middle of the channel, creating a parabolic laminar flow profile, see Fig. 5a, with concentration gradient due to diffusion. For an electroosmotic pump, electrical field is applied across the length of the open-ended channel. Most solid surfaces, including the wall of the microchannel, have an electric charge that attracts counter ions from the solution to form an electric double layer. These ions will move towards the opposite polarity electrode, pulling the body of liquid along towards the same direction. Unlike the parabolic profile that resulted from the pressure driven flow, the electroosmotic pump creates a simple laminar flow profile with a uniform velocity of the same frontline across the width of the channel, see Fig. 5b. Although dispersion and band broadening still remains, they are minimized. As compared to parabolic flow profile, this plug flow profile exhibits rather uniform diffusion. However, an important point to consider when working with enzyme is the possibility of protein adsorption onto the wall of the microchannel. Apart from depleting the amount of free enzyme, the portion of enzyme which is now immobilized by adsorption may alter the charge on the solid inner wall of the channel and result in changing the velocity of liquid flow.

The means to deliver and manipulate liquid flow, as well as the detection of the target substance produced within the microfluidic system, have continuously been the main research focuses. With the advance in digital technology, microfluidics have become a preferred approach in down scaling flow based chemical/biological analyses, including enzyme studies and those that utilize enzymatic reaction as part of the analyses. Microfluidic designs that offer concurrent multi-assay, accommodate incubation time, and enable long term monitoring of the reaction are desirable (Rho et al. 2016). Common main approaches, other than continuous flow homogeneous phase, include enzyme immobilization (heterogeneous phases) and droplets based microfluidics (aqueous segment in oil) which are described in the next sections.

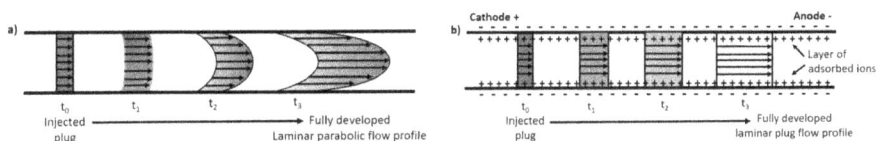

Figure 5. Laminar flow in a microchannel where there is a) a pressure driven parabolic gradient flow profile in which the direction of flow is due to the pressure pump, and b) an electroosmotic driven plug flow profile in which the direction of flow is due to the movement of a layer of adsorbed ions toward the oppositely charged electrode.

Common designs in microfluidic system

Designs of microchannel network usually depend on the application of a particular chemical reaction to be carried out, many times mimicking the process needed for batch-wise process. With the advance in technology such as 3D printers and laser engravers, various complicated designs can be made to serve specific purposes. Among the various novel proposals, some common microchannel designs for reagents introduction, reaction accommodation, and mixing and are shown in Fig. 6. Y- and T-junctions are the most widely used designs for introducing more than one reagent into the system either at the same time or at different times. At a glance, these two designs are not much different, both allow for the merging of two or more solutions. However, the difference in flow angles together with liquid properties such as viscosity and flow rate, have significant effect on liquid delivery through the microchannel. This aspect will be discussed more in the later section on droplets-based microfluidics. Serpentine design helps to fit a long channel within a short length platform. Two liquids, even though they are miscible, if introduced through two different inlets at a suitable flow rate usually can be controlled to flow in parallel after merging within a microchannel. Only diffusive mixing occurs at the liquid-liquid interface while flowing downstream. If more vigorous mixing is needed, the microchannel should be designed to disturb laminar flow and promote turbulence. An example of repetitive split and merge design helps to promote mixing of the reagents.

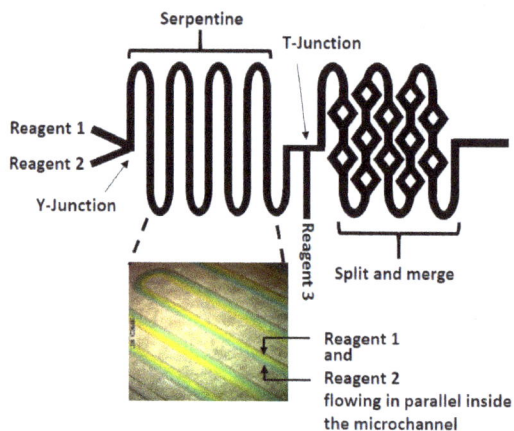

Figure 6. An example of microfluidic system with some common microchannel designs (modified from Siegesmund and Hartwell 2018, Ellerhorst and Hartwell 2019).

Flow based method with enzyme immobilization

In the conventional study of enzyme's kinetics, both enzyme and substrate are in solution or homogeneous phase. They have freedom of movement to interact from all angles, and if the orientation and energy is suitable, binding occurs. Kinetics measured in this homogeneous condition is known as *intrinsic* kinetics. After the analysis, the solution mixture, containing enzymatic product, free enzyme, and unreacted substrate, is usually discarded as these components cannot be easily separated for reuse. If the enzyme is high cost or low in availability, immobilization of enzyme on a solid surface, such as on a solid support (e.g., bead or monolith material packed in a column, or the inner wall of the capillary or microchannel), may be beneficial. Enzymatic reaction takes place when the substrate solution flows through the reactor with immobilized enzyme. Immobilizing enzyme molecules enables removal of products of the enzymatic reaction while keeping enzyme in place, and therefore, it is possible to reuse the reactor or amplify signal by introducing more substrate molecules. This in turn, lowers the consumption of costly enzymes. Asanomi et al. (2011) reviewed and summarized principles and advantages/disadvantages of various methods for immobilization of enzyme in a microchannel including particle entrapment, adsorption, and chemical crosslinking. The immobilized enzyme may also be reused to react with substrate at different concentrations in subsequent flow analyses. Apart from reusability, immobilization of enzyme can offer other benefits (Miyazaki et al. 2008). By keeping enzyme molecules in place, aggregation can be prevented which reduces the possibility of autolysis of some enzymes. Oriented immobilization of the enzyme to the solid support can help orient the binding sites outward for easy access by the substrate, and can also help to control and stabilize enzyme conformation. This can be done, for example, through a biotin-avidin binding, or through introduction of a functional group on the inner wall of the microchannel for covalent crosslinking with enzyme (Honda et al. 2005).

Nevertheless, it is important to keep in mind that in heterogeneous phase, substrate molecules in liquid phase have some mass transfer limitation in that they have to diffuse in order to reach the binding sites of the immobilized enzyme. If the enzyme turnover number is high, the rate of reaction between enzyme and substrate is faster than the diffusion rate of that substrate, then concentration gradient or partitioning of substrate will occur in the microporous environment. In other words, *observed* or *apparent* kinetics of the immobilized enzyme are masked by diffusional effect and may be distorted from the *intrinsic* kinetics in homogeneous phase. Therefore, an alternative kinetic model such as the Lilly-Hornby model was used, instead of the common Michaelis-Menten model, to evaluate apparent kinetic parameters (Lilly et al. 1966, Seong et al. 2003). The mathematical model of the *mass transfer limit* kinetics or *diffusional masked kinetics* usually yields K_m and V_{max} values different from the normal system. According to Webster (1983), decreasing enzyme activity within the unit volume, increasing substrate concentration, and decreasing pore size of the support material can minimize or eliminate diffusional effect to obtain *inherent* kinetics that is assumed to be the same as *intrinsic* kinetics and can be described using the normal Michaelis-Menten kinetics model.

Pore size and geometry as well as electrostatic charge of the support surface are important parameters (Webster 1983). Relatively small pore size, as compared to the size of substrate, can cause limitation in rotation of the large size substrate molecules and reduce the opportunity for proper interaction of substrate to the binding site of the immobilized enzyme. *Rotational masked* kinetic will cause kinetics distortion from the intrinsic condition, even with the absence of diffusional or concentration gradient effect. Figure 7 depicts intrinsic *vs* inherent kinetics to

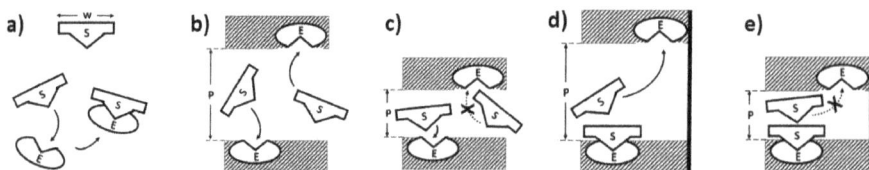

Figure 7. Different kinetic conditions a) intrinsic kinetics in homogeneous phase with no concentration gradient and no rotational nor diffusional limitation. b) an open large pore size relative to the size of substrate (P > W) yields inherent kinetics with no concentration gradient, and no rotational nor diffusion limitation c) an open small pore size relative to the size of substrate (P < W) yields rotational masked inherent kinetics d) a closed end large pore size relative to the size of substrate (P > W) yields diffusional masked inherent kinetics e) a closed end small pore size relative to the size of substrate (P < W) yields diffusional and rotational masked inherent kinetics. P is pore diameter. W is size of substrate molecule. S is substrate. E is enzyme. X shown on dotted arrows indicates the binding that cannot happen.

show the need of suitable geometry and pore size large enough for free rotation of substrate molecules in order to prevent concentration gradient, and rotational and diffusional masking.

To improve the efficiency of the packed-bed column, a single piece of monolith material has been used in place of small beads (He et al. 2010). Monolith material contains interconnected pore network that allows a more free flow of solution, with lower back pressure and better mass transfer. One last point to ponder when using immobilized enzyme reactor for kinetic study, reusing of enzyme may be affected by memory effect that causes deactivation or interferences from the previous run. Therefore, cleaning prior to the next analysis cycle is important to take care of the interferences due to the presence of unreacted substrate/product, etc., from the previous run. As the enzyme is generally used in excess, effect due to partial deactivation is generally insignificant.

Droplets based microfluidics

The demand for high throughput analysis has driven the development of microfluidic systems even further. Similar to using the conventional 96 well-plate with multi-tip micropipettes to carry out multiple reactions all at once in batch-wise process, multiple analyses in a microfluidic system has been a big goal. Droplet based microfluidics is the system that generates aqueous micro-droplets (dispersed phase) in an immiscible liquid (continuous phase). The droplet serves as an individual reactor where the chemical reaction takes place. Multiple droplets can be produced within a microchannel to accommodate and compare various reaction conditions, similar to the capability of a micro-well plate, but in a more automatic flow based system. Segmented flow, where slugs of liquid are separated by air segments, serve a similar purpose. However, the droplets platform is different in that the liquids in the droplet are encapsulated and protected from direct contact with the wall of the microchannel, whereas the slugs of liquids in a segmented flow system are exposed to the solid wall. Encapsulation of liquids inside the droplet and the immiscible nature of the carrier solution (continuous phase) will also prevent the reagents inside the droplets from dispersing and diffusing into the carrier steam. Minimizing loss of enzyme

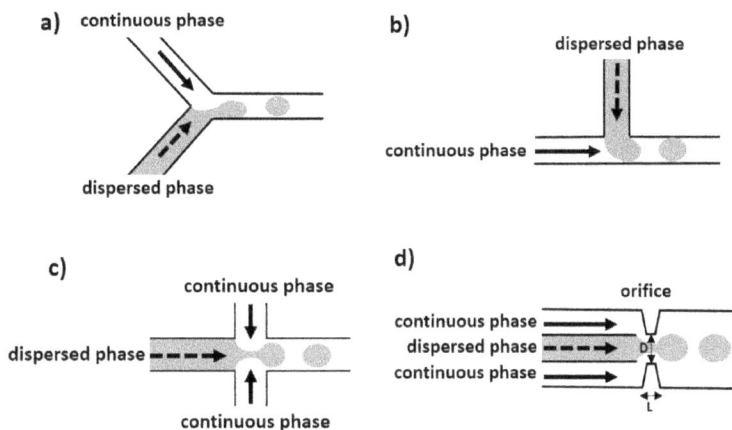

Figure 8. Some common droplets introduction designs a) Y-junction, b) T-junction, c) cross-flow focusing, and d) parallel flow focusing. D is width of orifice, and L is length of orifice.

or substrate can better ensure accuracy and precision of the kinetic measurements. Although many varieties of microchannel designs are possible, depending on the particular application, all usually incorporate some common inlet designs for generating droplets, namely the two inlets Y- and T-junctions, and the three inlets flow focusing in either cross junction or co-flow formats, see Fig. 8. Please note that the dispersed phase may be originating from two or more reagents (e.g., enzyme and substrate) merged together before intersecting with the continuous phase. The components in the dispersed phase may not completely mix when forming a droplet. The single shading shown in Fig. 8 is for simplicity in presentation, not necessarily representing a homogeneous mixture. For a complete reaction, various ways of droplet operations such as fusion, fission, and merging can be done as reviewed by Sohrabi et al. (2020).

The Y- and T-junctions, Figs. 8a and 8b, are popular due to the simplicity of the designs and fabrication. The critical region is the junction where the dispersed phase is pinched off due to the pressure and the stress applied when intersecting with the continuous phase. Droplet size, shape, and frequency of production can be varied by changing the geometry of the channel, i.e., height to width ratio of the channels (Wehking et al. 2014, Garstecki et al. 2006). Steegmans et al. (2009a, 2009b, and 2010) explained that droplet formation and detachment at a Y-junction occurs in one step by shear force between the two phases. In contrast, the process is two steps in T-junction configuration, where a droplet is formed by inertial forces of the dispersed phase before being detached by the dominant pressure of the continuous phase that overcomes the interfacial tension of the dispersed phase. Ushikubo et al. (2014) also found that droplet size depends mainly on viscosity and relative velocity of the two liquid phases, while interfacial forces is insignificant. They compared the Y- and T-junction performances on droplet production and found that the Y-junction is suitable for the high viscous continuous phase fluids, while T-junction is more feasible as it can be used with all types of fluids. T-junction is also reported to offer more uniform size droplets and its performance does not depend much on conditions such as velocity or flow rate.

In the flow-focusing configuration, unlike the Y- and T-junctions, shear force exerted by the continuous phase on the dispersed phase is symmetrical. Therefore, the generated droplets by the flow-focusing method are rather stable. The angle at the junction affects the hydrodynamic force in and around the region and changes characteristics of the droplets (Yu et al. 2019). In the cross junction design, Fig. 8c, the continuous phase flows in and focuses on the dispersed phase equally from both sides. The sum of vectors results in the droplet flowing forward. Another flow-focusing design, Fig. 8d, employs a narrow tubing (e.g., capillary) as a sub-channel to force the dispersed phase into the continuous phase flowing inside the main channel. Initially, the flows of both phases are parallel. The pressure increases at the narrow orifice and it overcomes surface tension of the dispersed phase, subsequently forcing droplets to form and detach (Nooranidoost et al. 2016). Physical parameters, e.g., droplet size and dispersion, impact directly the quality of the analysis results, e.g., throughput, dynamic working range, and accuracy (Rosenfeld et al. 2014). It has been demonstrated that the droplet size is directly related to orifice width and the distance of the orifice from the inlet, while it conversely relates to the orifice length up to a certain limit (Gupta et al. 2014). Apart from the sizes of the channels and orifice, the sizes of the droplets can be controlled by changing the flow rate of the continuous phase. It was found that droplet size is decreased by increasing flow rate of the continuous phase, or by using a lower ratio of dispersed flow rate to continuous flow rate (Nooranidoost and Kuma 2019).

Droplet-based microfluidics has become an attractive method for small volume enzyme kinetics because of important benefits including rapidity, biocompatible interfacial chemistry, and a dispersion-free system. Enzyme kinetics in the droplet-based platform can be carried out, for example, by introducing enzyme and substrate as a combined small volume into a continuous flow of the carrier fluid. Immiscibility of the droplets in the carrier helps to minimize dispersion and helps to retain enzymatic reagents within the droplet. Adsorption of enzyme and substrate onto the wall of the microchannel is also minimized. The mixing of content within the droplet (e.g., enzyme and substrate) occurs through diffusion, and can be enhanced with chaotic advection by using a winding microchannel such as serpentine design (e.g., downstream portion of Fig. 6) to promote stretch and fold movement of the liquid mixture (Song et al. 2006). The immiscible carrier fluid should be chosen so as to prevent side reactions as well as dispersion or partitioning of the enzyme-substrate products from the droplet. Aqueous phase enzyme-substrate droplets are usually carried through the microfluidic system with a surfactant or oil based liquid. Surfactants in the carrier fluid can help to lower the surface tension and adsorption at the liquid-liquid interface between the dispersed and the carrier phases (Dickinson 1991).

Ochoa et al. (2020) described a step-by-step procedure to design and fabricate a microfluidic chip for enzyme inhibition assay based on microfluidic droplets and using fluorescent image analysis. Some examples are included here to demonstrate how the droplet-based microfluidics system can mimic the batch-wise enzyme kinetics study such as those carried out in a micro-well plate. Song et al. (2006) also suggested that such systems as "preformed cartridge," proposed by Zheng and Ismagilov (2005), may be applied to identify the enzyme of a desired reactivity. In this application, droplets (plugs) of various enzymes in the preformed cartridge, flow and merge with a stream of a common substrate at a T-junction. The system should, therefore, be applicable to study the kinetic of a particular enzyme. For example, the preformed cartridge can be employed to contain a substrate of various concentrations which is then merged with a constant concentration stream of an enzyme.

On-chip dilution has also been reported for creating different concentrations of reagents. Song and Ismagilov (2003) utilized a three-inlet and flow of buffer in between two reagents (e.g., enzyme and substrate) to prevent direct mixing while varying the amount of reagents by controlling their flow rates. Later, Bui et al. (2011) proposed an on-chip dilution based on diffusive mixing. A Y-junction is utilized to produce laminar parallel flows of substrate and buffer solutions. Concentration gradients of substrate were created through diffusive mixing while the two solutions are flowing in parallel downstream, similar to the phenomenon shown in Fig. 6. The gradient along the liquid flows can be in linear, parabolic, or exponential fashion, depending on the concentration of the substrate, flow rate, viscosity, and channel length (Jeon et al. 2000). In Bui's work, the substrate linear gradient was merged with a constant concentration enzyme before entering the T-junction where droplets of enzyme-substrate were produced when intersecting with the immiscible continuous phase of oil solution. Due to the concentration gradient of the substrate, each droplet contains different substrate concentrations. Utilizing a multiple serpentine design to accommodate numerous droplets, a high throughput analysis of various enzymatic conditions can be carried out within a small device. Other methods for inducing concentration gradients have been reported, but diffusive mixing with laminar flow is the simplest and most stable.

Another example, by Jebrail and Wheeler (2010), demonstrates the use of electroosmotic flow to control the droplets. Electrical potentials are applied to an array of electrodes to control the mixing and movement of the droplets. The proposed microfluidics include both two plate format where liquid and droplets are accommodated in a closed microchannel, and one plate format where droplets are placed on the open channel. A hydrophobic insulator coated onto the electrodes isolated the droplets from direct contact with the electrode surfaces. Electrical charge was created on either side of the insulator when electrical potential was applied to the electrodes. These charges make it possible to manipulate liquid droplets to enter and leave the system, as well as to merge, mix, and split them as desired.

The applications of the droplet-based microfluidics have also been expanded to include encapsulation of rare cells or a single cell, e.g., bacteria, and DNA. This technological progress allows for the detection of a single enzymatic event and the screening of cells for a particular enzyme activity, which enables study of evolution of cells (Agresti et al. 2010), e.g., evaluation of the activity of enzyme expressed after repetitive cell mutation. Droplet fusion techniques (Joensson and Svahn 2012) can be very useful in adding enzyme, substrate, or other necessary reagents to the existing droplets that contain cells. Due to the large surface area to volume ratio of the micro-droplets, characteristics of the interface between the droplet's surface and the continuous phase (water/oil) may have great influence on the content inside the droplet. Therefore, compositions of aqueous and oil phases should be selected so as to minimize adverse phenomenon, e.g., nonspecific protein adsorption at the interface which can interfere with the enzymatic assay (Roach et al. 2005).

Detection

Image analysis based on fluorescence is a popular detection method in a microfluidic system while the UV-Visible region is not as widely used because of limited choices of the materials for microfluidics fabrication that do not absorb light in the range of UV-Visible detection wavelengths (Urban et al. 2006). By comparing fluorescence signals from images taken at the inlet and the outlet, rate of the enzymatic reaction can be determined. Fluorophores are usually attached to either the natural substrates (Joensson and Svahn 2012) or the enzyme (Girault et al. 2018). The fluorophores remain quenched until being released when substrates bind to the enzyme, as illustrated in Fig. 9, in which fluorescence intensity indicates the extent of enzymatic activity. In the droplet-based microfluidics, the observed fluorescence is usually the average signal of many droplets travelling through the detector during the particular set time duration, e.g., every 30s. However, the averaged responses cannot differentiate activity in each single droplet (Yin and Marshall 2012). In order to study enzyme at a single cell level, various attempts have been made for sorting of droplets (Beneyton et al. 2016, Vallejo et al. 2019), commonly known as Fluorescence Activated Droplets Sorting (FADS). For example, Vallejo et al. (2019) developed a microfluidic system to discover new enzyme variants of particular properties. Droplets containing single cell and substrate were collected for incubation, prior to being excited by a laser. The droplets that emit fluorescence upon exposure to the laser indicate the presence of the preferred enzyme variant. The emitted light activated the electrode, forcing the droplet containing the desired cell to move into a certain outlet.

Ultrahigh throughput analysis can be performed by designing a multi-microchannel to accommodate multi enzymatic reactions in parallel, and using more than one fluorogenic substrate. Example reports include screening for the particular cells that contain enzyme of specific characteristics of interest, e.g.,

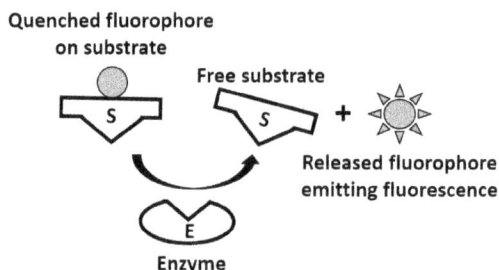

Figure 9. Illustration of the use of fluorogenic substrate as a tool to detect enzymatic activity.

with enantiospecificity, chemospecificity, and regiospecificity (Ma et al. 2018). Separate lasers are used to excite droplets in each microchannel to generate different emission wavelengths and therefore, avoid crosstalk of two fluorescence signals. This setup can also be applied to evaluate enzymatic activities toward two substrates simultaneously.

Paper based microfluidic devices

Despite the growth of microfluidics research, the fabrication process and related costs remain the important challenges, especially in places with limited resources (Ding et al. 2020). Even with advancement of technology, producing a polymer-based and glass-based microfluidic or Lab-on-Chip system is not a simple task and still requires well-trained personnel and some high cost equipment and/or reagents. Microfluidic paper based analytical devices (mPADs) have been gaining interest as simpler and more economic disposable microfluidic analysis systems (Noviana et al. 2021, Nishat et al. 2021).

The low cost, low nonspecific binding, and availability of paper (mostly cellulose or nitrocellulose) in a wide variety of thicknesses, porosities, pore sizes, and wicking properties make mPADs an attractive microfluidic platform. In addition, liquid absorption and capillary action allows for liquid flow on mPADs without the need for external power or pumps. Many reports have demonstrated simple yet effective means of creating microchannel designs with hydrophobic and hydrophilic zones through either physical patterning (e.g., laser engraving, and etching) or chemical patterning with hydrophobic inks (e.g., photolithography, wax printing, inkjet printing, and stamping). Nishat et al. (2021) has reviewed, summarized, and compared the pros and cons of some of these patterning methods, as well as the equipment needed. Two simple patterning methods, namely laser engraver and embedded crayon wax, are presented here as examples of physical and chemical patterning. As shown in Fig. 10, hydrophobic area can be easily created by manually applying crayon wax, similar to wax printing using a printer. Then, the paper was heated faced down on a

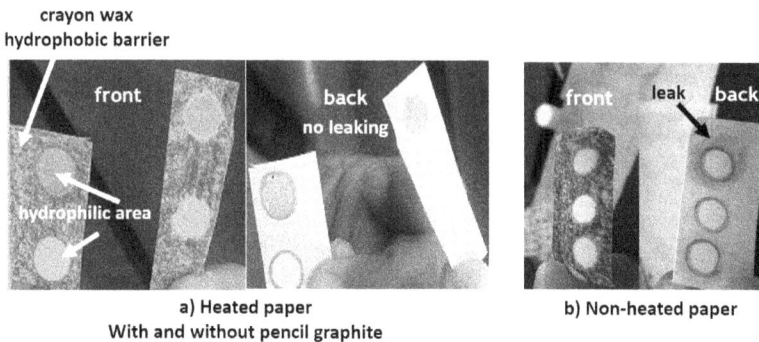

a) Heated paper
With and without pencil graphite

b) Non-heated paper

Figure 10. Waxed paper was created by applying crayon wax onto one side of a piece of filter paper. a) Heating of the waxed paper prevents liquid from leaking on both the front and back sides of the paper, and b) Non-heated waxed paper has the problem of liquid leaking from the hydrophilic region on the back side of the paper due to the lack of hydrophobic wax (modified from Nzobigeza and Hartwell 2019).

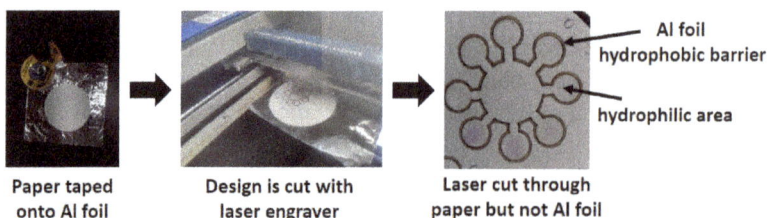

Figure 11. Creating a microfluidic pattern using a laser engraver. A piece of filter paper is taped onto aluminum foil using double sided tape. A laser engraver is used to cut the design imported from Corel Draw, creating a hydrophilic pattern, with the aluminum foil acting as hydrophobic barrier (modified from DeMott and Hartwell 2020).

hot plate. Heating is necessary for complete absorption of wax into the paper and to create a hydrophobic barrier on both sides of the paper. Without heating, liquid can leak out of the hydrophilic area on the back side which has not been directly coated with hydrophobic material. Adding a graphite barrier around the pattern can also better define the hydrophobic-hydrophilic boundary (Whitford et al. 2020). In order to create a more precise or complicated design, equipment with higher precision such as a laser engraver may be used. A common 60 W laser can cut through paper, but not the aluminum foil backing (Mahmud et al. 2016). As shown in Fig. 11, by taping the paper onto aluminum foil, the hydrophobic-hydrophilic pattern can be created and be kept intact. If backing is not needed, removable double sided tape will allow for easy detachment of the aluminum foil backing.

Despite many benefits of the mPADs, some drawbacks have been reported (Carrell et al. 2019), especially regarding poor detection limit, short shelf-life of the devices with pre-immobilized reagents, e.g., enzyme, and low reproducibility of the reaction. The main way to improve detection limit has been accomplished by integrating sensitive detection methods such as florescence, chemiluminescence, or electrochemical detection onto the mPADs. As mentioned earlier, a fluorescence substance is quenched until being released due to enzymatic reaction. As for chemiluminescence, the reaction is usually initiated by the oxidizing agent in the presence of a catalyst such as enzyme or metal ion (Li et al. 2019). The main advantage of chemiluminescence over the fluorescence detection method is owing to its simplicity with no requirement of excitation source. In addition, chemiluminescence offers a wide calibration range and low background signal. Nevertheless, most commercially available paper based biosensors utilize electrochemical detection with electrodes screen printed onto the paper because of their convenient mass production, cost-effectiveness, and ease of use (Caratelli et al. 2020).

Although enzymes can be immobilized on paper materials and V_{max} and K_m can be found similar to those of free enzymes, researchers have reported that enzymes tend to denature when being deposited and left to dry on paper for extended periods. Shelf-life of biological materials can be increased by refrigeration. However, while Ilacas and Gomez (2019) demonstrated successful storage of enzyme for 30 days at low temperature, many works report lowered or inhibition of enzyme activities on paper as compared to bulk solution. Mitchell et al. (2015) proposed a solvent-free reagent deposition method by transforming reagents into a solid form known as reagent pencil, in which the reagent to be deposited on the mPADs (e.g., enzyme) is mixed with polymer (e.g., polyethyleneglycol), graphite powder, water or acetone, and is pressed into the shape of a pencil core. By simply drawing onto the mPADs, enzymes such as glucose oxidase and horseradish peroxidase can be stabilized in this solid form at ambient conditions for more than 2 months. The method was demonstrated to work well with reagents with molecular weight in the range of 2000–6000 g/mol (Liu et al. 2017). The main challenge is the reproducibility of the amount of reagent being applied which depends largely on the pressure put on during application.

As discussed by Carrell et al. (2019), liquid flow and mixing on the paper surface is mainly due to capillary action which is highly dependent on the physical texture of the paper and often yields non-uniform and irreproducible mixing. Therefore, miscible liquids cannot be assumed to achieve complete mixing. The non-uniform distributions of liquid with aggregation of solutes around the outer edges of the test zones as solutions dry out on paper, known as coffee ring effect (Nishat et al. 2021), is another common problem affecting accuracy and precision of mPADs. Sedighi and Krull (2018) reported that pore size proportionally affects enzyme amplification in solution-phase where the smaller the pore size, the lower the amplification. On the other hand, the enzyme amplification increases with the smaller pore size in the surface-phase reaction where reagents were immobilized on the paper surface. Caratelli et al. (2020) developed a 3 layer electrochemical paper based microfluidic device for detection of enzymatic byproduct, in order to evaluate efficacy of drugs for treating Alzheimer's through an enzyme inhibition. It was pointed out that layers of paper origami mPADs act as diffusional barriers and cause the K_m value of the enzyme to be higher than the value found in bulk solution. This does not mean that the device cannot be used. It merely means that, no matter which enzymatic reactions and analysis methods are used, comparison of enzyme activities should be done using the same system (Razak et al. 2020). In order to improve reproducibility of the chemical reaction on mPADs, arrangement/design of the paper based device to enhance diffusive mixing of liquids (Osborn et al. 2010), as well as the application of an external force such as an acoustic wave (Rezk et al. 2012), have been reported.

Examples of flow based enzyme kinetics

Flow based techniques have been applied to countless chemical/biochemical analyses involving enzymatic reactions. Although most works may not directly use flow based techniques for studying enzyme's kinetics, the ability of these techniques to detect products of enzymatic reactions ensures that these flow based techniques can be used for kinetic study. Table 2 presents some selected examples of research related to employment of flow based systems to determine enzyme activity and/or kinetic parameters.

Table 2. Examples of flow based systems for determination of activities, kinetic parameters of enzymes, and their applications.

Flow based system, References	Enzyme	Application remarks
FIA, Kracke-Helm et al. 1991	β-galactosidase	On-line monitoring of intracellular enzyme production during cultivation of recombinant *E. coli*.
FIA, Becker et al. 1997	Phospholipase D (PLDs, EC 3.1.4.4)	Investigation of enzyme characteristics (K_m, V_{max}, and temperature dependence) with chemiluminescence detection of enzymatic product.
Stopped-flow FIA, Yerian et al. 1988	Penicillinase	Evaluation of penicillin sensor through characteristics of immobilized enzyme on cellulose (stability, response to different substrate, kinetic response, and effect of buffer) with flow Injection optosensing system.
SIA, Cedillo-Perez et al. 2017	β-Glucosidase	Determination of kinetic constants and behavior of free (homogeneous media) and immobilized enzyme on controlled pore glass (heterogeneous media).
SIA, Pinto et al. 2012	β-Glucosidase	Kinetic study of enzyme activity in sodium dodecylsulfate (SDS)/ionic liquid (IL) mixed micelles to investigate the influence of these potential reaction media for transglycosylation reactions.
SIA-LOV, Pereira et al. 2021	Cyclooxygenase	Automatic evaluation of cyclooxygenase 2 inhibition induced by metal-based anticancer compounds.
Microfluidics, Li et al. 2021	Horseradish peroxidase	Vibrating sharp-tip mixing of multiple streams of fluids for kinetic study, claimed to be the highest performance mixer to date in 3D printed micro-devices.
Microfluidics, Thomsen and Nidetzky 2008	thermophilic β-glycoside hydrolase	Comparison of kinetic parameters and activation energy of the free and immobilized enzymes. Microreactor is made of poly(methyl methacrylate) and poly(dimethyl-siloxane) (PMMA/PDMS). Enzyme immobilization is via cross-linking with glutardialdehyde onto the amino-silanized microstructured surface.
Droplets microfluidics, Hassan et al. 2016	Glucose oxidase	System is composed of PDMS microchannel with T-junction for droplets generation, and 3D printed flow cell for optical detection. Kinetic parameters were obtained by probing droplets at multiple points over time, with an example application of continuous glucose measurement.

Table 2 contd. ...

...Table 2 contd.

Flow based system, References	Enzyme	Application remarks
Droplets microfluidics, Ochoa et al. 2020	*P. aeruginosa* aryl sulfatase	Fabrication of PDMS microfluidic system using soft lithography method with fluorescent image analysis detection. Screening of potential inhibitor compounds by incorporating a potential inhibitor together with the enzyme and substrate inside the droplets generated at T-junction.
Functional paper-based microfluidics Kim 2016	Glucose oxidase and Horseradish peroxidase	Development of a computational model, depicting flow behavior in two different connected wicking materials, paper and hydrogel. The reaction kinetics of free enzymes and those entrapped in the polyacrylamide gel were compared.
Electrochemical paper based microfluidics Adkins et al. 2016	β-galactosidase	Microwire electrodes were embedded within the analysis channel between two paper layers. Measured enzyme kinetics were compared to those found in the literature.

Future trends

Microfluidics have progressed toward nanofluidic systems (van den Berg et al. 2010, Chen et al, 2021). This lower volume manipulation would require a precise reagent delivery system, and a sensitive detection method. Reduction of cost by using cheaper materials that enable the possibility of mass production of the microfluidic/ nanofluidic systems are also of interest. For portable and field analyses, reducing the number of fresh reagents would be of a great benefit. Therefore, long-term and stable storage of the device with pre-immobilized reagents is necessary, and reusability would be a plus. In many cases, the design of a device that can detect multiple analytes is also desirable. Finally, bio-compatible devices that would not cause after-use negative impact on health and environment would be welcome.

References

Adkins, J.A., E. Noviana and C.S. Henry. 2016. Development of a quasi-steady flow electrochemical paper-based analytical device. Anal. Chem. 88: 10639–10647.

Agresti, J.J., E. Antipov, A.R. Abate, K. Ahn, A.C. Rowat, J.-C. Baret, M. Marquez, A.M. Klibanov, A.D. Griffiths and D.A. Weitz. 2010. Ultrahigh-throughput screening in drop-based microfluidics for directed evolution. Proc. Natl. Acad. Sci. USA 107: 4004–4009.

Asanomi, Y., H. Yamaguchi, M. Miyazaki and H. Maeda. 2011. Enzyme-immobilized microfluidic process reactors. Molecules 16: 6041–6059.

Atkins, P. and J. de Paula. 2017. Homogeneous catalysis. pp. 295–297. *In*: Elements of Physical Chemistry. 7th edition. Oxford University Press. New York, USA.

Becker, M., U. Spohn and R. Ulbrich-Hofmann. 1997. Detection and characterization of phospholipase D by flow injection analysis. Anal. Biochem. 244: 55–61.

Beneyton, T., I.P.M. Wijaya, P. Postros, M. Najah, P. Leblond, A. Couvent, E. Mayot, A.D. Griffiths and A. Drevelle. 2016. High-throughput screening of filamentous fungi using nanoliter range droplet-based microfluidics. Sci. Reports 6: 27223.

Boyer, R. 2002. Chapter 6: Enzymes I, reactions, kinetics, and inhibition. pp. 137–1388. *In*: Concepts in Biochemistry. 2nd edition. John Wiley & Sons, Inc. New York. USA.

Bui, M.P.N., C.A. Li, K.N. Han, J. Choo, E.K. Lee and G.H. Seong. 2011. Enzyme kinetic measurements using a droplet-based microfluidic system with a concentration gradient. Anal. Chem. 83: 1603–1608.

Calabria, D., M.M. Calabretta, M. Zangheri, E. Marchegiani, I. Trozzi, M. Guardigli, E. Michelini, F. Di Nardo, L. Anfossi, C. Baggiani and M. Mirasoli. 2021. Recent advancements in enzyme-based lateral flow immunoassays. Sensors 21: 3358.

Caratelli, V., A. Ciampaglia, J. Guiducci, G. Sancesario, D. Moscone and F. Arduini. 2020. Precision medicine in Alzheimer's disease: An origami paper-based electrochemical device for cholinesterase inhibitors. Biosens. Bioelectron 165: 112411.

Carrell, C., A. Kava, M. Nguyen, R. Menger, Z. Munshi, Z. Call, M. Nussbaum and C. Henry. 2019. Beyond the lateral flow assay: A review of paper-based microfluidics. Microelectron. Eng. 206: 45–54.

Casavant, B.P., E. Berthier, A.B. Theberge, J. Berthier, S. Montanez-Sauri, L.L. Bischel, K. Brakke, C.J. Hedman, W. Bushman, N.P. Keller and D.J. Beebe. 2013. Suspended microfluidics. PNAS 110: 10111–10116.

Cedillo-Perez, V.V., V.A. Quintero-Lopez, U.R. Carrillo-Medrano and M.P. Canizares-Macias. 2017. β-Glucosidase activity by sequential injection analysis: Evaluation of the free and immobilized enzyme. Curr. Anal. Chem. 13: 128–136.

Chen, L., C. Yang, Y. Xiao, X. Yan, L. Hu, M. Eggersdorfer, D. Chen, D.A. Weitz and F. Ye. 2021. Millifluidics, microfluidics, and nanofluidics: Manipulating fluids at varying length scales. Materials Today Nano. 16: 100136.

DeMott, J. and S.K. Hartwell. 2020. An exploration of natural reagents for the ninhydrin reaction on a paper analytical device. Undergraduate Thesis. Chemistry Department, Xavier University, Cincinnati, Ohio, USA.

Dickinson, E. 1991. Chapter 9 Competitive adsorption and protein—surfactant interactions in oil-in-water emulsions. pp. 114–129. *In*: Microemulsions and Emulsions in Foods. ACS Symp. Ser. 448. American Chemical Society. USA.

Ding, Y., P.D. Howes and A.J. deMello. 2020. Recent advances in droplet microfluidics. Anal. Chem. 92: 132–149.

Ellerhorst, N.P. and S.K. Hartwell. 2019. Investigation of flow pattern in serpentine microchannel design. Undergraduate Thesis. Chemistry Department, Xavier University, Cincinnati, Ohio, USA.

Fisher, H.F. 2005. Transient-state kinetic approach to mechanisms of enzymatic catalysis. Acc. Chem. Res. 38: 157–166.

Garcia-Viloca, M., J. Gao, M. Karplus and D.G. Truhlar. 2004. How enzymes work: Analysis by modern rate theory and computer simulations. Science 303: 186–195.

Garstecki, P., M.J. Fuerstman, H.A. Stone and G.M. Whitesides. 2006. Formation of droplets and bubbles in a microfluidic T-junction—Scaling and mechanism of break-up. Lab Chip 6: 437–446.

Girault, M., T. Beneyton, D. Pekin, L. Buisson, S. Bichon, C. Charbonnier, Y. del Amo and J.-C. Baret. 2018. High-content screening of plankton alkaline phosphatase activity in microfluidics. Anal. Chem. 90: 4174–4181.

Gupta, A., H.S. Matharoo, D. Makkar and R. Kumar. 2014. Droplet formation via squeezing mechanism in a microfluidic flow-focusing device. Comput. Fluids 100: 218–226.

Hassan, S., A.M. Nightingalea and X. Niu. 2016. Continuous measurement of enzymatic kinetics in droplet flow for point-of-care monitoring. Analyst 141: 3266–3273.

Hartwell, S.K. and K. Grudpan. 2012. Flow-based systems for rapid and high precision enzyme kinetics studies. J. Anal. Methods Chem. Article ID 450716, 10 pages. doi:10.1155/2012/450716.

He, P., G. Greenway and S.J. Haswell. 2010. Development of enzyme immobilized monolith micro-reactors integrated with microfluidic electrochemical cell for the evaluation of enzyme kinetics. Microfluid. Nanofluid 8: 565–573.

Honda, T., M. Miyazaki, H. Nakamura and H. Maeda. 2005. Immobilization of enzymes on a microchannel surface through cross-linking polymerization. Chem. Commun. 5062–5064.

Ilacas, G. and F.A. Gomez. 2019. Microfluidic paper-based analytical devices (μPADs): Miniaturization and enzyme storage studies. Anal. Sci. 35: 379–384.

Jebrail, M.J. and A.R. Wheeler. 2010. Let's get digital: Digitizing chemical biology with microfluidics. Curr. Opin. Chem. Biol. 14: 574–581.

Jeon, N.L., S.K.W. Dertinger, D.T. Chiu, I.S. Choi, A.D. Stroock and G.M. Whitesides. 2000. Generation of solution and surface gradients using microfluidic systems. Langmuir 16: 8311–8316.

Joensson, H.N. and H.A. Svahn. 2012. Droplet microfluidics—A tool for single-cell analysis. Angew. Chem. Int. Ed. 51: 12176–12192.

Khongpet, W., P. Yanu, S. Pencharee, C. Puangpila, S.K. Hartwell, S. Lapanantnoppakhun, Y. Yodthongdee, A. Paukpol and J. Jakmunee. 2020. A compact multi-parameter detection system based on hydrodynamic sequential injection for sensitive determination of phosphate, nitrite, and nitrate in water samples. Anal. Methods 12: 855–864.

Khongpet, W., S. Pencharee, C. Puangpila, S.K. Hartwell, S. Lapanantnoppakhun and J. Jakmunee. 2018. Exploiting an automated microfluidic hydrodynamic sequential injection system for determination of phosphate. Talanta 177: 77–85.

Kim, S. 2016. Programming of wicking behavior of hydrogel and paper-based microfluidic device. M.S. Thesis. Bioengineering, University of Illinois at Urbana-Champaign, Illinois, USA.

Koshland, D.E. 1958. Application of a theory of enzyme specificity to protein synthesis. PNAS USA 44: 98–104.

Kracke-Helm, H.A., L. Brandes, B. Hitzmann, U. Rinas and K. Schiigerl. 1991. On-line determination of intracellular P-galactosidase activity in recombinant *Escherichia coli* using flow injection analysis (FIA). J. Biotechnol. 20: 95–104.

Kuddas, M. 2019. Chapter 1 Introduction to food enzymes. pp.1–18. *In*: Enzymes in Food Biotechnology: Production, Applications, and Future prospects. Elsevier Inc. New York. USA.

Li, F., J. Liu, L. Guo, J. Wang, K. Zhang, J. He and H. Cui. 2019. High-resolution temporally resolved chemiluminescence based on double layered 3D microfluidic paper-based device for multiplexed analysis. Biosens. Bioelectron 141: 111472.

Li, X., Z. He, C. Li and P. Li. 2021. One-step enzyme kinetics measurement in 3D printed microfluidics devices based on a high-performance single vibrating sharp-tip mixer. Anal. Chim. Acta 1172: 338677.

Lilly, M.D., W.E. Hornby and E.M. Crook. 1966. The kinetics of carboxymethylcellulose-ficin in packed beds. Biochem. J. 100: 718–723.

Liu, C.H., I.C. Noxon, L.E. Cuellar, A.L. Thraen, C.E. Immoos, A.W. Martinez and P.J. Costanzo. 2017. Characterization of reagent pencils for deposition of reagents onto paper-based microfluidic devices. Micromachines 8: 242.

Ma, F., M.T. Chung, Y. Yao, R. Nidetz, L.M. Lee, A.P. Liu, Y. Feng, K. Kurabayashi and G.Y. Yang. 2018. Efficient molecular evolution to generate enantioselective enzymes using a dual-channel microfluidic droplet screening platform. Nat. Commun. 9: 1030.

Mahmud, A., E.J.M. Blondeel, M. Kaddoura and B.D. MacDonald. 2016. Creating compact and microscale features in paper-based devices by laser cutting. Analyst 141: 6449–6454.

Mitchell, H.T., I.C. Noxon, C.A. Chaplan, S.J. Carlton, C.H. Liu, K.A. Ganaja, N.W. Martinez, C.E. Immoos, P.J. Costanzo and A.W. Martinez. 2015. Reagent pencils: A new technique for solvent-free deposition of reagents onto paper-based microfluidic devices. Lab Chip 15: 2213–2220.

Miyazaki, M., T. Honda, H. Yamaguchi, M.P.P. Briones and H. Maeda. 2008. Enzymatic processing in microfluidic reactors. Biotechnol. Genet. Eng. Rev. 25: 405–42.

Nishat, S., A.T. Jafry, A.W. Martinez and F.R. Awan. 2021. Paper-based microfluidics: Simplified fabrication and assay method. Sens. Actuators B: Chem. 336: 129681.

Nooranidoost, M. and R. Kuma. 2019. Geometry effects of axisymmetric flow-focusing microchannels for single cell encapsulation. Materials 12: 2811.

Nooranidoost, M., D. Izbassarov and M. Muradoglu. 2016. Droplet formation in a flow focusing configuration: Effects of viscoelasticity. Phys. Fluids 28: 123102.

Noviana, E., T. Ozer, C.S. Carrell, J.S. Link, C. McMahon, I. Jang and C.S. Henry. 2021. Microfluidic paper-based analytical devices: From design to applications. Chem. Rev. 121: 11835–11885.

Nzobigeza, W. and S.K. Hartwell. 2019. Paper-based assay of copper ion using egg white as natural reagent. Undergraduate Thesis. Chemistry Department, Xavier University, Cincinnati, Ohio, USA.

Ochoa, A., F. Trejo and L.F. Olguín. 2020. Chapter 14 Droplet-based microfluidics methods for detecting enzyme inhibitors. pp. 209–232. *In*: N.E. Labrou (ed.). Targeting Enzymes for Pharmaceutical Development Methods and Protocol. Humana Press. New York, USA.

Osborn, J.L., B. Lutz, E. Fu, P. Kauffman, D.Y. Stevens and P. Yager. 2010. Microfluidics without pumps: Reinventing the T-sensor and H-filter in paper networks. Lab Chip 10: 2659–2665.

Patton, C.J. and S.R. Crouch. 1986. Experimental comparison of flow-injection analysis and air-segmented continuous flow analysis. Anal. Chim. Acta 179: 189–201.

Pereira, S.A.P., F.D. Bobbink, P.J. Dyson and M.L.M.F.S. Saraiva. 2021. Automatic evaluation of cyclooxygenase 2 inhibition induced by metal-based anticancer compounds. J. Inorg. Biochem. 218: 111399.

Pereira, S.A.P., P.J. Dyson and M.L.M.F.S. Saraiva. 2020. Miniaturized technologies for high-throughput drug screening enzymatic assays and diagnostics: A review. Trends in Anal. Chem. 126: 115862.

Pinto, P.C.A.G., S.P.F. Costa, J.L.F.C. Lima and M.L.M.F.S. Saraiva. 2012. β-Galactosidase activity in mixed micelles of imidazolium ionic liquids and sodium dodecylsulfate: A sequential injection kinetic study. Talanta 96: 26–33.

Razak, N.N.A., S. Firmansyah and M.S.M. Annuar. 2020. Effects of microfluidization on kinetic parameter values of lipase hydrolysis reaction. Biocatal. Agric. Biotechnol. 27: 101660.

Rezk, A.R., A. Qi, J.R. Friend, W.H. Lib and L.Y. Yeo. 2012. Uniform mixing in paper-based microfluidic systems using surface acoustic waves. Lab Chip 12: 773–779.

Rho, H.S., A.T. Hanke, M. Ottens and H. Gardeniers. 2016. Mapping of enzyme kinetics on a microfluidic device. PLOS ONE, 1–14.

Roach, L.S., H. Song and R.F. Ismagilov. 2005. Controlling nonspecific protein adsorption in a plug-based microfluidic system by controlling interfacial chemistry using fluorous-phase surfactants. Anal. Chem. 77: 785–796.

Rosenfeld, L., T. Lin, R. Derda and S.K.Y. Tang. 2014. Review and analysis of performance metrics of droplet microfluidics systems. Microfluid. Nanofluid 16: 921–939.

Růžička, J. and E.H. Hansen. 1975. Flow injection analyses: Part I. A new concept of fast continuous flow analysis. Anal. Chim. Acta 78: 145–157.

Schilling, E. 2001. Basic microfluidic concepts. https://faculty.washington.edu/yagerp/microfluidicstutorial/basicconcepts/basicconcepts.htm. Accessed April 27, 2022.

Scott, J.E. and K.P. Williams. 2012. Validating identity, mass purity and enzymatic purity of enzyme preparations. pp. 81–92. *In*: S. Markossian (ed.). Assay Guidance Manual (updated 2021). Eli Lilly & Company and the National Center for Advancing Translational Sciences. USA.

Sedighi, A. and U.J. Krull. 2018. Enzymatic amplification of oligonucleotides in paper substrates. Talanta 186: 568–575.

Seong, G.H., J. Heo and R.M. Crooks. 2003. Measurement of enzyme kinetics using a continuous-flow microfluidic system. Anal. Chem. 75: 3161–3167.

Siegesmund, E. and S.K. Hartwell. 2018. Microfluidics: Design and assembly. Undergraduate Thesis. Chemistry Department, Xavier University, Cincinnati, Ohio, USA.

Silvestre, C., P. Pinto, M.A. Segundo, M.L. Saraiva and J. Lima. 2011. Enzyme based assays in a sequential injection format: A review. Anal. Chim. Acta 689: 160–177.

Sohrabi, S., N. Kassir and M.K. Moraveji. 2020. Droplet microfluidics: Fundamentals and its advanced applications. RSC Adv. 10: 27560–27574.

Song, H. and R.F. Ismagilov. 2003. Millisecond kinetics on a microfluidic chip using nanoliters of reagents. J. Am. Chrm. Soc. 125: 14613–14619.

Song, H., D.L. Chen and R.F. Ismagilov. 2006. Reactions in droplets in microfluidic channels. Angew. Chem. Int. Ed. 45: 7336–7356.

Steegmans, M.L.J., C.G.P.H. Schroën and R.M. Boom. 2009b. Generalised insights in droplet formation at T-junctions through statistical analysis. Chem. Eng. Sci. 64: 3042–3050.

Steegmans, M.L.J., J. De Ruiter, K.G.P.H. Schroën and R.M. Boom. 2010. A descriptive force-balance model for droplet formation at microfluidic Y-junctions. AIChE. J. 56: 2641–2649.

Steegmans, M.L.J., K.G.P.H. Schroën and R.M. Boom. 2009a. Characterization of emulsification at flat microchannel Y junctions. Langmuir 25: 3396–3401.

Strelow, J., W. Dewe, P.W. Iversen, H.B. Brooks, J.A. Radding, J. McGee and J. Weidner. 2012. Mechanism of action assays for enzymes. pp. 105–126. *In*: S. Markossian (ed.). Assay Guidance Manual (updated 2021). Eli Lilly & Company and the National Center for Advancing Translational Sciences. USA.

Thomsen, M.S. and B. Nidetzky. 2008. Microfluidic reactor for continuous flow biotransformations with immobilized enzymes: The example of lactose hydrolysis by a hyperthermophilic β-glycoside hydrolase. Eng. Life Sci. 8: 40–48.

Tkachenko, E., E. Gutierrez, M.H. Ginsberg and A. Groisman. 2009. An easy to assemble microfluidic perfusion device with a magnetic clamp. Lab Chip 9: 1085–1095.

Tsao, C. and D.L. DeVoe. 2009. Bonding of thermoplastic polymer microfluidics. Microfluid. Nanofluid. 6: 1–16.

Urban, P.L., D.M. Goodall and N.C. Bruce. 2006. Enzymatic microreactors in chemical analysis and kinetic studies. Biotechnol. Adv. 24: 42–57.

Ushikubo, F.Y., F.S. Birribilli, D.R.B. Oliveira and R.L. Cunha. 2014. Y- and T-junction microfluidic devices: Effect of fluids and interface properties and operating conditions. Microfluid. Nanofluid 17: 711–720.

Vallejo, D., A. Nikoomanzar, B.M. Paegel and J.C. Chaput. 2019. Fluorescence-activated droplet sorting for single-cell directed evolution. ACS Synth. Biol. 8: 1430–1440.

van den Berg, A., H.G. Craighead and P. Yang. 2010. From microfluidic applications to nanofluidic phenomena. Chem. Soc. Rev. 39: 899–900.

van Staden, J.F. and S.I. Raluca. 2002. New horizons in sequential injection kinetic analysis. Anal. Bioanal. Chem. 374: 3–12.

Vasella, A., G.J. Davies and M. Böhm. 2002. Glycosidase mechanisms. Curr. Opin. Chem. Biol. 6: 619–629.

Webster, A. 1983. Intrinsic, inherent, and observed kinetic data with immobilized enzymes: The concept of rotational masking. Biotechnol. Bioeng. 25: 2479–2484.

Wehking, J.D., M. Gabany, L. Chew and R. Kumar. 2014. Effects of viscosity, interfacial tension, and flow geometry on droplet formation in a microfluidic T-junction. Microfluid. Nanofluid 16: 441–453.

Whitford, E., W. Nzobigeza and S.K. Hartwell. 2020. Paper based assay of copper (II) ion using egg white as a natural chromogenic reagent. Anal. Lett. 53: 2465–2480.

Yerian, T.D., G.D. Christian and J. Ruzicka. 1988. Flow injection analysis as a diagnostic tool for development and testing of a penicillin sensor. Anal. Chem. 60: 1250–1256.

Yin, H. and D. Marshall. 2012. Microfluidics for single cell analysis. Curr. Opin. Biotechol. 23: 110–119.

Yu, W., X. Liu, Y. Zhao and Y. Chen. 2019. Droplet generation hydrodynamics in the microfluidic cross-junction with different junction angles. Chem. Eng. Sci. 203: 259–284.

Zheng, B. and R.F. Ismagilov. 2005. A microfluidic approach for screening submicroliter volumes against multiple reagents by using preformed arrays of nanoliter plugs in a three-phase liquid/liquid/gas flow. Angew. Chem. Int. Ed. Engl. 44: 2520–2523.

Chapter 2

Chemoproteomics

An Extremely Powerful Kit in Drug Discovery Toolbox

Anupama Binoy,[1,#] *Revathy Sahadevan*[1,#] and *Sushabhan Sadhukhan*[1,2,*]

Introduction

The human genome project (HGP) has revolutionized the scientific world by providing a blueprint of genes encoding the entire human proteome and one cannot overstate its importance for modern pharmacology. This allowed us to understand the information present on approximately 20,000 protein-coding genes (Legrain et al. 2011). However, there is a huge knowledge gap on various functional and redundant proteins in the human proteome, their cellular localization, and distribution, their abundance as well as their interaction with small molecules. Additionally, the existence of splice variants and the post-translational modifications (PTMs) make the human proteome more complicated than it seems. In 2009, the human proteome project was launched to prepare a human proteome map with a gene-centric approach (A Gene-Centric Human Proteome Project: HUPO--the Human Proteome Organization, 2010). The intention was to generate publicly accessible information resources for further exploration by the scientific community. The word 'proteome' was coined quite early by Mark Wilkins in 1994 (Wilkins and Gooley 1997) describing it as the

---

[1] Department of Chemistry, Physical & Chemical Biology Laboratory, Indian Institute of Technology Palakkad, Kerala 678 623, India.

[2] Department of Biological Sciences & Engineering, Indian Institute of Technology Palakkad, Kerala 678 623, India.

Emails: anupamabinoy@iitpkd.ac.in; 202004003@smail.iitpkd.ac.in

* Corresponding authors: sushabhan@iitpkd.ac.in

# Authors contributed equally.

study of a whole set of proteins present in a whole organism, or a cell or a tissue, and also encompassing the modifications that happen on the proteins due to several signaling and environmental factors. Over the years, proteomics has played vital roles in drug discovery processes, particularly in understanding the mechanism of action of a given molecule of interest such as drugs or inhibitors. It has enabled identification of biomarkers specific to the disease which is useful in early detection, and prognosis as well as in monitoring the progression of several diseases (Aslam et al. 2017).

However, along with the launch of the human proteome project, the limelight has been shifted towards the emergence of chemoproteomics and probe-based tools for identifying the global interacting proteome set in an organism. This also brings in a tremendous opportunity for the scientific world for the discovery of new and more target-specific approaches to drug discovery. Chemoproteomics presents a robust platform for the discovery of new first-in-class medicine. It is a rapidly growing field in chemical biology with an arsenal of techniques to understand protein-small molecule interactions. It has paved the way for a proteome-wide evaluation of identifying the targets by characterizing interactions between small molecules and their protein targets (Spradlin et al. 2021). This technology includes a growing set of techniques like activity-based protein profiling (ABPP) (Wang et al. 2018) and targeted protein degradation using proteolysis-targeting chimeras (PROTACs) (Belcher et al. 2021, Qi et al. 2021).

Chemoproteomics provides a strong foundation for studying the interaction between small molecules and proteins, often with the help of some chemical probes that are synthesized based on the structure of the molecules of interest. These probes are also known as bioorthogonal probes as they function without interfering with the native physiology of the living system when used in the context of *in vivo* conditions. They are bolstered with a functional group to label or pulldown substrate proteins it interacts with which is most of the time the proteins of interest. They enable probing through selective covalent bond formations with the proteins (often in the active site). As some of the classic examples, chemoproteomics has been successfully applied to elucidate the binding partners of Class I and II HDACs in the native cellular state (Salisbury and Cravatt 2007), identify *in situ* molecular targets of polyphenols (Wang et al. 2016), study various protein PTMs (Erkan et al. 2020), etc. Some of these are discussed in detail in the respective sections below. Identifying the protein targets through chemoproteomics also has its limitations as sometimes the structural modification of the target leads to perturbation in molecular interactions and biological functions (Piazza et al. 2020). Here, in this chapter, we will focus on chemical reactions for the development of chemoproteomics tools and the applications of the same in the field of biotechnology; in particular the drug discovery and identification of disease-associated biomarkers.

Structural elements in the probes used in chemoproteomics

Chemoproteomics use small molecule probes to profile proteins/enzymes based on their activity within the native cellular environment. These probes are often referred to as activity-based probes (ABP) (Berger et al. 2004) and are designed to be bioorthogonal in nature, i.e., they should act without interfering with the native physiology of live cells (Chan et al. 2021). It is generally comprised of three components, (i) a warhead or reactive group, (ii) a reporter tag, and (iii) a linker connecting the first two, i.e., the warhead and the reporter tag (Fig. 1). A reactive group (also known as warhead) is designed to enable the covalent interaction between the small molecule and its target protein. It can be the entire molecule of interest itself or a part of it with the desired functional group which can selectively interact with amino acids in the active site of enzymes. Several reactive functional groups such as sulphonate ester (Adam et al. 2001, Adam et al. 2002a), epoxide (Greenbaum et al. 2000, Chen et al. 2003, Marques et al. 2017), fluorophosphonate (Liu et al. 1999, Field et al. 2020, Faucher et al. 2020), peptide (acyloxy)methyl ketone (Sexton et al. 2006, van de Plassche et al. 2020), etc., have been widely used to design and synthesize reactive warheads. A reporter tag is used to visualize, pulldown, or enrich the target protein for identification and further validation studies. Fluorescent or radiolabeled tags are generally used for visualization, and biotin tags are used for the purification and mass spectrometric identification of target proteins. A linker group is introduced to connect the reporter tag with the reactive warhead. They are designed to control the steric hindrances and specificity of the attached groups (Meng et al. 2017). Linkers can be long alkyl chains, or poly ethylene glycol (PEG), or short peptides. To study the small molecule-protein non-covalent interactions, photo-cross linkers such as diazirine, benzophenone groups, etc., have been introduced to the linker region of the probe (Guo and Li 2017). Upon photo-irradiation, these photo-crosslinkers yield reactive carbenes which are capable of capturing non-covalently interacting proteins or adjacent proteins that reside near the active site of the target protein. This in turn helps in the identification of the mode of action of the small molecules by elucidating if the small molecule is working through any off-targets. There exists another type of probe which has been designed by introducing a terminal alkyne or azide functional group to the molecule of interest. These kinds of probes exert the least steric hindrance in terms of their interaction with bio-macromolecules. The reporter tags can be introduced in the downstream process *via* alkyne-azide Click reactions with an alkyne/azide derivative of fluorescent counterpart or biotin (Meldal et al. 2008).

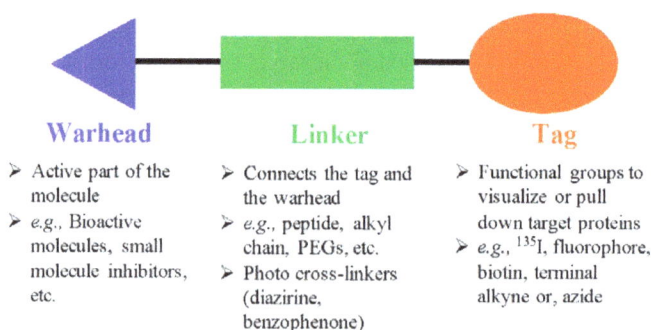

Warhead
- Active part of the molecule
- *e.g.,* Bioactive molecules, small molecule inhibitors, etc.

Linker
- Connects the tag and the warhead
- *e.g.,* peptide, alkyl chain, PEGs, etc.
- Photo cross-linkers (diazirine, benzophenone)

Tag
- Functional groups to visualize or pull down target proteins
- *e.g.,* ^{135}I, fluorophore, biotin, terminal alkyne or, azide

Figure 1. Structural components of bioorthogonal probe routinely used in chemoproteomics.

Importance of chemoproteomics in drug discovery

The discovery of new in-class active drugs for various existing and emerging diseases has always been a challenging task due to the lack of sufficient knowledge on the biological functions of protein(s) involved. The expression and function of proteins inside the body are dependent on multiple factors including but not limited to post-transcriptional processes, PTMs, cellular localization, signaling factors, protein turnover, environmental cues, etc. The major problem encountered by the scientific community is in determining the specific molecular and cellular functions of many proteins encoded by the genome of an organism and that has posed a severe limitation in the discovery of new medicines. Advancements in biochemical, molecular, and cellular techniques have enabled the functional understanding of certain proteins through *in vitro* experiments. However, the proteins will not function quite the same in isolation as they do under physiological conditions. Further, in the era of genetic methods dominion over pharmacology, genetic disruption of given protein resulted in its altered expression for the rest of the organism's life, and that held a major limitation in its application in pharmaceutical sectors. The development of high-throughput screening technologies has provided a platform to study the interaction of proteins with millions of small molecules. Despite this, these advancements were unable to bring the expected momentum in the discovery of new first-in-class medicines. The screening of drugs is time-consuming, expensive, and encompasses a lot of risk factors. The probability of a drug molecule entering the Phase 1 clinical trials after a lot of strenuous preclinical testing is only around 10% (Dowden and Munro 2019). The major reason for this failure lies in the limited drug efficacy, on and off-targets (specific and non-specific targets), heterogeneity among various diseases, and its impact on individuals per se. In addition, identifying the undruggable proteome has remained the bottleneck in modern drug discovery as there is a lack of knowledge about well-defined binding pockets for a large fraction of the proteome and their interactions with small molecules (Dang et al. 2017). The aforementioned issues could be resolved largely, if not fully, by exploiting chemoproteomics wherein various probe molecules (of the ligand of interest) can be used in conjunction with advanced mass spectrometry-based proteomics. The active site-directed chemoproteomics strategy leverages the covalent attack mechanism used by most of the active sites of enzymes to react with its substrate. The chemoproteomics tools are synthesized by tagging a suicidal substrate which serves both as an inhibitor for the respective enzyme as well as ABPs (Activity-based Probes). In addition to site-directed proteome profiling, multiple enzyme classes can be targeted concurrently. Chemoproteomics enables the identification of protein target and mechanism of action of a small molecule simultaneously, unlike the typical way where the efficacy of small molecules is tested first and their targets and mode of actions are studied later.

Chemoproteomics to identify molecular targets of small molecules

A complex cellular system often makes it difficult to decipher the mechanism of action of small molecules, their interaction with protein targets, and overall pharmacological effects. This is actually the most critical step in the drug discovery process, and chemoproteomics encompasses versatile tools for the identification of molecular targets that are the key players behind the activity of a given small molecule. The added advantage is that chemoproteomics does this in a very initial stage of the drug development, which certainly helps in making the drug discovery process more cost-effective. Identification of the therapeutic targets uses chemical probes whose mechanism can be well explained through affinity and activity-based chemoproteomics, and also with the recently developed label-free technologies (Drewes and Knapp 2018).

The mechanism of action of a small molecule depends upon their interaction (both covalent or non-covalent) with intracellular protein targets and this lays the basis for the discovery of new drug products with effective pharmacological activity. Therefore, target identification becomes a critical step in the development of new therapeutics (Schenone et al. 2013, Comess et al. 2018). However, numerous studies have time and again stated that most drugs interact with multiple proteins which makes the identification of primary targets substantially more difficult. In the last two decades, chemoproteomics has unraveled methods that can comprehensively reveal multiple targets of small molecules. Several other methods also exist such as gene expression signatures to connect the compound with its targets, chemical genomics approaches, and yeast two-hybrid methods (Caligiuri et al. 2006, Hillenmeyer et al. 2008, Lamb et al. 2006, Zon and Peterson 2005). However, these strategies showed multiple interferences, off-targets, and had narrow applicability.

Chemoproteomics recruits interdisciplinary approaches including chemical synthesis, cellular and protein biology for comprehensively fishing out multiple targets of small molecules and their enrichment, followed by mass spectrometric analysis to identify the proteins of interest (Beroza et al. 2002). Drug discovery process mostly uses chemoproteomics in two different ways: (i) compound centric chemical proteomics (CCCP) which originates from the classical drug affinity chromatography (where the molecule is immobilized on a matrix) followed by proteomics analysis, and (ii) activity-based protein profiling where the parent molecule of interest is modified into an activity-based probe with special attention to its ability to enrich the protein targets by binding (Rix and Superti-Furga 2009). In both methods, it is very important to retain the biological activity of the small molecules of interest throughout the process. Here in this chapter, we will shed light on routinely used chemoproteomics strategies, particularly in drug discovery processes with some classical examples.

1. Immobilization-based probe-driven chemoproteomics:

This approach is based on the immobilization of the molecule of interest on inert and biocompatible support resins such as agarose or other macroscopic beads.

The enrichment of the target proteins is done through affinity chromatography. Here, the compound is immobilized on solid support through a linker arm, and the matrix is used to fish out the proteins interacting with the molecule of interest from the cellular or tissue extract (Fig. 3). The recent advances in mass spectrometric techniques in terms of sensitivity and high throughput nature have further enhanced the power of immobilization-based probes for the identification of unrecognized potential targets of small molecules (Godl et al. 2003). Immobilized affinity probes were mostly utilized in identifying the specific cellular targets for kinase inhibitors. Immobilization of an analog of SB203580 (Godl et al. 2003), a mitogen-activated protein kinase p38 inhibitor on chromatographic beads led to the identification of several unknown high-affinity targets such as cyclin G-associated kinase (GAK) and casein kinase 1 (CK1) as well as RICK (Rip-like interacting caspase-like apoptosis-regulatory protein kinase/Rip2/CARDIAK) apart from its known targets. SU6668, an indolinone compound, which was initially known to selectively inhibit the receptor tyrosine kinase involved in tumor angiogenesis was subjected to immobilization-based chemoproteomics for the affinity capture of other potential cellular targets. In that study, the researcher could identify previously unknown targets of SU6668 such as Aurora kinases and TANK-binding kinase 1 (Godl et al. 2005). Graves et al. (2002), conducted a study to explain the mechanistic role of quinolines such as 4-aminoquinoline chloroquine and the quinolinemethanol mefloquine in the treatment of malaria. They screened several quinoline drugs against the purine binding proteome based on their structural similarity by exploiting displacement affinity chromatography using target capturing through gamma-phosphate-linked ATP-Sepharose. Aldehyde dehydrogenase 1 (ALDH1) and quinone reductase 2 (QR2) were the only two human proteins specific to human red blood cell purine binding proteome which were identified as potential targets of these quinoline drugs (Graves et al. 2002). In another study, Schreiber's group constructed a FK-506 affinity matrix using its amino derivative to identify its target protein, FK-506-binding protein (FKBP) with a cis-trans peptidyl-prolyl isomerase activity (Harding et al. 1989). It was also observed that FK-506 (a macrolide) and cyclosporine A (a cyclic undecapeptide), despite both serving as immunosuppressants by inhibiting T-cell activation, had very different molecular targets. Cyclosporine A targeted cyclosporin A-binding protein, and cyclophilin whereas FK-506 targeted FK-506-binding protein (FKBP) (Siekierka et al. 1989a, Siekierka et al. 1989b). Despite its significant contribution to the drug discovery process, it is important to note that this method is applicable to cellular/tissue extracts and not directly to the native cellular environment and that itself represents a major flaw of this strategy. Another major limitation of immobilization-based chemical proteomics lies in the steric hindrance which might influence the interaction between the true protein targets and the interacting small molecule.

2. Activity-based probe chemoproteomics:

To overcome the steric interference caused by immobilization-based chemoproteomics, activity-based probes were conceptualized. Activity-based protein profiling (ABPP) is based on the covalent interaction between small molecules (such as inhibitors, ligands, drugs, etc.) and the reactive amino acid side chains present in the active site of the enzymes; unlike other techniques which mainly focus on non-covalent interactions. ABPP is mostly utilized under physiological conditions and it potentially reflects true drug-target interactions taking place inside the native cellular environment. In principle, the activity-based probes should not differ significantly from the parent molecule in terms of its biological activities. As discussed in above, there are two important components in an activity-based probe: (i) the reactive group (in the compound of interest) for binding or covalently modifying the active site of the target proteins and (ii) a reporter tag for the detection and enrichment of the target proteins. Often, a linker is placed in between the molecule of interest and the reporter group to avoid any possible steric hindrance. The probes are incubated with the live cells and allowed to physiologically interact with the cellular components before they are lysed, followed by an affinity enrichment of the target protein using the reporter group. Biotin is the most frequently used reporter owing to its strong affinity towards avidin, where enrichment is performed using either avidin or streptavidin (an avidin analog with a more convenient binding constant with biotin) beads. While, in the case of fluorescent reporters, efficient and rapid detection, as well as visualization of proteins, can be done through an in-gel fluorescence experiment (Fig. 3). ABPP probes have been utilized to study different enzyme families like kinases (Yee et al. 2005, Liu et al. 2005), phosphatases (Kumar et al. 2004), oxidoreductases (Adam et al. 2001, Adam et al. 2002b, 2002a), glycosidases (Hekmat et al. 2005, Vocadlo and Bertozzi 2004), proteases (Liu et al. 1999, Kidd et al. 2001), etc. This strategy involved the covalent labeling of the active site of enzymes in an activity-based manner which helped to distinguish between functional enzymes from their inactive zymogens (Kidd et al. 2001, Liu et al. 1999, Adam et al. 2001).

ABPP has also been successfully applied in identifying protein targets of various natural products/drug molecules. For example, Yee et al. (2005) attached fluorescent labels (BODIPY and rhodamine) to Wortmannin enabling the identification of its binding partners from HEK 293T cell lysates. This led to the identification of PI-3 kinases and other associated proteins as the major targets of Wortmannin (Yee et al. 2005). With a similar strategy, Liu and co-workers identified other kinases such as Polo-like kinase 1 and 3 (Plk1and Plk3) as the cellular targets of Wortmannin (Liu et al. 2007). Klaic et al. (2012) developed an affinity probe of biotinylated celestrol to elucidate its potential targets as annexin II, β-tubulin, and eukaryotic elongation factor 1A (Klaic et al. 2012). Greenbaum et al. (2000), developed ABPP probes based on epoxysuccinyl peptide with different reporter groups such as the radionuclide 125I, biotin, and the BODIPY (fluorophore). They were able to validate cathepsin as the major target of the E-64, an inhibitor of cysteine protease (Greenbaum et al. 2000). Cravatt's group successfully profiled the global serine hydrolases in the complex proteome by synthesizing a biotinylated fluorophosphonate (FP-biotin) which showed affinity toward several members of the serine hydrolase family (Liu et al. 1999). They also synthesized a variant of FP-biotin by replacing the hydrophobic alkyl chain linker with a more hydrophilic polyethylene glycol (PEG) moiety to obtain FP-PEG-biotin. Both the probes showed a similar ability to fish out multiple members of serine hydrolases like serine peptidases, lipases, and esterases simultaneously (Kidd et al. 2001).

3. Click-able probes:

The application of biotinylated probes poses problems due to their large sizes which can interfere with the way they interact with the biomacromolecules as compared to the parent analogs of the probes. Advancements in ABPP technology alleviated steric hindrances associated with a bulky reporter like biotin or fluorescent tags by adapting a two-step strategy that took advantage of bioorthogonal reactions. Bioorthogonal reactions refer to the chemical reactions that can happen within the living system without interfering its native biological processes (Sletten and Bertozzi 2009, Prescher et al. 2004, Prescher and Bertozzi 2005). This method involves the functionalization of the molecule of interest with a small functional group (such as a terminal alkyne or azide) that does not interfere with the native physiological activities of a living system, which ultimately can yield the profiling of enzyme activities *in vitro* as well as *in vivo* systems (Fig. 3). As depicted in Fig. 2, the identification of target proteins can be achieved by the downstream addition of reporter tags through bioorthogonal reactions between the complements such as alkyne with an azide moiety (Vasilyeva et al. 2016, Zhou and Fahrni 2004), alkene with tetrazine (Wu and Devaraj 2018), or Staudinger ligation (Sletten and Bertozzi 2011, Bednarek et al. 2020).

Due to the minimal steric hindrance and bio-orthogonal nature of these probes, they pose the least interferences to the pharmacological activity of the drug

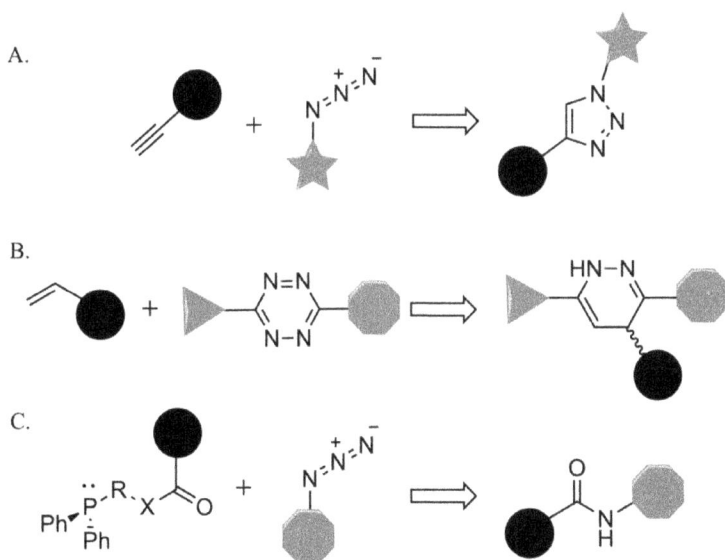

Figure 2. Bioorthogonal reactions used in chemoproteomics. (A) Cu(I)-catalyzed alkyne-azide Click reaction, (B) alkene-tetrazine reaction, and (C) Staudinger ligation.

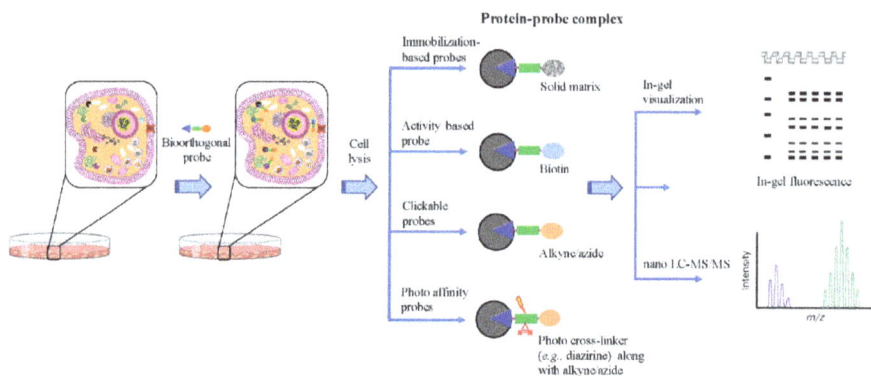

Figure 3. Illustrated representation of various chemoproteomics-based tools and their mode of action for identification and validation of protein targets.

molecules. Click chemistry is described as a group of chemical reactions with high selectivity and high yield along with its ability to be conducted under simple reaction conditions. Due to these unique features of Click chemistry, it has emerged as a highly recommended tool in biomedical applications, especially in the field of drug discovery (Hein et al. 2008). Several research groups have utilized Click chemistry-based probes for the identification of potential protein targets of various natural products (Gersch et al. 2012, Yang and Liu 2015, Ziegler et al. 2013, Wright and Sieber 2016). For example, Bottcher and Sieber (2010) developed a showdomycin probe by modifying its 5'-OH group with an alkyne moiety. The probe resulted in the detection of a series of essential enzymes which belongs to the oxidoreductase and transferase families in different bacterial systems as a target of showdomycin. The identification of the molecular targets MurA1 and MurA2, which are essential enzymes for bacterial cell-wall biosynthesis, was of particular interest in their study (Böttcher and Sieber 2010). Qin et al. (2020) synthesized itaconate–alkyne (ITalk) as a bioorthogonal probe for itaconate to quantify the site-specific interaction of itaconate in a cellular extract of inflammatory macrophages (Qin et al. 2020). Further to understand the functional role of itaconate in host-pathogen interaction, they synthesized a series of itaconate-based bioorthogonal probes which enabled quantitative and site-specific profiling of itaconate proteins in *Salmonella*. Through this study, they were able to demonstrate that itaconate acts as a covalent inhibitor for isocitrate lyase, a key enzyme involved in the glyoxylate cycle (Zhang et al. 2021). Similarly, the development of bioorthogonal azide or alkyne probes for studying the protein PTMs such as palmitoylation, and myristoylation has been well exploited in the last decade by several groups (Hannoush and Arenas-Ramirez 2009, Charron et al. 2009, Hang et al. 2007, Kostiuk et al. 2008, Martin and Cravatt 2009). An FDA-approved anti-obesity drug, tetrahydrolipstatin (Orlistat) with a potential antitumor ability was modified into a chemical probe by conjugating it with an alkyne moiety. This alkyne-bearing analog of Orlistat was able to identify eight new targets apart from the previously known fatty acid synthase enzyme (Yang et al. 2010).

4. *Photoaffinity labeling*

It is relatively easier to identify the proteins which covalently interact with the small molecules as the resulting complex is stable enough for downstream processing and mass-spectrometric analysis. On the contrary, there exists a significant number of biologically active molecules which interact non-covalently (*via* hydrophobic-hydrophobic, van der Walls interaction, ionic interaction, or hydrogen bonding, etc.) with the proteins and this results in very labile protein-small molecule complexes which are prone to degradation as soon as cells are lysed. In this kind of situation, photoaffinity labeling comes to the rescue. It is a unique and unbiased strategy that utilizes photoactivation of the probe that is bearing a photo-crosslinker. These groups remain inactive under normal chemical and biological conditions but get activated when they are irradiated at a specific wavelength to yield highly reactive transient species (e.g., carbenes) that crosslink the probe to its binding partners (Fig. 3). This is followed by a Click reaction which enables the downstream enrichment of captured proteins. Benzophenones, aryl azides, and alkyl or aryl diazirines are commonly used photo crosslinkers in this context (Mishra et al. 2020). Hulce et al. (2013) utilized photoaffinity probes for the global identification of cholesterol-binding proteins (Hulce et al. 2013). They synthesized a set of structurally similar probes made of sterol moiety connected to the photoactivatable diazirine group. HeLa cells were incubated with this probe and further photoactivated through exposure to 365 nm UV light to covalently cross-link the probe to its binding proteins. Later, Click reaction with the biotin azide followed by streptavidin enrichment led to the identification of approximately 250 cholesterol-binding proteins. Suberoylanilide hydroxamic acid (SAHA) was the first histone deacetylase (HDAC) inhibitor to get the FDA approval. Salisbury and Cravatt (2007) developed SAHA-BPyne as a probe for suberoylanilide hydroxamic acid (SAHA), the first FDA-approved HDAC inhibitor to identify the proteins that come together (non-covalently) to form the active complex for histone deacetylases (HDACs). SAHA was modified using benzophenone moiety (a photo-crosslinker) to covalently capture the HDAC-associated proteins like MTA2, CoREST, and methyl CpG binding protein 3 (MBD3) (Salisbury and Cravatt 2007).

Applications of chemoproteomics for small molecule library screening, biomarker discovery and *in vivo* imaging of enzymes

Chemoproteomics has also been explored to screen libraries of small molecule inhibitors in a high-throughput manner. Previously, we have discussed different probe-based identification of protein targets of small molecules where the probes are designed to target a specific class of enzymes *via* installing known functional groups and/or binding groups into their structures. These directed probes can covalently modify the enzyme active sites and target a large set of closely related enzymes with multiple isoforms (Cravatt et al. 2008, Nomura et al. 2010). It is widely used in drug screening to identify the selectivity of a particular target (Fig. 4A). On the other hand, in non-directed ABPP method, the action of an inhibitor can be monitored

A.

B.

C.

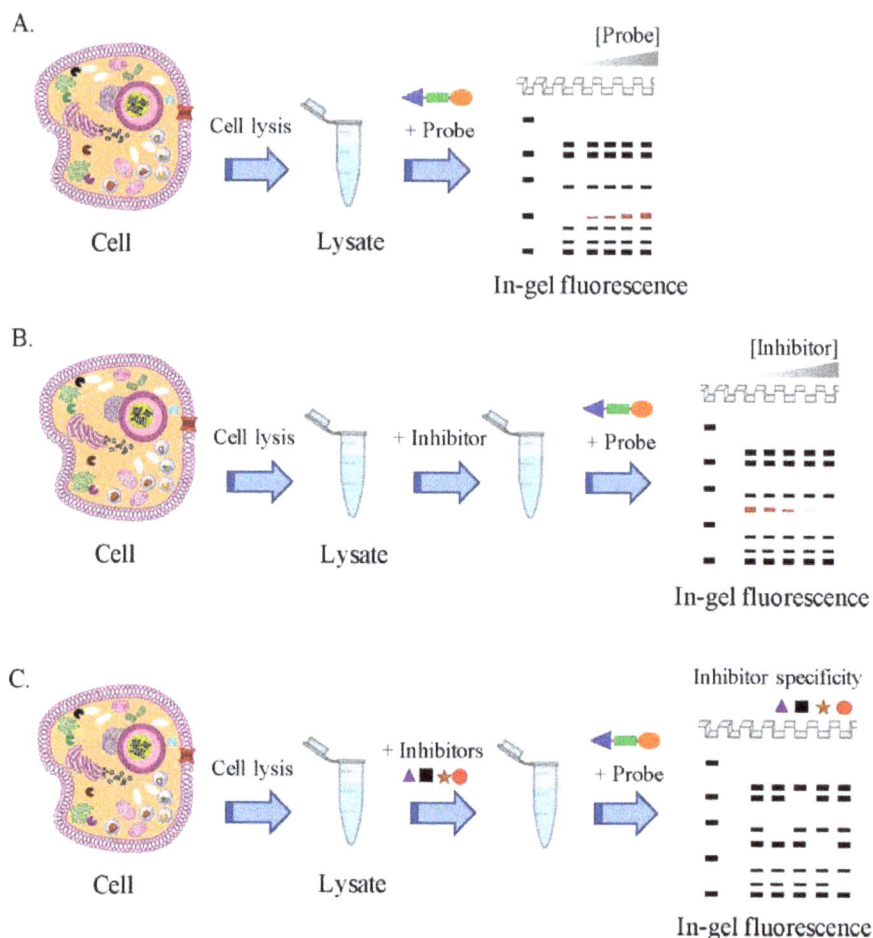

Figure 4. Schematic representation of application of ABPP in drug screening. (A) Direct labeling of proteins with probes to identify the targets, (B) Competitive labeling and (C) Screening of library of inhibitors for their specificity.

via chemoproteomics and identify the target proteins (Fig. 4B), or a combinatorial library of probes can also be synthesized and screened against many enzymes to identify specific protein labeling events happening in cell (Fig. 4C) (Saghatelian and Cravatt 2005). This method has been proven to be effective as an important tool in the discovery of both reversible and irreversible inhibitors.

1. Screening of small molecule library for drug discovery

The advantages of using chemoproteomics (using activity-based probes) in drug discovery is two-fold. One that it can evaluate multiple enzymes at once in their very native environment and second, it needs minimum structural information about the target proteins to do so. On a similar note, inhibitors can also be developed for the enzymes (orphan enzymes) that lack any information about their activities or substrates. Selective inhibitors can be readily identified from the non-specific inhibitors and this is particularly of immense importance for the ones that belong to a large family of enzymes having multiple isoforms such as kinases or hydrolases. Greenbaum et al. (2002) used this approach to identify the targets of cysteine proteases using a library of irreversible inhibitors containing primary tripeptide skeleton linked with an epoxide electrophile. The inhibitory action was visualized with radiolabeled active site-directed probe, 125I-DCG-04 treatment for proteases followed by SDS-PAGE and phosphorimaging. This method allowed them to identify multiple targets of each inhibitor in a single gel-based assay without any optimization of substrate and kinetic conditions (Greenbaum et al. 2002). A similar strategy was used by Leung et al. (2003) to screen the serine hydrolase inhibitors from a library of compounds containing electrophilic ketone group. The protein-inhibitor complex was visualized through competitive assay with fluorophosphonate rhodamine (FP-Rhodamine). They found that fatty acid amide hydrolase (FAAH), an endocannabinoid-degrading enzyme and triacylglycerol hydrolase are the most important targets of the serine hydrolase inhibitors (Leung et al. 2003). In another study, Bonavia et al. (2011) found that carbamoyl-phosphate synthetase 2, and aspartate transcarbamylase and dihydroorotase (CAD) complex are the targets of isoxazole- pyrazole and their proline derivatives that are effective against respiratory syncytial virus (Bonavia et al. 2011).

2. Identification of biomarkers for human diseases

The non-directed ABPP is also effective in identifying the biomarkers of various human diseases (Jones and Neubert 2017). Biomarkers are measurable and quantifiable molecular signatures of biological status that can serve as indicators of diseases or targets of drugs in clinical diagnosis and therapy. Biomarkers can be genes, proteins, or even small molecule metabolites. Among many tools to identify the protein biomarkers, chemoproteomics stands out with the aid of strategically designed molecular probes. These probes must be selective and sensitive to detect a particular protein from a complex biological system (Zhang et al. 2017). For example, a fluorescent fluorophosphonate chemical probe (FP-TAMRA) was used to identify the enzymes plasmin and kallikrein as biomarkers of calcium, potassium, and sodium homeostasis (Navarrete et al. 2013). In another study by Oikonomopoulou and co-workers, a specific serine hydrolase, kallikrein-related peptidase 6 (KLK6) was identified as a biomarker using a biotinylated probe containing a phosphonate ester warhead in the ascites fluid from ovarian cancer patients (Oikonomopoulou et al. 2008).

3. In vivo imaging of enzymes

In addition to PTMs and catalytic activities, the temporal expression and subcellular distribution of proteins also play an important role in the function of that proteins. Thus, visualizing the proteins in their native environment, i.e., their localization inside the cell or a whole organism is of utmost significance and can reveal important information about their function. Among the many applications of chemoproteomics, the most fascinating one is using these bioorthogonal probes for live imaging of molecules in living systems such as live cells or whole organisms. For these applications, the probe molecule requires certain properties such as it should be biocompatible, cell-permeable, should not involve any toxic metal catalysts and most importantly the labeling resulting from the protein-probe interaction should be rapid (Lukinavicius et al. 2014). In this context, Pan et al. (2017) developed a series of alkyne-diaziridine-containing probes, targeting bromo domain-containing protein (BRD4) that recognizes acetylated lysine residues located on histones. BRD4 is an important enzyme involved in many cellular processes including mitosis, angiogenesis, inflammation, etc. These probes were clicked *in vivo* with fluorophores which allowed the visualization of BRD4 located throughout the nucleus of HepG2 cells (Pan et al. 2017). In another study to visualize cytoskeleton proteins tubulin and actin, fluorogenic probes were developed namely SiR-tubulin and SiR-actin. These probes were able to bind with the microtubules more efficiently as compared to bovine serum albumin. The increased fluorescence intensities imparted by these probes when bound to microtubules further justified the finding. These probes were also efficient for live cell imaging (Lukinavičius et al. 2014). Sherratt et al. (2017) replaced methionine of the culture media with homopropargylglycine, harboring an alkyne functionality, which led them to perform bioorthogonal reactions with fluorescent azide to visualize live bacteria. This allowed rapid screening and identification of living pathogenic organisms (Sherratt et al. 2017).

Future of chemoproteomics and Conclusions

Preceding discussion clearly shows that chemoproteomics-based tools were mainly designed to target enzymes that have defined catalytic sites. The widely targeted enzyme superfamily belongs to kinases which hold special places in almost all signaling pathways. But if we carefully look into the molecular and biochemical pathways associated with diseases progression, the first set of protein family affected in pathogenic conditions are membrane proteins like receptor proteins, ion channels, etc. (Sigismund et al. 2018, Niemeyer et al. 2001). Therefore, we think that there should be a paradigm shift for the application of chemoproteomics from enzymes to other types of proteins such as ion channels, receptor, and transport proteins, chaperons, etc., which play an instrumental role in the pathogenesis of several dreadful diseases like cancers, viral infections like COVID 19, HIV, etc. Additionally, it is very important to consider all the physiological conditions prevailing in a whole-cell while trying to understand the function of a disease-associated protein target or its interaction with any small molecule drug. To date, ABPs or chemoproteomic tools have been synthesized mostly focusing on a single protein PTM at a time but in the native physiological cellular environment, a particular function is modulated or affected through heterologous modification and expression systems. Therefore, it is very important to consider multiple protein modifications at the same time while synthesizing a chemoproteomic probe to get a comprehensive understanding on the PTM crosstalk. Currently, available chemoproteomic approaches seem to be limited to the enzyme classes with multiple isoforms like DUBs, DHHCs, kinases, proteases, glutathione S-transferases, etc. The future advancement in chemoproteomics should certainly be directed towards the identification of other classes of enzymes responsible for the pathogenesis of diseases. Simultaneously, these approaches should be advanced through interdisciplinary methods in order to utilize this strategy towards theranostic application along with identification and validation of molecular targets.

In this chapter, we have tried to review the advancement in the field of chemoproteomics towards understanding the interaction between a small molecule drug and its protein targets. The different techniques utilized to profile the interacting protein target of small molecule inhibitor are explained and discussed with appropriate examples. However, these techniques also have the limitation of missing out low abundant proteins and restricted tissue distributed proteins. The advent of chemoproteomics has enabled us to understand the complete mechanistic action of the inhibitors by not only identifying their on-targets but also picking out the off-targets, some of which might be associated with toxicity and other side effects. Chemoproteomics have reduced the overall expense, both in terms of time and money, which were usually incurred for the development of a drug molecule and bringing them to clinical trials. This approach has enabled the identification of novel disease-modifying targets that were not considered druggable earlier. This further opens up opportunities and ideas for the development of newer chemical probes and highlights the potential of collaborative work between chemistry and biology disciplines in biotechnology.

Acknowledgments

The authors are thankful for the financial support obtained from the Indian Institute of Technology Palakkad and Department of Science and Technology-Science & Engineering Research Board (DST-SERB), Govt. of India for providing all the infrastructure for writing this Book Chapter. AB is supported by the National Post-doctoral fellowship granted by DST-SERB (PDF/2020/001950), Govt. of India, RS is supported by scholarship from Council of Scientific & Industrial Research (CSIR), Govt. of India (09/1282(0003)/2019-EMR-I). We also acknowledge the financial support from DST-SERB (ECR/2017/002082) and CSIR (02(0434)/21/EMR-II) to SS.

References

A gene-centric human proteome project: HUPO-the human proteome organization. 2010. Mol. Cell. Proteomics 9: 427–429.

Adam, G.C., B.F. Cravatt and E.J. Sorensen. 2001. Profiling the specific reactivity of the proteome with non-directed activity-based probes. Chem. Biol. 8: 81–95.

Adam, G.C., E.J. Sorensen and B.F. Cravatt. 2002a. Proteomic profiling of mechanistically distinct enzyme classes using a common chemotype. Nat. Biotechnol. 20: 805–809.

Adam, G.C., E.J. Sorensen and B.F. Cravatt. 2002b. Trifunctional chemical probes for the consolidated detection and identification of enzyme activities from complex proteomes. Mol. Cell. Proteomics 1: 828–835.

Aslam, B., M. Basit, M.A. Nisar, M. Khurshid and M.H. Rasool. 2017. Proteomics: Technologies and their applications. J. Chromatogr. Sci. 55: 182–196.

Bednarek, C., I. Wehl, N. Jung, U. Schepers and S. Bräse. 2020. The Staudinger ligation. Chem. Rev. 120: 4301–4354.

Belcher, B.P., C.C. Ward and D.K. Nomura. 2021. Ligandability of E3 ligases for targeted protein degradation applications. Biochemistry (in press, DOI: 10.1021/acs.biochem.1c00464).

Berger, A.B., P.M. Vitorino and M. Bogyo. 2004. Activity-based protein profiling. applications to biomarker discovery, *in vivo* imaging and drug discovery. Am. J. PharmacoGenomics 4: 371–381.

Beroza, P., H.O. Villar, M.M. Wick and G.R. Martin. 2002. Chemoproteomics as a basis for post-genomic drug discovery. Drug Discov. Today 7: 807–14.

Bonavia, A., M. Franti, E.P. Keaney, K. Kuhen, M. Seepersaud, B. Radetich et al. 2011. Identification of broad-spectrum antiviral compounds and assessment of the druggability of their target for efficacy against respiratory syncytial virus (RSV). Proc. Natl. Acad. Sci. U.S.A. 108: 6739–6744.

Böttcher, T. and S.A. Sieber. 2010. Showdomycin as a versatile chemical tool for the detection of pathogenesis-associated enzymes in bacteria. J. Am. Chem. Soc. 132: 6964–6972.

Caligiuri, M., L. Molz, Q. Liu, F. Kaplan, J.P. Xu, J.Z. Majeti et al. 2006. MASPIT: Three-hybrid trap for quantitative proteome fingerprinting of small molecule-protein interactions in mammalian cells. Chem. Biol. 13: 711–722.

Chan, W.C., S. Sharifzadeh, S.J. Buhrlage and J.A. Marto. 2021. Chemoproteomic methods for covalent drug discovery. Chem. Soc. Rev. 50: 8361–8381.

Charron, G., M.M. Zhang, J.S. Yount, J. Wilson, A.S. Raghavan, E. Shamir et al. 2009. Robust fluorescent detection of protein fatty-acylation with chemical reporters. J. Am. Chem. Soc. 131: 4967–4975.

Chen, G., A. Heim, D. Riether, D. Yee, Y. Milgrom, M.A. Gawinowicz et al. 2003. Reactivity of functional groups on the protein surface: Development of epoxide probes for protein labeling. J. Am. Chem. Soc. 125: 8130–8133.

Comess, K.M., S.M. McLoughlin, J.A. Oyer, P.L. Richardson, H. Stockmann, A. Vasudevan et al. 2018. Emerging approaches for the identification of protein targets of small molecules—A practitioners' perspective. J. Med. Chem. 61: 8504–8535.

Cravatt, B.F., A.T. Wright and J.W. Kozarich. 2008. Activity-based protein profiling: From enzyme chemistry to proteomic chemistry. Annu. Rev. Biochem. 77: 383–414.

Dang, C.V., E.P. Reddy, K.M. Shokat and L. Soucek. 2017. Drugging the 'undruggable' cancer targets. Nat. Rev. Cancer. 17: 502–508.

Dowden, H. and J. Munro. 2019. Trends in clinical success rates and therapeutic focus. Nat. Rev. Drug Discov. 18: 495–496.

Drewes, G. and S. Knapp. 2018. Chemoproteomics and chemical probes for target discovery. Trends Biotechnol. 36: 1275–1286.

Erkan, H., D. Telci and O. Dilek. 2020. Design of fluorescent probes for bioorthogonal labeling of carbonylation in live cells. Sci. Rep. 10: 7668.

Faucher, F., J.M. Bennett, M. Bogyo and S. Lovell. 2020. Strategies for tuning the selectivity of chemical probes that target serine hydrolases. Cell Chem. Biol. 27: 937–952.

Field, S.D., W. Lee, J.K. Dutra, F.S.F. Serneo, J. Oyer, H. Xu et al. 2020. Fluorophosphonate-based degrader identifies degradable serine hydrolases by quantitative proteomics. ChemBioChem. 21: 2916–2920.

Gersch, M., J. Kreuzer and S.A. Sieber. 2012. Electrophilic natural products and their biological targets. Nat. Prod. Rep. 29: 659–82.

Godl, K., J. Wissing, A. Kurtenbach, P. Habenberger, S. Blencke, H. Gutbrod et al. 2003. An efficient proteomics method to identify the cellular targets of protein kinase inhibitors. Proc. Natl. Acad. Sci. U.S.A. 100: 15434–15439.

Godl, K., O.J. Gruss, J. Eickhoff, J. Wissing, S. Blencke, M. Weber et al. 2005. Proteomic characterization of the angiogenesis inhibitor SU6668 reveals multiple impacts on cellular kinase signaling. Cancer Res. 65: 6919–6926.

Graves, P.R., J.J. Kwiek, P. Fadden, R. Ray, K. Hardeman, A.M. Coley et al. 2002. Discovery of novel targets of quinoline drugs in the human purine binding proteome. Mol. Pharmacol. 62: 1364–1372.

Greenbaum, D., K.F. Medzihradszky, A. Burlingame and M. Bogyo. 2000. Epoxide electrophiles as activity-dependent cysteine protease profiling and discovery tools. Chem. Biol. 7: 569–581.

Greenbaum, D.C., W.D. Arnold, F. Lu, L. Hayrapetian, A. Baruch, J. Krumrine et al. 2002. Small molecule affinity fingerprinting: A tool for enzyme family subclassification, target identification, and inhibitor design. Chem. Biol. 9: 1085–1094.

Guo, H. and Z. Li. 2017. Developments of bioorthogonal handle-containing photo-crosslinkers for photoaffinity labeling. Med. Chem. Commun. 8: 1585–1591.

Hang, H.C., E.J. Geutjes, G. Grotenbreg, A.M. Pollington, M. Jose Bijlmakers and H.L. Ploegh. 2007. Chemical probes for the rapid detection of fatty-acylated proteins in mammalian cells. J. Am. Chem. Soc. 129: 2744–2745.

Hannoush, R.N. and N. Arenas-Ramirez. 2009. Imaging the lipidome: ω-alkynyl fatty acids for detection and cellular visualization of lipid-modified proteins. ACS Chem. Biol. 4: 581–587.

Harding, M.W., A. Galat, D.E. Uehling and S.L. Schreiber. 1989. A receptor for the immunosuppressant FK506 is a cis-trans peptidyl-prolyl isomerase. Nature 341: 758–760.

Hein, C.D., X.M. Liu and D. Wang. 2008. Click chemistry, a powerful tool for pharmaceutical sciences. Pharm. Res. 25: 2216–2230.

Hekmat, O., Y.W. Kim, S.J. Williams, S. He and S.G. Withers. 2005. Active-site peptide 'fingerprinting' of glycosidases in complex mixtures by mass spectrometry. Discovery of a novel retaining beta-1,4-glycanase in *Cellulomonas fimi*. J. Biol. Chem. 280: 35126–35135.

Hillenmeyer, M.E., E. Fung, J. Wildenhain, S.E. Pierce, S. Hoon, W. Lee et al. 2008. The chemical genomic portrait of Yeast: Uncovering a phenotype for all genes. Science 320: 362–365.

Hulce, J.J., A.B. Cognetta, M.J. Niphakis, S.E. Tully and B.F. Cravatt. 2013. Proteome-wide mapping of cholesterol-interacting proteins in mammalian cells. Nat. Methods 10: 259–264.

Jones, L.H. and H. Neubert. 2017. Clinical chemoproteomics - opportunities and obstacles. Sci. Transl. Med. 9: 1–7.

Kidd, D., Y. Liu and B.F. Cravatt. 2001. Profiling serine hydrolase activities in complex proteomes. Biochemistry 40: 4005–4015.

Klaić, L., R.I. Morimoto and R.B. Silverman. 2012. Celastrol analogues as inducers of the heat shock response. design and synthesis of affinity probes for the identification of protein targets. ACS Chem. Biol. 7: 928–937.

Kostiuk, M.A., M.M. Corvi, B.O. Keller, G. Plummer, J.A. Prescher, M.J. Hangauer et al. 2008. Identification of palmitoylated mitochondrial proteins using a bio-orthogonal azido-palmitate analogue. FASEB J. 22: 721–32.

Kumar, S., B. Zhou, F. Liang, W.Q. Wang, Z. Huang and Z.Y. Zhang. 2004. Activity-based probes for protein tyrosine phosphatases. Proc. Natl. Acad. Sci. U.S.A. 101: 7943–7948.

Lamb, J., E.D. Crawford, D. Peck, J.W. Modell, I.C. Blat, M.J. Wrobel et al. 2006. The connectivity map: Using gene-expression signatures to connect small molecules, genes, and disease. Science 313: 1929–1935.

Legrain, P., R. Aebersold, A. Archakov, A. Bairoch, K. Bala, L. Beretta et al. 2011. The human proteome project: current state and future direction. Mol. Cell. Proteomics 10: M111.009993.

Leung, D., C. Hardouin, D.L. Boger and B.F. Cravatt. 2003. Discovering potent and selective reversible inhibitors of enzymes in complex proteomes. Nat. Biotechnol. 21: 687–691.

Liu, Y., K.R. Shreder, W. Gai, S. Corral, D.K. Ferris and J.S. Rosenblum. 2005. Wortmannin, a widely used phosphoinositide 3-kinase inhibitor, also potently inhibits mammalian polo-like kinase. Chem. Biol. 12: 99–107.

Liu, Y., M.P. Patricelli and B.F. Cravatt. 1999. Activity-based protein profiling: The serine hydrolases. Proc. Natl. Acad. Sci. U.S.A. 96: 14694–14699.

Liu, Y., N. Jiang, J. Wu, W. Dai and J.S. Rosenblum. 2007. Polo-like kinases inhibited by wortmannin: labelling sites and downstream effects. J. Biol. Chem. 282: 2505–2511.

Lukinavičius, G., L. Reymond, E. D'Este, A. Masharina, F. Göttfert, H. Ta et al. 2014. Fluorogenic probes for live-cell imaging of the cytoskeleton. Nat. Methods 11: 731–733.

Marques, A.R.A., L.I. Willems, D.H. Moro, B.I. Florea, R. Scheij, R. Ottenhoff, C.P.A.A. van Roomen et al. 2017. A specific activity-based probe to monitor family GH59 galactosylceramidase, the enzyme deficient in Krabbe disease. ChemBioChem. 18: 402–412.

Martin, B.R. and B.F. Cravatt. 2009. Large-scale profiling of protein palmitoylation in mammalian cells. Nat. Methods 6: 135–138.

Meldal, M. and C.W. Tornøe. 2008. Cu-catalyzed azide-alkyne cycloaddition. Chem. Rev. 108: 2952–3015

Meng, F., Y. Liu, J. Niu and W. Lin. 2017. Novel alkyl chain-based fluorescent probes with large Stokes shifts used for imaging the cell membrane and mitochondria in different living cell lines. RSC Adv. 7: 16087–16091.

Mishra, P.K., C.M. Yoo, E. Hong and H.W. Rhee. 2020. Photo-crosslinking: An emerging chemical tool for investigating molecular networks in live cells. ChemBioChem. 21: 924–932.

Navarrete, M., J. Ho, O. Krokhin, P. Ezzati, C. Rigatto, M. Reslerova et al. 2013. Proteomic characterization of serine hydrolase activity and composition in normal urine. Clin. Proteomics 10: 1–11.

Niemeyer, B.A., L. Mery, C. Zawar, A. Suckow, F. Monje, L.A. Pardo et al. 2001. Ion channels in health and disease. 83rd Boehringer ingelheim onds Iinternational titisee conference. EMBO Reports 2: 568–573.

Nomura, D.K., M.M. Dix and B.F. Cravatt. 2010. Activity-based protein profiling for biochemical pathway discovery in cancer. Nat. Rev. Cancer 10: 630–638.

Oikonomopoulou, K., K.K. Hansen, A. Baruch, M.D. Hollenberg and E.P. Diamandis. 2008. Immunofluorometric activity-based probe analysis of active KLK6 in biological fluids. Biol. Chem. 389: 747–756.

Pan, S., S.Y. Jang, D. Wang, S.S. Liew, Z. Li, J.S. Lee and S.Q. Yao. 2017. A suite of 'minimalist' photo-crosslinkers for live-cell imaging and chemical proteomics: Case study with BRD4 inhibitors. Angew. Chem. Int. Ed. Engl. 56: 11816–11821.

Piazza, I., N. Beaton, R. Bruderer, T. Knobloch, C. Barbisan, L. Chandat et al. 2020. A machine learning-based chemoproteomic approach to identify drug targets and binding sites in complex proteomes. Nat. Commun. 11: 4200.

Prescher, J.A. and C.R. Bertozzi. 2005. Chemistry in living systems. Nat. Chem. Biol. 1: 13–21.

Prescher, J.A., D.H. Dube and C.R. Bertozzi. 2004. Chemical remodelling of cell surfaces in living animals. Nature 430: 873–877.

Qi, S.M., J. Dong, Z.Y. Xu, X.D. Cheng, W.D. Zhang and J.J. Qin. 2021. PROTAC: An effective targeted protein degradation strategy for cancer therapy. Front. Pharmacol. 12: 692574.

Qin, W., Y. Zhang, H. Tang, D. Liu, Y. Chen, Y. Liu et al. 2020. Chemoproteomic profiling of itaconation by bioorthogonal probes in inflammatory macrophages. J. Am. Chem. Soc. 142: 10894–10898.

Rix, U. and G. Superti-Furga. 2009. Target profiling of small molecules by chemical proteomics. Nat. Chem. Biol. 5: 616–624.

Saghatelian, A. and B.F. Cravatt. 2005. Assignment of protein function in the postgenomic era. Nat. Chem. Biol. 1: 130–142.

Salisbury, C.M. and B.F. Cravatt. 2007. Activity-based probes for proteomic profiling of histone deacetylase complexes. Proc. Natl. Acad. Sci. U.S.A. 104: 1171–1176.

Schenone, M., V. Dancík, B.K. Wagner and P.A. Clemons. 2013. Target identification and mechanism of action in chemical biology and drug discovery. Nat. Chem. Biol. 9: 232–240.

Sexton, K.B., D. Kato, A.B. Berger, M. Fonovic, S.H.L. Verhelst and M. Bogyo. 2006. Specificity of aza-peptide electrophile activity-based probes of Caspases. Cell Death Differ. 14: 727–732.

Sherratt, A.R., Y. Rouleau, C. Luebbert, S. Bidawid, N. Corneau, J. Paul et al. 2017. Rapid screening and identification of living pathogenic organisms via optimized bioorthogonal non-canonical amino acid tagging. Cell Chem. Biol. 24: 1048–1055.e3.

Siekierka, J.J., M.J. Staruch, S.H. Hung and N.H. Sigal. 1989a. FK-506, a potent novel immunosuppressive agent, binds to a cytosolic protein which is distinct from the Cyclosporin A-binding protein, Cyclophilin. J. Immunol. 143: 1580–1583.

Siekierka, J.J., S.H.Y. Hung, M. Poe, C.S. Lin and N.H. Sigal. 1989b. A cytosolic binding protein for the immunosuppressant FK506 has peptidyl-prolyl isomerase activity but is distinct from Cyclophilin. Nature 341: 755–757.

Sigismund, S., D. Avanzato and L. Lanzetti. 2018. Emerging functions of the EGFR in cancer. Mol. Oncol. 12: 3–20.

Sletten, E.M. and C.R. Bertozzi. 2009. Bioorthogonal chemistry: Fishing for selectivity in a sea of functionality. Angew. Chemie Int. Ed. 48: 6974–6998.

Sletten, E.M. and C.R. Bertozzi. 2011. From mechanism to mouse: A tale of two bioorthogonal reactions. Acc. Chem. Res. 44: 666–676.

Spradlin, J.N., E. Zhang and D.K. Nomura. 2021. Reimagining druggability using chemoproteomic platforms. Acc. Chem. Res. 54: 1801–1813.

van de Plassche, M.A.T., M. Barniol-Xicota and S.H.L. Verhelst. 2020. Peptidyl acyloxymethyl ketones as activity-based probes for the main protease of SARS-CoV-2*. Chembiochem. 21: 3383–3388.

Vasilyeva, S.V., V.V. Filichev and A.S. Boutorine. 2016. Application of Cu(I)-catalyzed azide–alkyne cycloaddition for the design and synthesis of sequence specific probes targeting double-stranded DNA. Beilstein J. Org. Chem. 12: 1348–1360.

Vocadlo, D.J. and C.R. Bertozzi. 2004. A strategy for functional proteomic analysis of glycosidase activity from cell lysates. Angew. Chemie. 43: 5338–5342.

Wang, J., J. Zhang, C.J. Zhang, Y.K. Wong, T.K. Lim, Z.C. Hua et al. 2016. *In situ* proteomic profiling of curcumin targets in HCT116 colon cancer cell line. Sci. Rep. 6: 22146.

Wang, S., Y. Tian, M. Wang, M. Wang, G.B. Sun and X.B. Sun. 2018. Advanced activity-based protein profiling application strategies for drug development. Front. Pharmacol. 9: 353.

Wilkins, M.R. and A.A. Gooley. 1997. Protein identification in proteome projects. pp. 35–64. *In*: M.R. Wilkins, K.L. Williams, R.D. Appel and D.F. Hochstrasser (eds.). Proteome Research: New Frontiers in Functional Genomics. Principles and Practice. Springer, Berlin, Heidelberg.

Wright, M.H. and S.A. Sieber. 2016. Chemical proteomics approaches for identifying the cellular targets of natural products. Nat. Prod. Rep. 33: 681–708.

Wu, H. and N.K. Devaraj. 2018. Advances in tetrazine bioorthogonal chemistry driven by the synthesis of novel tetrazines and dienophiles. Acc. Chem. Res. 51: 1249–1259.

Yang, P. and K. Liu. 2015. Activity-based protein profiling: recent advances in probe development and applications. Chembiochem. 16: 712–24.

Yang, P.Y., K. Liu, M.H. Ngai, M.J. Lear, M.R. Wenk and S.Q. Yao. 2010. Activity-based proteome profiling of potential cellular targets of Orlistat - an FDA-approved drug with anti-tumor activities. J. Am. Chem. Soc. 132: 656–666.

Yee, M.C., S.C. Fas, M.M. Stohlmeyer, T.J. Wandless and K.A. Cimprich. 2005. A cell-permeable, activity-based probe for protein and lipid kinases. J. Biol. Chem. 280: 29053–29059.

Zhang, L., S. Wan, Y. Jiang, Y. Wang, T. Fu, Q. Liu et al. 2017. Molecular elucidation of disease biomarkers at the interface of chemistry and biology. J. Am. Chem. Soc. 139: 2532–2540.

Zhang, Y., W. Qin, D. Liu, Y. Liu and C. Wang. 2021. Chemoproteomic profiling of itaconations in Salmonella. Chem. Sci. 12: 6059–6063.

Zhou, Z. and C.J. Fahrni. 2004. A fluorogenic probe for the Copper(I)-catalyzed azide-alkyne ligation reaction: modulation of the fluorescence emission via 3(n, Π*)-1(π, Π*) inversion. J. Am. Chem. Soc. 126: 8862–8863.

Ziegler, S., V. Pries, C. Hedberg and H. Waldmann. 2013. Target identification for small bioactive molecules: Finding the needle in the haystack. Angew. Chemie. 52: 2744–2792.

Zon, L.I. and R.T. Peterson. 2005. *In vivo* drug discovery in the Zebrafish. Nat. Rev. Drug Discov. 4: 35–44.

Chapter 3

Combinatorial Chemistry and High Throughput Screening Methods

Vaishnavi Puppala,[#] Shivcharan Prasad[#] and Ipsita Roy*

Introduction

The basic aim of combinatorial chemistry is to synthesize a large number of molecules and then identify target-specific binders. This is carried out most commonly by high throughput screening (HTS) with the help of computational methods. A large number of structurally diverse molecules can be synthesized within a short period of time which can be submitted for pharmacological studies. There are two phases in a combinatorial approach; the first is to develop chemical libraries. This is followed by target-based identification of active ingredients. Each library has a set of chemicals which is associated with information such as chemical structure, physiochemical characteristics, purity and quantity of the compound. A brief history of the development of combinatorial synthesis is shown in Fig. 1.

Combinatorial chemistry and high throughput screening approaches have crossed the "hype" and "disillusionment" stages of their life cycle (Grygorenko et al. 2020). A more nuanced approach has emerged. The high attrition rate of the conventional drug discovery process and the success of HTS in drug repurposing have reaffirmed the faith that researchers had reposed in this technique. This chapter aims to describe different approaches of creating combinatorial libraries, followed by the advances in techniques to screen them for biological activity.

Department of Biotechnology, National Institute of Pharmaceutical Education and Research, Sector 67, S.A.S. Nagar, Punjab 160062, India.
* Corresponding author: ipsita@niper.ac.in
[#] These authors contributed equally.

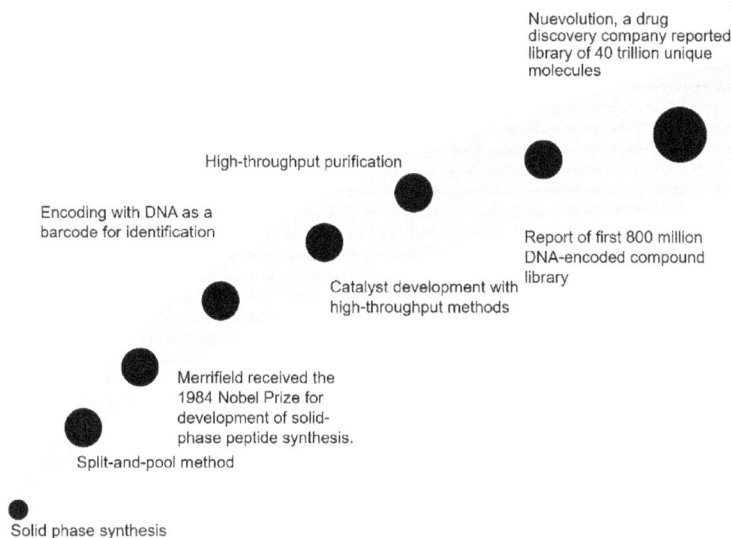

Figure 1. A brief history of developments in combinatorial chemistry.

Different approaches to generate combinatorial libraries

Two approaches are generally adopted to generate combinatorial libraries. These are (i) parallel chemical synthesis, and (ii) combinatorial synthesis. Both techniques are used to generate a number of synthetic analogs, but which technique is used depends on when it is utilized, how the steps are executed and the state of the final product. Some major characteristics of these two methodologies are highlighted in Table 1.

Table 1. Comparison of parallel and combinatorial synthetic methods.

Properties	Parallel	Combinatorial
Types of products	Individual compound	Mixtures of compounds
Sample size	Limited as it depends on availability of reactor vessel	Large number of compounds in a single process (up to millions)
Method	One-bead-one compound	Solid-phase or tethered synthesis
Synthetic flow	Linear	Split and mix
Product identity	Mostly known	Requires a deconvolution procedure to identify individual compounds from mixtures
Compound identification	By physical position within parallel platform or vessel	Chemically, by tagging, color or bar coding,
Used for	Accelerating discovery of new compounds and optimization of optimal process conditions	Screening of properties

Parallel synthesis

This technique helps to speed up the discovery of new compounds and screening for optimal process conditions. It is a time saving method and allows one to run multiple experiments simultaneously. For example, in the pharmaceutical industry, parallel synthesis is used to synthesize libraries with diverse chemical structures that can be screened for potential biological activity. It is based on the 'one bead, one compound' (OBOC) approach and may be applied at different stages of drug discovery, e.g., lead generation, lead optimization and screening for optimal process conditions. Optimization of the process provides a better understanding of the effect of reaction variables, solvent systems, optimal temperatures and concentrations to be used for the reaction (Aube 2014). In this approach, organic synthesis is carried out using linear sequences on individual substrates to provide a single end product in separate vessels or microtiter plates. Instead of converting one starting material into one final product, a set of similar starting materials are processed simultaneously.

The unique setup of rows and columns provides flexibility to researchers to organize the building blocks to be combined and an easy way to identify compounds in a particular well. The advantage of this approach is its 'neatness'; every pathway is clearly identified and can be monitored. A systematic arrangement of reaction sequences minimizes formation of byproducts while generating an adequate amount of diversity (Aube 2014). In parallel synthesis, the number of compounds that can be synthesized depends on the availability of reaction vessels. As the diversity of compounds increases, the number of reaction vessels too increases. Most parallel synthesis hardware platforms contain less than 200 reaction sites (6–96 well setups being the most common) which is much lower than the demand of the final aim. The approach requires very careful setup of chemical design principles so that the maximum chemical diversity can be generated in a given library. The pioneering work of Merrifield in designing solid-phase synthesis addressed this problem (discussed in the next section) (Merrifield 1963).

Combinatorial approach

Solid phase synthesis

In this approach, molecules are transiently bound to a solid support and progressive synthesis of product molecules is performed via the strategy of protecting and deprotecting key functional groups. It was initially applied to the production of peptides (Merrifield 1963). An amino acid residue is attached to a resin (200-mesh polymer beads) with the help of a linker. The attached amino acid is blocked at one end to inhibit it from participating in any further reaction. The byproducts are removed by filtration and recrystallization of intermediates. Separation and purification are accomplished simply by filtering and washing the beads with appropriate solvents. The reagents for the next peptide bond formation are then added, and the purification steps repeated. A chain is built up by a series of peptide bond coupling steps and the final product is cleaved from the resin. Some examples of commonly used resins and linkers in solid phase synthesis are listed in Table 2.

Table 2. An illustrative list of commonly used resins and linkers in solid phase synthesis.

Resin	Linker
TentaGel®	1–2% Divinyl benzene
Polyamide	Alcohol linkers
Cross-linked polystyrene	Carboxylic acid linkers
Aminomethylated polystyrene	Amide linkers
ChemMatrix®	Carboxamide and other linkers

Split-and-mix approach

In the initial phase of development of combinatorial chemistry, much emphasis was given to solid phase synthesis (Furka et al. 1991, Furka 2002). Árpád Furka (Eötvös Loránd University, Hungary) soon realized the futility of using the Merrifield approach to develop a genuine combinatorial library and provided a theoretical solution called "synthetic back searching strategy" which he notarized in 1982. Later on, his laboratory showed the practical application of this approach by (i) linking different monomers to small portions of the resin, (ii) mixing these coupled matrix portions, (iii) adding a second monomer to this mixed portion and repeating this process (Fig. 2). The method allowed solid-supported synthesis to expand the number of compounds that can be prepared in a given number of vessels (Furka et al. 1989a, b). This later on became famous as the split-mix or portioning-mixing method of combinatorial synthesis.

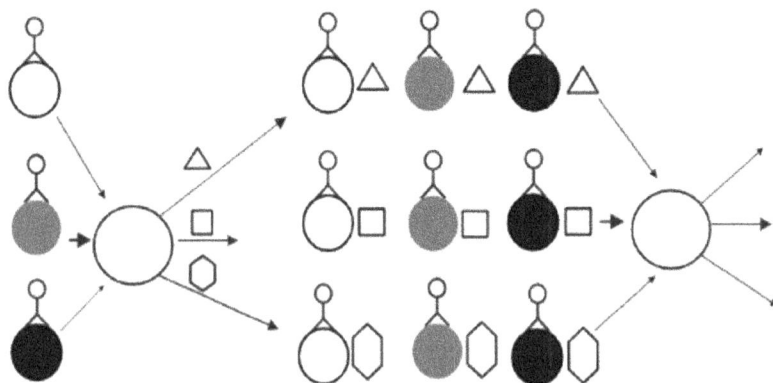

Figure 2. Schematic diagram of split-pool synthesis.

The advantage of this scheme is its suitability for multi-step reactions; isolation of intermediates at every step is not necessary. Moreover, purification is easier as unutilized/excess reagents can be removed, with control over the location of reactions (Grygorenko et al. 2020). However, the initial approach did not allow for identification of products formed at each step. The yield was also quite low. These drawbacks were overcome by the approach of string synthesis adopted by Advanced ChemTech Inc. who used untagged support matrices and monitored each step of synthesis by a computer.

This approach is largely based on solid-based synthetic methods, and a mixture of similar type of compounds is produced in the same reaction vessel (Fig. 2). On the other hand, solution-based synthesis has limited applicability, even though it provides flexibility, especially for larger number of chemical reactions. The major problem with solution-based synthesis is that a large number of reagents are taken together in a reaction vessel or microtiter well plate. This results in several side reactions which makes it difficult to keep track of the main product(s). Moreover, as compounds are not attached to a solid support, isolation of product is a big challenge. To make extraction process easier, newer strategies have been applied, such as use of ion exchange resins. These methods eliminate byproducts so that the need for aqueous workup is not required, resulting in easier extraction of final product (Aube 2014).

Alternate approaches

The screening of specific ligands against the desired target is of prime importance in a new drug discovery program. HTS and fragment-based discovery (FBD) procedures are generally used to validate a large number of small molecules (ranging between thousand to million) against desired targets of interest (Jorgensen 2004, Folmer 2016). Exploring the full potential of HTS requires individual synthesis and testing of hundreds to thousands of chemical compounds, which demands significant resources (Blay et al. 2020, Lloyd 2020). A major challenge with the split-and-pool strategy is the associated difficulty of determining the identity of the functional moiety after biological screening. This led to the development of alternate approaches, such as protein ligand discovery. For example, encoded combinatorial antibody libraries typically contain billions of unique structures and routinely provide binders against virtually any protein target of interest (Canal-Martín and Pérez-Fernández 2020, Sioud 2019). In 1992, Brenner and Lerner proposed encoded chemical libraries, in which chemical molecules immobilized on bead matrices are conjugated to unique DNA sequences which function as barcodes and store information about their identity (Brenner and Lerner 1992, Gironda-Martínez et al. 2021, Lerner and Neri, 2020). Further modifications in the methodology led to direct coupling of chemical entities to double stranded DNA fragments, without the use of beads (Decurtins et al. 2016, Flood et al. 2020, Halford 2017, Neri and Lerner 2018). DNA-encoded chemical libraries (DEL) have emerged as a powerful tool for hit identification in the pharmaceutical industry and in academia (Sunkari et al. 2022). DEL is a collection of small molecules, individually coupled with DNA tags that have unique information about the identity and structure of each library member (Fig. 3). They have been used to screen for antimicrobials (Cochrane et al. 2021). The advantage of DEL, generated by the amalgamation of synthetic and biological components, is that millions, billions and even trillions of chemical compounds can be screened in a single and simple experiment (Halford 2017). The approach leads to saving of storage space while keeping track (of identity) of each member of the library.

Using this approach, a drug discovery company, Nuevolution, came up with a huge library of 40 trillion unique molecules in 2017. Newer approaches allow DELs to identify binding partners in complex biological milieu (Huang et al. 2022). In a

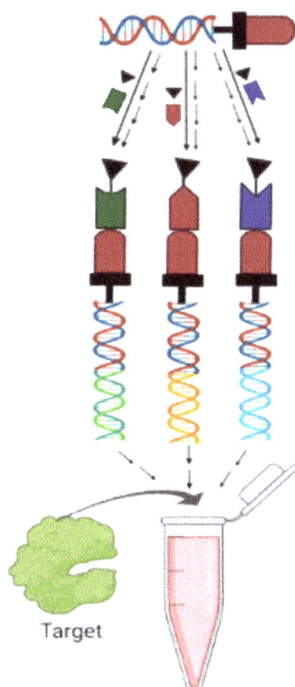

Figure 3. Schematic diagram of synthesis of DNA-encoded libraries.

novel approach, DEL (containing RNA-binding natural product molecules) has been combined with a labeled RNA fold library to isolate a molecule which is bound to an oncogenic primary miRNA and inhibited proliferation of MDA-MB-231 triple negative breast cancer cells (Benhamou et al. 2022). A few examples of lead molecules developed using screening of alternate libraries and those presently in clinical trials are shown in Table 3.

Table 3. An illustrative list of therapeutic leads screened using alternate libraries.

Molecule	Sponsor	Status	Library used	Application	Reference
GSK2982772	GSK	Phase IIa	DNA encoded	Psoriasis, rheumatoid arthritis, ulcerative colitis	Halford 2017
GSK'481	GSK	Phase IIa	DNA encoded	Inflammation	Harris et al. 2016
LLP2A-Ale	California Institute for Regenerative Medicine	Phase I	OBOC peptidomimetics	Osteopenia	Guan et al. 2012
X-165	X-CHEM	Phase 1	DNA encoded	Idiopathic pulmonary fibrosis	Cuozzo et al. 2020
2.45	Roche/ Chugai	Lead	DNA encoded	Alport syndrome/ Chronic kidney disease	Richter et al. 2019

Technological advancements

Traditional tools of HTS have been comprehensively discussed earlier (Reymond and Babiak 2007). Over time, advancements in the division of the omics studies such as genomics, metabolomics, and proteomics has provided impetus to the development of high throughput screening (HTS) techniques (Zhou et al. 2021a). As described above, HTS is a lateral process of analyzing and identifying specific binders among thousands to millions of molecules to the target. Earlier, mixtures were examined mostly in 96-well titer plates (Attene-Ramos et al. 2014). Nowadays, most of the HTS methods have embraced robotic/automated systems for screening millions of samples in order to prevent human error, save time, and allow extensive data analysis through robust software (Murray and Wigglesworth 2016). HTS acts as an initial point in the drug discovery process by providing further optimized hits to increase its efficiency. It bypasses the use of structure-based design in drug discovery. However, it can be, and is, performed in parallel with other strategies like computational techniques and fragment-based drug design (Janzen 2014).

HTS methods are classified as cell-based assays, including cell viability, reporter gene-based, second messenger, high throughput microscopy assay; biophysical assays including fluorescence polarization, anisotropy, FRET, TR-FRET, and fluorescence timeline image analysis. Other HTS methods include binding-affinity-based NMR, SPR, mass spectroscopy, and DSF (Blay et al. 2020, Wu et al. 2013). A few of these are described below.

Biophysical assays

Classically, biophysical or biochemical assays require a purified target protein of interest against which binding of ligands or inhibition of activity is examined (Fang 2012). The assays are performed competitively. A compound should displace a known ligand or substrate. The assays are typically conducted in 96- or 384-well plates, providing a negotiable sample volume and reducing advanced instrumentation (Stoddart et al. 2016).

Over a decade, quick shift in techniques from radioactive screening to optical assays has occurred in the platforms of both biotechnology and pharmaceutical companies. The assays have the advantages of being economical, safe to handle, and generate no radioactive waste (Owicki 2000). Compared to filtration assays, fluorescence polarization assays do not require any physical separation of bound and free ligands in a homogenous solution, ultimately decreasing the number of processing steps (Blay et al. 2020). Roehrl et al. identified small organic molecules as therapeutic molecules targeted for intracellular protein-protein interaction (Roehrl et al. 2004). Through fluorescence polarization assay, they reported the discovery of INCA-6 peptide inhibitor from a library of 16,320 compounds. The Oregon Green tagged VIVIT peptide, which has high affinity for calcineurin, was used as the fluorophore. The screened molecule was able to inhibit the NFAT (nuclear factor of activated T cells, a substrate of calcineurin phosphatase)–calcineurin interaction by displacing the VIVIT peptide from the phosphatase and acted as an anti-inflammatory agent in Cl.7W2 T cells (Roehrl et al. 2004). In an effort to reduce the usage of labels, Pu et al. have extended the use of infrared matrix-assisted laser desorption electrospray ionization (IR-MALDESI), a label-free system, to high throughput screening (Pu et al. 2021). The authors developed assays for three enzymes by analyzing a range of analytes like small molecule metabolites, lipids and short peptides and showed the proof-of-concept of the technique with one of them for high throughput lead identification. Screening of 3,588 compounds provided a good correlation between IC_{50} against isocitrate dehydrogenase 1 using IR-MALDESI-MS and fluorescence assays. Modifying the mode of triggering led to more than two-fold reduction in assay time per well (Pu et al. 2021). Thus, conventional assays, which have high sensitivity but are otherwise give low throughput, can be repurposed for high throughput screening by tweaking relevant parameters.

Förster or fluorescence resonance energy transfer (FRET) works on the principle of non-radiative energy, where the energy emitted by a donor molecule is absorbed by the acceptor. The phenomenon works between two fluorophores distanced within 1–10 nm scale and have overlapping emission and absorption spectra. The FRET-based assay can be manipulated, either induced or abolished, through catalytic activities, intramolecular tension, or conformational changes produced by protein-protein interactions. Han et al. demonstrated the FRET-based screening of anticancer drugs, such as MDM2 (mouse double minute 2 homolog) inhibitors, which prevented the binding of p73 to MDM2 (Han et al. 2021). The authors utilized the interaction between intrinsic fluorescence of MDM2 (due to Trp, Q18W) and 1-naphthylalanine tagged to p73 which could be disrupted by the addition of the MDM2 inhibitor Nutlin-3. This enabled the authors to develop a high throughput screening method for anti-cancer drugs. The FRET assays can be employed not only for screening of drug candidates but also for the identification of the therapeutic target. Liu et al. used fluorescence lifetime imaging coupled with FRET (FLIM-FRET) to identify novel epigenetic biomarkers in breast cancer (Liu et al. 2019). The authors screened eleven epigenetics-relevant markers which were present in close proximity to ER-dependent genes and monitored their interaction with ERa using FLIM-FRET. Potential hits were validated by inhibiting the enzyme involved in their epigenetic modification. This led to improved therapeutic response in MCF7 cells and mice xenograft models when combined with a traditional anti-cancer molecule like tamoxifene (Liu et al. 2019). Using the fluorescence lifetime plate reader platform, small molecules capable of reducing FRET interaction between two huntingtin fragments tagged with donor and acceptor fluorophores were identified (Lo et al. 2020). Six molecules were shortlisted for a library of 1280 compounds (LOPAC) screened in the high throughput format which could slow down protein aggregation and dissociate preformed fibrils in cell models. The same approach and the library had previously been used to identify molecules which could inhibit the interaction between cardiac sarcoplasmic reticulum Ca-ATPase (SERCA2a) and phospholamban (Stroik et al. 2018). Inhibition of this interaction, which had earlier been classified as "undruggable", is a therapeutic target in calcium-dependent cardiac contractile dysfunction, and demonstrates the power of fluorescence-based tools to isolate bioactive molecules.

Surprisingly, core biophysical tools have also been adapted to high throughput screening platforms. The fusion of different mass spectrometric techniques with existing platforms has been discussed above. Determination of enthalpy of binding (ΔH) between the target and the analyte was shown to be a reliable measure of screening binder molecules (Baggio et al. 2017). Incorporation of an 'anchoring fragment' generated focused positional scanning libraries (ƒPOS), which allowed detection of binding events when measured with isothermal titration calorimetry. Using this approach, the authors identified a tetrapeptide against the BIR3 domain of XIAP, an anti-apoptotic protein with the highest ΔH value (−12.2 kcal/mol) which showed high affinity (K_d) for the target, thus validating the results observed with enthalpy measurements (Baggio et al. 2017). The establishment of 'Structure Activity Relationship' by NMR (SAR by NMR) identifies shorter sequences of limited affinity to the target which are then fused to generate potent bidentate ligands (Shuker et al. 1996). Baggio et al. have combined SAR by NMR with ƒPOS to create an HTS platform (Baggio et al. 2018). They screened a library of 1,00,000 compounds, each containing the metal-chelating hydroxamic moiety at the C-terminal, using smaller lots of mixtures. Each mixture was tested to bind the metalloproteinase MMP-12 using 1D and 2D HMQC (heteronuclear multiple quantum coherence) correlation spectroscopy and the preferred element at each position was identified. The highest-ranking elements were individually synthesized. They finally identified a compound (ƒ17-ƒ9-ƒ28-CONHOH) which showed low IC_{50} value (nanomolar) against MMP12 enzyme activity and negligible inhibition of other MMPs (Baggio et al. 2018). Affinity-based mass spectroscopy has been developed as a high throughput screening tool in conjunction with size exclusion (SEC) and reverse phase (RP) chromatographies, which the authors have referred to as ALIS (Automated Ligand Identification System) (Annis et al. 2004). SEC is used to separate the binders from the non-binders while the bound ligand is separated from the target using the RP column and loaded onto the mass spectrometer for analysis. Screening of 2,500 compounds could be accomplished within ten minutes and yielded a novel inhibitor of dihydrofolate reductase (Annis et al. 2004). The method does not require any prior knowledge about the identity of the target and can screen over 2,50,000 compounds per day. A more sophisticated version of affinity-based mass spectroscopy has now been developed to determine binding affinity of molecules for a membrane-bound target (Lu et al. 2019). Membrane-bound targets, such as G protein-coupled receptors (GPCRs), are notoriously difficult to work with due to their frequent loss of conformation during purification. The cost of purification too adds to the cost of HTS. Preliminary screening of a library of 4,800 compounds against purified adenosine A_{2A} receptor led to the identification of 22 initial hits out of which 15 had no reported history of A_{2A} receptor activity (Lu et al. 2019). Increasing the complexity of the library to 20,000 compounds still identified the same compounds as in the initial library, demonstrating the specificity of binding.

Cell based assay systems

Cell-based assays are used in every step of the drug discovery process, from hit identification and validation, followed by primary screening, lead identification, lead optimization, safety, and toxicological screening (Macarrón and Hertzberg 2009). Screening of drugs is mainly based on their ability to induce a cell- or organism-relevant phenotype (Zheng et al. 2013). The assay system consists of target-based and phenotypic cell-based approaches that identify lead molecules. Screening of multiple targets in cells is an important advantage of cell-based assays. An important preliminary step in any screening procedure is to monitor the effect of the molecule(s) on cell survival. Cell viability assays also help to identify compounds with anti-proliferative activity against cancer cells or pathogens and monitor toxicity in organs such as the liver (Adan et al. 2016, McKim 2010). The major disadvantage is the lack of knowledge about the cellular target(s). Cell viability can be assessed using various dyes such as ruthenium, Alamar blue and tetrazolium compounds (Rajalingham 2016). 2,3-Bis-(2-methoxy4-nitro-5-sulphophenyl) (XTT) endpoint assay was employed to screen the antiproliferative activity of a wide variety of natural plant extract active compounds against differential cancer lines (Cortelo et al. 2021). Out of these, successive rounds of antiproliferative assay led to the selection of seventeen extracts for bioactivity-guided cell-based screening assays. The most active moiety was isolated from the hexane extract of *Salacia crassifolia* root wood. This was active against sixty different cancer cell lines. Long et al. have adopted the paired cell line approach to target the highly metastatic osteosarcoma (Long et al. 2021). Starting with 1,54,002 extracts of natural product origin from different living systems such as plant, marine, microbial extracts, they screened the extracts for their ability to inhibit the growth of highly metastatic MG63.3 cells while sparing MG63 cells which have low metastaticity. The authors proposed that this strategy would allow identification of metastatic pathway-specific molecules (Long et al. 2021). Two active natural products, lovastatin from the extract of *Monascus ruber* and limonoid toosendanin from *Melia toosendan*, were isolated through MG63/63.3 pairing assay. Although toosendanin has hepatotoxic activity, it may provide a scaffold for drug development. The results led the authors to screen commercially available statins against osteosarcoma metastasis. They suggested cerivastatin as a potential repurposed drug based on its ability to activate caspase-3 (Long et al. 2021) although this requires further validation.

Spherical nucleic acids (SNAs) are nanoparticles functionalized with oligonucleotide ligands (DNA or RNA) (Cutler et al. 2012). As far as cancer therapeutics is concerned, their major advantages are cell membrane permeability and nuclease resistance. Yamankurt et al. synthesized a library of SNAs that varied in their nanoparticle core, oligonucleotide composition and model antigens based on pre-defined criteria (Yamankurt et al. 2019). They screened > 950 SNA structures for their ability to activate TLR9 signaling and estimated the amount of secreted embryonic alkaline phosphatase in the culture medium of RAW-Blue macrophages using SAMDI (self-assembled monolayers for MALDI) (Yamankurt et al. 2019). This study did not claim to identify and optimize an immunotherapeutic candidate for any specific disease; the purpose was to explore the structure-activity relationships and rules to design SNA libraries. Finally, with the aid of machine learning approach, about 1,000 variations of the SNA structure were identified which could activate and generate immune response. Kitakaze et al. developed an HTS-based cellular assay using a versatile UPR (unfolded protein response) reporter output to monitor protein folding in the endoplasmic reticulum (Kitakaze et al. 2019). They identified 2-phenylimidazo[2,1-b]benzothiazole (IBT) backbone through an initial high throughput screening of > 2,00,000 compounds using tunicamycin as an endoplasmic reticulum stress inducer (Kitakaze et al. 2019). This backbone was present in four of the ten best compounds and functioned as a chemical chaperone. The chaperones are validated by biochemical and chemical biology approaches showing inhibition of protein aggregation, directly binding to the unfolded or misfolded proteins. The molecule IBT21 successfully prevented cell death caused due to chemical-induced endoplasmic reticulum stress and proteotoxin and had a significantly lower IC_{50} value as compared to azoramide, a UPR modulator and 4-phenylbutyrate, a chemical chaperone (Kitakaze et al. 2019).

The continually observed increase in antibiotic resistance and the need to develop newer antibiotics has re-emphasized to the need to identify the mechanism of action of antibiotics. A novel scaffold was synthesized where *RFP* gene, coding for red fluorescent protein, was cloned under the SOS-inducible *sulA* promoter and responded to DNA damage, and *katushka2S* gene coding for far-red fluorescent protein, was cloned downstream of the tryptophan attenuator by replacing tryptophan codons with those of alanine. p*sulA* is activated by ribosome-stalling compounds (Osterman et al. 2016). This multiplex reporter system, called pDualrep2, with two distinct fluorescent signatures, permitted screening and functionality checks of antibiotic molecules from natural and synthetic origin at the same time on an agar plate in the high throughput mode. A study using pDualrep2 screened a library of 47,000 natural and synthetic molecules from research institute of chemical diversity collection for inhibiting DNA biosynthesis and translation (Ivanenkov et al. 2020). Differentially substituted 1,2,3,4-tetrahydrocarbazole derivates were identified as DNA biosynthesis inhibitors although MIC (minimum inhibitory concentration) value was significantly higher than the standard molecule, erythromycin.

Microfluidics

In the course of modernization, microfluidics appeared to be advantageous for HTS due to miniaturization, with lower consumption of sample/reagents, low operating costs, and time saving as compared to standard techniques (Macarron et al. 2011). Development in microfluidics can be followed in three modes, which are reciprocal to each other, including perfusion flow, droplet-based and microarray. The technical challenges associated with miniaturization have been addressed using various modifications, some of which are briefly discussed in this section. The evolution of high throughput screening using the standard microtiter technology has been fraught with several issues, such as manual pipetting and errors, loss of solvent by evaporation, and high reagent usage (Liu and Li 2018). Even after introduction of modifications, several challenges have limited its miniaturization and portability options. Considering the difficulties faced by these systems, several portable liquid handling techniques with or without pumping systems were introduced. Li et al. developed a microfluidic array reactor with a capillary action and degassed PDMS (polydimethylsiloxane) pumping system in combination (Li et al. 2012). This is an automated system and can perform multiple functions in parallel. The device was designed with no external power supply, easy handling, and no complex supporting equipment. The modular strategy introduced flexibility and increased the application base of the approach. Practical usage of this power-free device, without the need for multiple injections, was demonstrated by optimizing the crystallization conditions of lysozyme (Liu and Li 2018).

Wu et al. have attempted to overcome the requirement of sophisticated facilities of liquid/plate handling while carrying out HTS by developing a bench-top microarray platform (Wu et al. 2011). The arrayed PDMS [poly(dimethylsiloxane)] posts microarray platform delivered 320 natural compounds from a library to segregated PEG microwells containing MCF-7 human breast cancer cells. Cell viability was examined by detecting the fluorescence produced by calcein AM using a microarray scanner (Wu et al. 2011). 9-Methoxy-camptothecin, a molecule with known anti-cancer activity, was identified as a potential anti-tumor drug. The authors reported that in a 72-hour period, 2,100 assays could be carried out with a single device, including cell culture time. Using this set up, drug-drug interaction could also be monitored by incubating members of the library with verapamil, a vasodilator and a P-gp inhibitor (Wu et al. 2011). The above arrangement was modified by regulating the release of compounds from photocrosslinked PEGDA (polyethylene glycol diacrylate) hydrogel arrays and to monitor apoptosis and necrosis using Sytox Orange and allophycocyanin-conjugated annexin V (Kwon et al. 2011). Du et al. have developed the sequential operation droplet array (SODA) technique by which droplets could be processed using a capillary linked to a syringe pump (Du et al. 2013, 2018). This allowed all functions of cell culture to be carried out simultaneously in an oil-covered droplet of 500 nl size and permitted delivery of drug molecules as per a pre-decided schedule. Using A549 small lung cancer cells, the authors monitored the effect of drug combinations of flavopiridol, 5-fluorouracil and paclitaxel with scheduled dosing and were able to determine the optimized combination of drugs to achieve maximum inhibition of cell growth (Du et al. 2013). This approach allowed up to 1,000-fold reduction in drug concentration as compared to conventional drug screening approaches. A device for 3D cell culture based combinatorial drug screening array has been developed on a microfluidic system in combination with pneumatic valves (Chang et al. 2019). Using MDA-MB-231 and MCF-7 breast cancer cells as illustration, combination drug screening using doxorubicin and paclitaxel was performed.

Combinatorial chemistry and drug discovery

In the race to discover efficient and specific drugs, many techniques or approaches have been designed. The hunt for potent drugs against targets includes conventional methods that follow structure-based identification of pharmacophore, followed by synthesis of analogs of a hit molecule and optimization of analogues/hits by structure-activity relationship (SAR) evaluation (Gaurav and Gautam 2014). These classical methods are associated with disadvantages in terms of cost and time investment. A comparison between chemical and biological screening methods is summarized in Fig. 4.

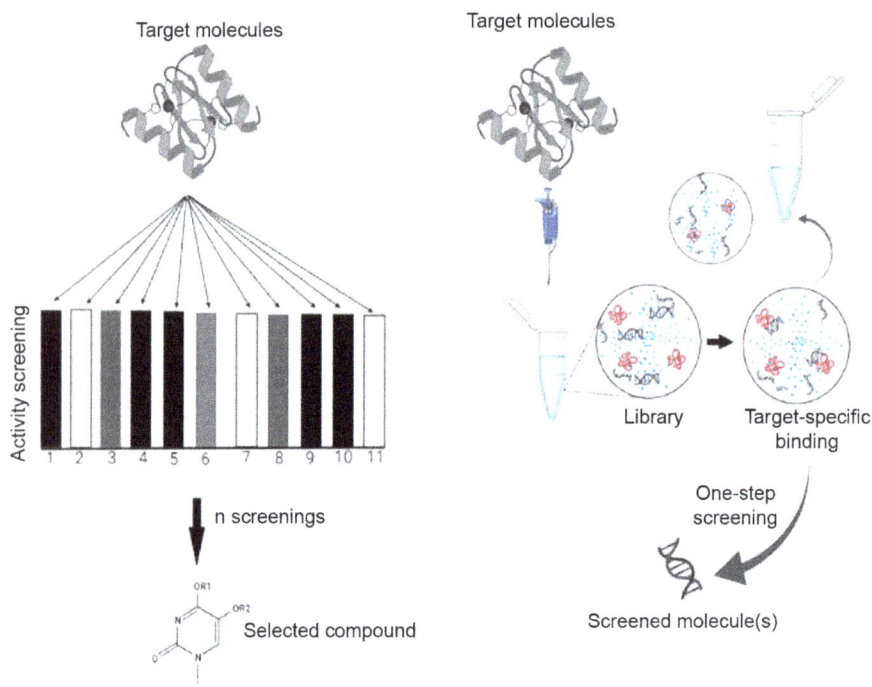

Figure 4. Schematic diagram of combinatorial v/s biological screening.

Novel strategies have evolved to overcome the limitations of conventional drug discovery by generating a dynamic combinatory library (Li et al. 2013). Dynamic combinatorial library (DCL) is a part of Dynamic combinatorial chemistry (DCC), increasing the scope of drug discovery (Aldib et al. 2012, Mondal and Hirsch 2015). Structurally diverse compound libraries can be generated by combining different building blocks in a one-pot reaction through a combinatorial chemistry strategy (Cheng et al. 2018) as has been described above. DCC is classified as supramolecular chemistry (Frei et al. 2019) and the library is formed by building blocks with complementary functional groups which undergo reversible covalent or non-covalent reactions (Mondal and Hirsch 2015). The term 'dynamic' refers to the possibility of interconversion between library members via reversible covalent or non-covalent interactions. The library make-up is determined by the equilibrium stability of the members under reaction conditions. This interconverting equilibrium shifts towards the ligand exhibiting 'true' binding towards the target, which can then be amplified. The system works by generation of the library and affinity screening in parallel (Niedbała and Jurczak 2020). The study utilizes molecular recognition based on a controlled thermodynamic system (Frei et al. 2019). A thermodynamically controlled system works on Le Chatelier's principle. A shift in synthetic equilibria is achieved in a mixture of compounds possessing stable binding affinity, excluding non-binders and amplification (concentration) of the product. The success of DCC was apparent in the recognition of novel ligands and inhibitors for various protein targets (Huang and Leung 2016). The binding molecules were identified by an analytical tool such as LC-MS or HPLC, which are capable enough to detect the minute changes in the overall library. Some examples are discussed below.

The activity of pseudocholinesterase (butyrylcholinesterase) remains unchanged or is increased during the progression of Alzheimer's disease (Geula and Darvesh 2004) and other disease conditions (Li et al. 2008, Sato et al. 2014). A polymer-based-DCL was developed through reversible acyl hydrazone formation reaction; firstly, an effective binding site was identified and later, information about the amplified product was utilized for synthesizing a specific multivalent potent inhibitor against the tetrameric butyrylcholinesterase enzyme using multiple receptor/ligand interactions (Zhao et al. 2021). The resultant multivalent inhibitor, APG5b, displayed significant selectivity for butyrylcholinesterase over acetylcholinesterase and displayed weak cytotoxicity against tested cancer cell lines.

In the rapid drug discovery process, protein-directed dynamic combinatorial chemistry (PD-DCC) has emerged as a high-powered tool for identification of ligands as a tool (Huang and Leung 2016). The PD-DCC aims at selecting and synthesizing prime binders while escaping the unwanted binders. Even though a valuable and useful tool, the challenges faced by the system include an increase in the equilibration time due to reduced protein activity or inhibition and limited availability of analytical detectors for screening larger libraries. In contrast to PD-DCC, the kinetic target-guided synthesis undergoes an irreversible reaction between the building blocks, requires a near-stoichiometric amount of protein and accelerates the reaction rate, resulting in the stabilization of a ternary complex. The kinetic and thermodynamic targeted studies allow the exploration of unpredicted protein conformations as they do not interfere with the protein's flexibility (Qiu et al. 2019). This microflow-based system succeeded in identifying a previously unreported inhibitor, ethyl octadecenoate, against bovine serum albumin tested as a model protein. The elements required in this approach are the protein template, building blocks, biocompatible alterable chemistry, and a measurable output.

Dynamic combinatorial chemistry has also evolved to modulate protein-protein interaction with the discovery of small-molecule stabilizers against the family of 14-3-3 proteins which have over 500 interaction partners (Hartman et al. 2020). Among the various isoforms of the protein, the authors explored the screening of binders targeting the zeta isoform. The reversible formation of the acylhydrazone linkage was used as the starting point for DCC. Binding affinities were established by surface plasmon resonance and led to the identification of A1H3 and A2H3 acyl hydrazone compounds as potent ligands. Addition of the best binder to a complex of the 29-mer fragment of synaptopodin with 14-3-3 led to reduction in signal by fluorescence polarization, validating the method (Hartman et al. 2020). A similar polymer-based dynamic combinatorial library was successful in producing multivalent inhibitors against the pentameric *E. coli* heat-labile enterotoxin B subunit (Xu et al. 2020). The addition of a template protein as the external stimulus to the library amplified a specific side chain which was used to produce the APG4a inhibitor containing only the amplified ligand. The benefits of multivalent interactions in biochemical systems, combining several weak interactions resulting in high-affinity and robust binding, has been discussed earlier (Fasting et al. 2012).

Serine is required for *de novo* synthesis of purine and deoxythymidine, which supports the proliferation of cancer cells. The enzyme that catalyzes the rate-limiting step of biosynthesis of serine, 3-phosphoglycerate dehydrogenase (PHGDH), has emerged as a potential anti-cancer target (Jing et al. 2015). Screening an in-house library comprising of 2,860 compounds, Zhou et al. identified H-G6 as a potent inhibitor which was able to inhibit PHGDH activity by more than half at 1 μM concentration (Zhou et al. 2021b). In search of more potent inhibitors, the authors employed activity-directed combinatorial chemical synthesis strategy to generate analogs of H-G6. Of these, b36 was found to cause selective cytotoxicity in several cancer cell lines, sparing normal cells (Zhou et al. 2021b).

Staphylococcus aureus belongs to the class of gram-positive bacteria responsible for causing many nosocomial infections. The treatment against the organism is complex because of acquired resistance to different antibiotics, as in the case of methicillin-resistant *S. aureus* (MRSA), and has underlined the need for discovery of targeted drugs through a combinatorial chemistry approach (Rakesh et al. 2018). Enzymes secreted by the bacteria cause structural modifications of the drug at the amino and hydroxyl groups and perturb its functional properties (Mingeot-Leclercq et al. 1999). Starting with resorcinol derivatives and aryloxy acetonitriles, Yu et al. synthesized a library of 148 distinct biarylhydroxyketones (Yu et al. 2014). Of these, 24 molecules were successful in inhibiting hemolysis of rabbit erythrocytes when challenged with MRSA. The presence of a bulky hydrophobic group on either or both of the aromatic rings was deemed essential for hemolytic activity of the antiviral agent (Yu et al. 2014).

Garcia et al. reported the identification of ligands for a bio-macromolecular template by adding a template molecule to perturb the composition of a dynamic combinatorial library (Garcia et al. 2018). The study found a high-affinity binder against the bromodomain-containing protein of *Trypanosoma cruzi* which causes Chagas disease. Significantly, the molecule exhibited low cytotoxicity against Vero cells. Thus, rational analysis of DCL can provide helpful information about the properties of library members against a perturbing stimulus (Karageorgis et al. 2014).

Reversible and irreversible myeloperoxidase inhibitors as anti-inflammatory molecules were identified using a combinatorial library of aldehyde and hydrazine derivatives (Soubhye et al. 2017). Targeted development was carried out by incubating the enzyme with mixtures of aromatic aldehydes (first group) and hydrazine/hydrazide derivatives or aliphatic aldehydes and hydrazine/hydrazide derivatives (second group). Individually, neither the aromatic nor the aliphatic aldehydes showed any significant inhibitory activity against the enzyme while the hydrazines (at 1 μM) showed ~ 60% inhibition. However, incubation with the mixed libraries resulted in 100% inhibition with the first group and ~ 96% inhibition with the second group. The recurring building blocks in the strongest inhibitors were identified and coupled to generate molecules with the maximum enzyme inhibitory activity. The lowest IC_{50} value against the enzyme was 79 nM. Finally, evaluation of the inhibitors in a carrageenan-induced inflammation model in rats suggested that a single dose of irreversible inhibitors could lower the activity of the enzyme released upon activation of neutrophils to a level similar to those in control animals (Soubhye et al. 2017).

Conclusion

Although combinatorial chemistry provides an exciting and rational approach to identifying lead compounds, the strategy has succeeded in identifying only three marketed molecules (where synthetic origins are known) till date (Newman and Cragg 2020). These have been approved as new chemical entities: sorafenib from Bayer as an antitumor compound approved by FDA in 2005, vemurafenib approved by FDA in 2011, and translarna approved in the EU in 2014. One reason for the early failure of HTS was the limited diversities of libraries. Another issue, which has frequently cropped in bulletin boards on combinatorial chemistry in 1990s, has been the purity of the product. With approaches such as dynamic combinatorial chemistry being developed, the numbers being handled have become staggering. Bar coding with DNA has allowed trillions of molecules to be handled in a small microtube. Purity of the molecule during development is also not a concern any longer. Machine learning approaches have been incorporated for understanding structure-activity relationships in combinatorial polymer synthesis (Gormley and Webb 2021). Advances in synthetic biology have permitted the development of combinatorial libraries and engineering of microbes to synthesize metabolites and macromolecules (Sarnaik et al. 2020). Thus, a more realistic and nuanced expectation may succeed in a more fruitful integration of HTS approach with new age combinatorial tools in the drug discovery process.

References

Adan, A., Y. Kiraz and Y. Baran. 2016. Cell proliferation and cytotoxicity assays. Curr. Pharm. Biotechnol. 14: 1213–1221.

Aldib, I., J. Soubhye, K. Zouaoui Boudjeltia, M. Vanhaeverbeek, A. Rousseau, P.G. Furtmuller et al. 2012. Evaluation of new scaffolds of myeloperoxidase inhibitors by rational design combined with high-throughput virtual screening. J. Med. Chem. 55: 7208–7218.

Annis, D., J. Athanasopoulos, P. Curran, J. Felsch, K. Kalghatgi, W. Lee et al. 2004. An affinity selection-mass spectrometry method for the identification of small molecule ligands from self-encoded combinatorial libraries—Discovery of a novel antagonist of *E. coli* dihydrofolate reductase. Int. J. Mass Spectrum. 238: 77–83.

Attene-Ramos, M.S., C.P. Austin and M. Xia. 2014. High Throughput Screening. Elsevier, Amsterdam, The Netherlands.

Aube, J., S.A. Rogers and C. Santini. 2014. Enabling technologies in high throughput chemistry. pp. 1–27. *In*: P. Knochel (ed.). Comprehensive Organic Synthesis (Second Edition). Elsevier.

Baggio, C., L. Cerofolini, M. Fragai, C. Luchinat and M. Pellecchia. 2018. HTS by NMR for the identification of potent and selective inhibitors of metalloenzymes. ACS Med. Chem. Lett. 9: 137–142.

Baggio, C., P. Udompholkul, E. Barile and M. Pellecchia. 2017. Enthalpy-based screening of focused combinatorial libraries for the identification of potent and selective ligands. ACS Chem. Biol. 12: 2981–2989.

Benhamou, R.I., B.M. Suresh, Y. Tong, W.G. Cochrane, V. Cavett, S. Vezina-Dawod et al. 2022. DNA-encoded library versus RNA-encoded library selection enables design of an oncogenic noncoding RNA inhibitor. Proc. Natl. Acad. Sci. U.S.A. 119: e2114971119.

Blay, V., B. Tolani, S.P. Ho and M.R. Arkin. 2020. High-throughput screening: Today's biochemical and cell-based approaches. Drug Discov. Today 10: 1807–1821.

Brenner, S. and R.A. Lerner. 1992. Encoded combinatorial chemistry. Proc. Natl. Acad. Sci. U.S.A. 89: 5381–5383.

Canal-Martin, A. and R. Perez-Fernandez. 2020. Protein-directed dynamic combinatorial chemistry: An efficient strategy in drug design. ACS Omega 41: 26307–26315.

Chang, H.C., C.H. Lin, D.S. Juang, H. Wu, C.Y. Lee, C. Chen et al. 2019. Multilayer architecture microfluidic network array for combinatorial drug testing on 3D-cultured cells. Biofabrication 11: 035024, 1–19.

Cheng, CW., Y. Zhou, WH. Pan, S. Dey, C.Y. Wu, W.L Hsu et al. 2018. Hierarchical and programmable one-pot synthesis of oligosaccharides. Nat. Commun. 9: 5202, 1–9.

Cochrane, W.G., P.R. Fitzgerald and B.M. Paegel. 2021. Antibacterial discovery via phenotypic DNA-encoded library screening. ACS Chem. Biol. 16: 2752–2756.

Cortelo, P.C., D.P. Demarque, R.G. Dusi, L.C. Albernaz, R. Braz-Filho, E.I. Goncharova et al. 2021. A molecular networking strategy: High-throughput screening and chemical analysis of Brazilian cerrado plant extracts against cancer cells. Cells 10: 691, 1–13.

Cuozzo, J.W., M.A. Clark, A.D. Keefe, A. Kohlmann, M. Mulvihill, H. Ni et al. 2020. Novel autotaxin inhibitor for the treatment of idiopathic pulmonary fibrosis: A clinical candidate discovered using DNA-encoded chemistry. J. Med. Chem. 63: 7840–7856.

Cutler, J.I., E. Auyeung and C.A. Mirkin. 2012. Spherical nucleic acids. J. Am. Chem. Soc. 134: 1376–1391.

Decurtins, W., M. Wichert, R.M. Franzini, F. Buller, M.A. Stravs, M.A. Zhang et al. 2016. Automated screening for small organic ligands using DNA-encoded chemical libraries. Nat. Protoc. 4: 764–780.

Du, G.S., J.Z. Pan, S.P. Zhao, Y. Zhu, J.M.J. den Toonder and Q. Fang. 2013. Cell-based drug combination screening with a microfluidic droplet array system. Anal. Chem. 85: 6740–6747.

Du, G.S., J.Z. Pan, S.P. Zhao, Y. Zhu, J.M.J. den Toonder and Q. Fang. 2018. A microfluidic droplet array system for cell-based drug combination screening. Methods Mol. Biol. 1771: 203–211.

Fang, Y. 2012. Ligand-receptor interaction platforms and their applications for drug discovery. Expert Opin. Drug Discov. 7: 969–988.

Fasting, C., C.A. Schalle, M. Weber, O. Seitz, S. Hecht, B. Koksch et al. 2012. Multivalency as a chemical organization and action principle. Angew. Chem. Int. Ed. 51: 10472–10498.

Flood, D.T., C. Kingston, J.C. Vantourout, P.E. Dawson and P.S. Baran. 2020. DNA encoded libraries: A visitor's guide. Isr. J. Chem. 60: 268–280.

Folmer, R.H.A. 2016. Integrating biophysics with HTS-driven drug discovery projects. Drug Discov. Today 3: 491–498.

Frei, P., R. Hevey and B. Ernst. 2019. Dynamic combinatorial chemistry: A new methodology comes of age. Chem. Eur. J. 25: 60–73.

Furka, A. 2002. Combinatorial chemistry: 20 years on.... Drug Discov. Today 7: 1–4.

Furka, A., F. Sebestyen, M. Asgedom and G. Dibo. 1991. General method for rapid synthesis of multicomponent peptide mixtures. Int. J. Pept. Protein Res. 37: 487–493.

Furka, A., F. Sebestyen, M. Asgedom and G. Dibo. 1989a. Abstracts of the 10th International Symposium on Medicinal Chemistry. Elsevier. Amsterdam.

Furka, A., F. Sebestyen, M. Asgedom and G. Dibo. 1989b. Abstracts of the 14th International Congress on Biochemistry. Berlin.

Garcia, P., V.L. Alonso, E. Serra, A.M. Escalante and R.L.E. Furlan. 2018. Discovery of a biologically active bromodomain inhibitor by target-directed dynamic combinatorial chemistry. ACS Med. Chem. Lett. 9: 1002–1006.

Gaurav, A. and V. Gautam. 2014. Structure-based three-dimensional pharmacophores as an alternative to traditional methodologies. J. Recept. Channel Res. 7: 27–38.

Geula, C. and S. Darvesh. 2004. Butyrylcholinesterase, cholinergic neurotransmission and the pathology of Alzheimer's disease. Drugs Today (Barc.) 40: 711–721.

Gironda-Martinez, A., E.J. Donckele, F. Samain and D. Neri. 2021. DNA-encoded chemical Libraries: A comprehensive review with successful stories and future challenges. ACS Pharmacol. Transl. Sci. 4: 1265–1279.

Gormley, A.J. and M.A. Webb. 2021. Machine learning in combinatorial polymer chemistry. Nat. Rev. Mater. 6: 642–644.

Grygorenko, O.O., D.M. Volochnyuk, S.V. Ryabukhin and D.B. Judd. 2020. The symbiotic relationship between drug discovery and organic chemistry. Chem. Eur. J. 26: 1196–1237.

Guan, M., W. Yao, R. Liu, K.S. Lam, J. Nolta, J. Jia et al. 2012. Directing mesenchymal stem cells to bone to augment bone formation and increase bone mass. Nat. Med. 18: 456–462.

Halford, B. 2017. How DNA-encoded libraries are revolutionizing drug discovery. Chem. Eng. News 25: 28–33.

Han, A.R., T. Durgannavar, D. Ahn and S.J. Chung. 2021. A FRET-based fluorescent probe to screen anticancer drugs, inhibiting p73 binding to MDM2. ChemBioChem 22: 830–833.

Harris, P.A., B.W. King, D. Bandyopadhyay, S.B. Berger, N. Campobasso, C.A. Capriotti et al. 2016. DNA-encoded library screening identifies benzo [b][1, 4] oxazepin-4-ones as highly potent and monoselective receptor interacting protein 1 kinase inhibitors. J. Med. Chem. 5: 2163–2178.

Hartman, A.M., W.A. Elgaher, N. Hertrich, S.A. Andrei, C. Ottmann and A.K. Hirsch. 2020. Discovery of small-molecule stabilizers of 14-3-3 protein–protein interactions via dynamic combinatorial chemistry. ACS Med. Chem. Lett. 11: 1041–1046.

Huang, Y., Y. Li and X. Li. 2022. Strategies for developing DNA-encoded libraries beyond binding assays. Nat. Chem. 14: 129–140.

Huang, R. and I.K.H. Leung 2016. Protein-directed dynamic combinatorial chemistry: A guide to protein ligand and inhibitor discovery. Molecules 21: 910.

Ivanenkov, Y.A., I.A. Osterman, E.S. Komarova, A.A. Bogdanov, P.V. Sergiev, O.A. Dontsova et al. 2020. Tetrahydrocarbazoles as novel class of DNA biosynthesis inhibitors in bacteria. Anti-Infective Agents 18: 121–127.

Janzen, W.P. 2014. Screening technologies for small molecule discovery: The state of the art. Chem. Bio. 21: 1162–1170.

Jing, Z., W. Heng, L. Xia, W. Ning, Q. Yafei, Z. Yao et al. 2015. Downregulation of phosphoglycerate dehydrogenase inhibits proliferation and enhances cisplatin sensitivity in cervical adenocarcinoma cells by regulating Bcl-2 and caspase-3. Cancer Biol. Ther. 16: 541–548.

Jorgensen, W.L. 2004. The many roles of computation in drug discovery. Science 303: 1813–1818.

Karageorgis, G., S. Warriner and A. Nelson. 2014. Efficient discovery of bioactive scaffolds by activity-directed synthesis. Nat. Chem. 6: 872–876.

Kitakaze, K., S. Taniuchi, E. Kawano, Y. Hamada, M. Miyake, M. Oyadomari et al. 2019. Cell-based HTS identifies a chemical chaperone for preventing ER protein aggregation and proteotoxicity. eLife 8: e43302.

Kwon, C.H., I. Wheeldon, N.N. Kachouie, S.H. Lee, H. Bae, S. Sant et al. 2011. Drug-eluting microarrays for cell-based screening of chemical-induced apoptosis. Anal. Chem. 83: 4118–4125.

Lerner, R.A. and D. Neri. 2020. Reflections on DNA-encoded chemical libraries. Biochem. Biophys. Res. Commun. 527: 757–759.

Li, B., E.G. Duysen and O. Lockridg. 2008. The butyrylcholinesterase knockout mouse is obese on a high-fat diet. Chem. Biol. Interact. 175: 88–91.

Li, G., Y. Luo, Q. Chen, L. Liao and J. Zhao. 2012. A "place n play" modular pump for portable microfluidic applications. Biomicrofluidics 6: 014118-1-014118–16.

Li, J., P. Nowak and S. Otto. 2013. Dynamic combinatorial libraries: From exploring molecular recognition to systems chemistry. J. Am. Chem. Soc. 135: 9222–9239.

Liu, W., Y. Cui, W. Ren and J. Irudayaraj. 2019. Epigenetic biomarker screening by FLIM-FRET for combination therapy in ER+ breast cancer. Clin. Epigenetics 11: 16, 1–9.

Liu, Y. and G. Li. 2018. A power-free, parallel loading microfluidic reactor array for biochemical screening. Sci. Rep. 8: 13664, 1–9.

Lloyd, M.D. 2020. High-throughput screening for the discovery of enzyme inhibitors. J. Med. Chem. 63: 10742–10772.

Lo, C.H., N.K. Pandey, C.K.W. Lim, Z. Ding, M. Tao, D.D. Thomas et al. 2020. Discovery of small molecule inhibitors of huntingtin exon 1 aggregation by FRET-based high-throughput screening in living cells. ACS Chem. Neurosci. 11: 2286–2295.

Long, S.A., S. Huang, A. Kambala, L. Ren, J. Wilson, M. Goetz et al. 2021. Identification of potential modulators of osteosarcoma metastasis by high-throughput cellular screening of natural products. Chem. Biol. Drug Des. 97: 77–86.

Lu, Y., S. Qin, B. Zhang, A. Dai, X. Cai, M. Ma et al. 2019. Accelerating the throughput of affinity mass spectrometry-based ligand screening toward a G protein-coupled receptor. Anal. Chem. 91: 8162–8169.

Macarron, R. and R.P. Hertzberg. 2009. Design and implementation of high-throughput screening assays. Methods Mol. Biol. 565: 1–32.

Macarron, R., M.N. Banks, D. Bojanic, D.J. Burns, D.A. Cirovic, T. Garyantes et al. 2011. Impact of high-throughput screening in biomedical research. Nat. Rev. Drug Discov. 10: 188–195.

McKim, J.M. 2010. Building a tiered approach to in vitro predictive toxicity screening: A focus on assays with *in vivo* relevance. Comb. Chem. High Throughput Screen 13: 188–206.

Merrifield, R.B. 1963. Solid phase peptide synthesis. I. The synthesis of a tetrapeptide. J. Am. Chem. Soc. 85: 2149–2154.

Mingeot-Leclercq, M.P., Y. Glupczynski and P.M. Tulkens. 1999. Aminoglycosides: Activity and resistance. Antimicrob. Agents Chemother. 43: 727–737.

Mondal, M. and A.K.H. Hirsch. 2015. Dynamic combinatorial chemistry: A tool to facilitate the identification of inhibitors for protein targets. Chem. Soc. Rev. 44: 2455–2488.

Murray, D. and M. Wigglesworth. 2016. HTS methods: Assay design and optimisation. pp. 1–15. *In*: High throughput Screening Methods: Evolution and Refinement. The Royal Society of Chemistry, London.

Neri, D. and R.A. Lerner. 2018. DNA-encoded chemical libraries: A selection system based on endowing organic compounds with amplifiable information. Annu. Rev. Biochem. 87: 479–502.

Newman, D.J. and G.M. Cragg. 2020. Natural products as sources of new drugs over the nearly four decades from 01/1981 to 09/2019. J. Nat. Prod. 83: 770–803.

Niedbała, P. and J. Jurczak. 2020. One-pot parallel synthesis of unclosed cryptands-searching for selective anion receptors via static combinatorial chemistry techniques. ACS Omega 5: 26271–26277.

Osterman, I.A., E.S. Komarova, D.I. Shiryaev, I.A. Korniltsev, I.M. Khven, D.A. Lukyanov et al. 2016. Sorting out antibiotics' mechanisms of action: A double fluorescent protein reporter for high-throughput screening of ribosome and DNA biosynthesis inhibitors. Antimicrob. Agents Chemother. 60: 7481–7489.

Owicki, J.C. 2000. Fluorescence polarization and anisotropy in high throughput screening: Perspectives and primer. J. Biomol. Screen 5: 297–306.

Pu, F., A.J. Radosevich, J.W. Sawicki, D. Chang-Yen, N.N. Talaty, S.M. Gopalakrishnan et al. 2021. High-throughput label-free biochemical assays using infrared matrix-assisted desorption electrospray ionization mass spectrometry. Anal. Chem. 93: 6792–6800.

Qiu, C., Z. Fang, L. Zhao, W. He, Z. Yang, C. Liu et al. 2019. Microflow-based dynamic combinatorial chemistry: A microscale synthesis and screening platform for the rapid and accurate identification of bioactive molecules. React. Chem. Eng. 4: 658–662.

Rajalingham, K. 2016. Cell-based assays in high-throughput mode (HTS). BioTechnologia 97: 227–234.

Rakesh, K.P., M.H. Marichannegowda, S. Srivastava, X. Chen, S. Long, C.S. Karthik et al. 2018. Combating a master manipulator: *Staphylococcus aureus* immunomodulatory molecules as targets for combinatorial drug discovery. ACS Comb. Sci. 20: 681–693.

Reymond, J.L. and P. Babiak. 2007. Screening systems. Adv. Biochem. Eng. Biotechnol. 105: 31–58.

Richter, H., A.L. Satz, M. Bedoucha, B. Buettelmann, A.C. Petersen, A. Harmeier et al. 2019. DNA-encoded library-derived DDR1 inhibitor prevents fibrosis and renal function loss in a genetic mouse model of Alport syndrome. ACS Chem. Biol. 14: 37–49.

Roehrl, M.H.A., S. Kang, J. Aramburu, G. Wagner, A. Rao and P.G. Hogan. 2004. Selective inhibition of calcineurin-NFAT signaling by blocking protein-protein interaction with small organic molecules. Proc. Natl. Acad. Sci. U.S.A. 101: 7554–7559.

Sarnaik, A., A. Liu, D. Nielsen and A.M. Varman. 2020. High-throughput screening for efficient microbial biotechnology. Curr. Opin. Biotechnol. 64: 141–150.

Sato, K.K., T. Hayashi, I. Maeda, H. Koh, N. Harita, S. Uehara et al. 2014. Serum butyrylcholinesterase and the risk of future type 2 diabetes: The Kansai Healthcare Study. Clin. Endocrinol. 80: 362–367.

Shuker, S.B., P.J. Hajduk, R.P. Meadows and S.W. Fesik. 1996. Discovering high-affinity ligands for proteins: SAR by NMR. Science 274: 1531–1534.

Sioud, M. 2019. Phage display libraries: From binders to targeted drug delivery and human therapeutics. Mol. Biotechnol. 61: 286–303.

Soubhye, J., M. Gelbcke, P. Van Antwerpen, F. Dufrasne, M.Y. Boufadi, J. Neve et al. 2017. From dynamic combinatorial chemistry to *in vivo* evaluation of reversible and irreversible myeloperoxidase inhibitors. ACS Med. Chem. Lett. 8: 206–210.

Stoddart, L.A., C.W. White, K. Nguyen, S.J. Hill and K.D.G. Pfleger. 2016. Fluorescence and bioluminescence-based approaches to study GPCR ligand binding. Br. J. Pharmacol. 173: 3028–3037.

Stroik, D.R., S.L. Yuen, K.A. Janicek, T.M. Schaaf, J. Li, D.K. Ceholski et al. 2018. Targeting protein-protein interactions for therapeutic discovery via FRET-based high-throughput screening in living cells. Sci. Rep. 8: 12560, 1–13.

Sunkari, Y.K., V.K. Siripuram, T.L. Nguyen and M. Flajolet. 2022. High-power screening (HPS) empowered by DNA-encoded libraries. Trends Pharmacol. Sci. 43: 4–15.

Wu, B., Z. Zhang, R. Noberini, E. Barile, M. Giulianotti, C. Pinilla et al. 2013. HTS by NMR of combinatorial libraries: A fragment-based approach to ligand discovery. Chem. Biol. 20: 19–33.

Wu, J., I. Wheeldon, Y. Guo, T. Lu, Y. Du, B. Wang et al. 2011. A sandwiched microarray platform for benchtop cell-based high throughput screening. Biomaterials 32: 841–848.

Xu, J., S. Zhang, S. Zhao and L. Hu. 2020. Identification and synthesis of an efficient multivalent *E. coli* heat labile toxin inhibitor—A dynamic combinatorial chemistry approach Bioorg. Med. Chem. 28: 115436, 1–8.

Yamankurt, G., E.J. Berns, A. Xue, A. Lee, N. Bagheri, M. Mrksich et al. 2019. Exploration of the nanomedicine-design space with high-throughput screening and machine learning. Nat. Biomed. Eng. 3: 318–327.

Yu, G., D. Kuo, M. Shoham and R. Viswanathan. 2014. Combinatorial synthesis and *in vitro* evaluation of a biaryl hydroxyketone library as antivirulence agents against MRSA. ACS Comb. Sci. 16: 85–91.

Zhao, S., J. Xu, S. Zhang, M. Han, Y. Wu, Y. Li et al. 2021. Multivalent butyrylcholinesterase inhibitor discovered by exploiting dynamic combinatorial chemistry. Bioorg. Chem. 108: 104656, 1–8.

Zheng, W., N. Thorne and J.C. McKew. 2013. Phenotypic screens as a renewed approach for drug discovery. Drug Discov. Today 18: 1067–1073.

Zhou, M., A. Varol and T. Efferth. 2021a. Multi-omics approaches to improve malaria therapy. Pharmacol. Res. 167: 105570, 1–14.

Zhou, X., Y. Tan, K. Gou, L. Tao, Y. Luo et al. 2021b. Discovery of novel inhibitors of human phosphoglycerate dehydrogenase by activity-directed combinatorial chemical synthesis strategy. Bioorg. Chem. 115: 105159, 1–14.

Chapter 4

Fundamentals and Applications of Solid/Gas Biocatalysis

Sergio Huerta-Ochoa,[1,*] *Carlos Omar Castillo-Araiza,*[2]
Itza Nallely Cordero-Soto,[1] *Yahir Alejandro Cruz-Martínez,*[1]
Lilia Arely Prado-Barragán[1,*] and *Olga Miriam Rutiaga Quiñones*[3]

Introduction

Solid/gas (SG) biocatalysis is a non-conventional catalytic process, which is carried out by immobilizing enzymes or whole-cells on a solid support through which the substrates flow in the gas phase using the ability of some enzymes to catalyze reactions with vaporized substrates (Kulishova and Zharkov 2017). To the best of our knowledge, the early SG biocatalytic systems were developed by the research groups of Professors Marie-Dominique Legoy (Pulvin et al. 1986) and Alexander M. Klibanov (Bárzana et al. 1987) in the mid-1980s. In their application, Pulvin et al. (1986) used horse liver alcohol dehydrogenase [E.C. 1.1.1.1] along with the coenzyme for the aldehyde production, coimmobilized onto albumin-glutaraldehyde porous particles. The vapors of acetaldehyde and butanol flowed through the reactor by a carrier gas saturated with water vapor. One year later, Bárzana et al. (1987) implemented another SG biocatalysis system, using alcohol oxidase (EC 1.1.3.13) as biocatalyst from *Pichia pastoris*, catalase (EC 1.11.1.6) from *Aspergillus niger*, and peroxidase (EC 1.11.1.7) from horseradish roots to oxidase ethanol vapors with molecular oxygen.

---

[1] Biotechnology Department, Universidad Autónoma Metropolitana Iztapalapa, México City, MEXICO.
[2] Laboratory of Catalytic Reactor Engineering Applied to Chemical and Biological Systems. Departamento de IPH, Universidad Autónoma Metropolitana-Iztapalapa, México City, MEXICO.
[3] Department of Chemistry-Biochemistry, Tecnológico Nacional de México/Instituto Tecnológico de Durango, Durango, MEXICO.
* Corresponding authors: sho@xanum.uam.mx; lapb@xanum.uam.mx

Technical characteristics of SG biocatalysis consist of enhanced biocatalyst stability, and reduced mass transfer resistances, thus offering technological, environmental, and economic advantages. These result in SG biocatalysis system becoming a white biotechnology process. Therefore, these aspects have led to the development of three potential applications: (a) the analysis of gaseous compounds (biosensors) in a diversity of sectors, among them medicine, food industry, environmental monitoring, metabolism, agriculture, military, and security (Bankole et al. 2022); (b) the development of SG bioreactors for the synthesis of high value volatile organic compounds of interest for the pharmaceutical, cosmetic and food industries (Cordero-Soto et al. 2020, García-Martínez et al. 2022); and (c) the degradation of toxic volatile organic compounds used in industry and agriculture as herbicides, insecticides, solvents or intermediates for chemical synthesis (Erable et al. 2005, Erable et al. 2009).

Finally, in this chapter, the effects of thermodynamics, kinetics, and transport aspects on the performance of the biocatalyst are analyzed in terms of engineering fundamentals, which are relevant for their conceptual design and operation of the SG for various applications (Cordero-Soto et al. 2020, García-Martínez et al. 2022).

Definition and description of the SG biocatalysis

SG biocatalysis definition

The relevant research findings in the 1980s about enzymatic catalysis in organic solvents led to the tempting possibility to carry it with reactants in the gas phase which involved attractive reactions and applications of industrial interest (Pulvin et al. 1986, Bárzana et el. 1987). SG biocatalysis has been defined as the use of lyophilized biocatalysts (purified enzyme or cells), non-supported or immobilized in a solid phase, in which substrates in the gaseous phase react selectively to produce the desired compounds. Due to the SG system, it is possible to control thermodynamic activities of the species involved during the reactions such as substrates, products, and water. Thus the SG system becomes a powerful tool to investigate the influence of enzyme microenvironment on catalysis (Bousquet-Dubouch et al. 2001, Dunn and Daniel 2004).

Main factor influencing SG biocatalysis

Water activity

During the SG biocatalytic reaction, the biocatalyst is active even at low hydration levels below the monolayer coverage. Water plays an important role during the operation of SG biocatalytic system such that water molecules can be adsorbed on the enzyme as part of the protein structure, modify the microenvironment of the active site or even participate as a reactant during the substrate transformation (Fig. 1). Moreover, the hydration degree of the enzymes in water limited systems is an important operational factor since it impacts their flexibility, stability, structure, dynamics, and function (Yang and Russell 1996a, b, Kurkal et al. 2005, Bellissent-

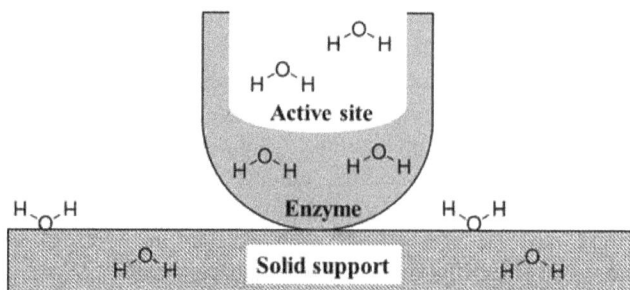

Figure 1. Water molecules in low water SG biocatalytic system.

Funel et al. 2016). Counter-ions strongly associate with charges in the protein or its substrates in low-dielectric media provoking large effects on enantioselectivity (Halling 2000). Additionally, Dimoula et al. (2009) observed that water adsorption, during the acetophenone reduction in continuous SG reactor, decreased the amount of adsorbed acetophenone, by nearly 25%, at a a_w of 0.54 impacting the reaction rate. On the other hand, water plays an important role in affecting the thermodynamic parameters of the reaction and influencing the kinetics of the reactions performed (Graber et al. 2003, Lamare et al. 2004). Besides, lower hydration level enhances the thermostability of the biocatalyst and increases the optimum temperature of the gas-phase reaction (Bárzana et al. 1989a, Lamare et al. 2004).

Operational pressure

Operational pressure in the SG bioreactor is an important parameter since reduced pressure has interesting technical advantages. Among these are, increases the use of substrates with higher boiling points, raises substrate concentration, allows the control of the thermodynamic activity around the enzyme as it varies the partial pressure of the substrates in the gas phase, decrease the carrier gas consumption, and improve the efficiency of the product condensation (Mizobuchi and Nagayama 2015). Debeche et al. (2005) observed a low-pressure drop during the passage of reactants through a reactor packed using two different fiber supports (a glass fiber and a carbon fiber) using *Psedomonas cepaciae* lipase PS to catalyze hydrolysis and transesterification reactions.

Supports

SG biocatalysis can be carried out with non-supported or immobilized biocatalysts; however, at higher water activities, denaturation and aggregation of the catalysts could take place when they are non-supported. Therefore, catalyst immobilization is a useful technique for SG biocatalysis as there is no desorption of the catalysts from the support during the process. The main desirable characteristics of the supports are high mechanical, thermal and chemical stability, and high surface area. The hydrophilic/hydrophobic properties of the support influence the outcome of the immobilization process and thereby of the biocatalytic reaction. The most widely used are porous supports that present a high surface area allowing high protein loading.

For instance, *Candida antarctica* lipase B from Novozym® (Csanádi et al. 2012); and *Candida antarctica* lipase B from Roche (Bousquet-Dubouch et al. 2001) were immobilized on the macro-porous acrylic resin with specific surface areas (m² g⁻¹) of about 130 and 4–6, respectively. Nevertheless, Alcohol dehydrogenase (ADH) from *Parvibaculum lavamentivorans* (Sigma-Aldrich) was supported on non-porous glass beads (Mizobuchi and Nagayama 2015) with a protein load of 0.8–4.0 mg protein (g support)⁻¹. Other types of support have also been studied. For instance, Debeche et al. (2005) observed that the reaction rate of ethyl acetate hydrolysis catalyzed by *Psedomonas cepaciae* was around ten times higher when the lipase PS was adsorbed onto carbon fiber support as compared to glass fiber support. Both supports have similar specific surface areas and a protein load in the range of 1–20 mg protein (g support)⁻¹ but differ in their hydrophilic/hydrophobic properties. While the glass fibers are highly hydrophilic, the carbon fibers are hydrophobic.

Other aspects

Although the SG system can work as a continuous process removing substrate and products from the packed bed with the carrier gas, inhibition of enzyme activity by the product has been reported. For instance, gas-phase oxidation of ethanol vapors with molecular oxygen catalyzed by an alcohol oxidase adsorbed on DEAE-cellulose was inhibited by the product hydrogen peroxide. This inhibition can be reduced by addition of catalase or peroxidase to the dry formulation (Bárzana et al. 1987). In another example, the gas-phase dehalogenation reaction of 1-Chlorobutane using the haloalkane dehalogenase from *Rhodococcus erythropolis* NCIMB 13064 was inhibited by HCl production. This inhibition was reduced using a triethylamine volatile buffer for pH control (Erable et al. 2005).

Technological, environmental, and economic advantages of the SG system

The introduction of enzymes or cells into an SG system where the thermodynamic constraints are minimized offers important advantages (Table 1A). Furthermore, SG biocatalysis is an emerging technology for basic research and the development of new cleaner industrial processes (Lamare et al. 2004, Cordero-Soto et al. 2020) presenting environmental and economic advantages (Table 1B). These characteristics widens the potential industrial applications of biocatalysts. Among these applications biosensors, bioconversion processes, and remediation of flue gases are widely reported in the literature.

SG Biosensors

The biosensors technology development and commercialization started around the 60s, the first device developed was the enzymatic sensor to determine the glucose concentration in blood through the catalyzed glucose oxidase reaction (Patel 2002). The first biosensor was a glucose analyzer developed by Clark and Lyons in 1962 and marketed in 1975 by Yellow Springs Instrument Company. This biosensor was named "enzyme electrode" and consisted of a glucose oxidase enzyme linked to an oxygen electrode. However, the term biosensor began to be used in 1977

Table 1. Various advantages of the SG system.

Advantages	References
(A) Technological advantages	
There are no problems of solubility of substrates and products due to the absence of solvents in the system.	Goubet et al. 2002, Kulishova and Zharkov 2017
Mass transfer limitations are negligible as diffusivities of reactants and products in the gaseous phase are orders of magnitude higher than in water	Bárzana et al. 1989a, Goubet et al. 2002, Kulishova and Zharkov 2017
Diffusivities increase in proportion to the temperature at a power of 1.5, which is higher than those in water	Goubet et al. 2002, Kulishova and Zharkov 2017
Continuous processes are easily designed	Goubet et al. 2002, Kulishova and Zharkov 2017
The low a_w brings higher temperature stability of the biocatalysts, avoids microbial contamination, and allows to switch equilibrium constant to synthesis instead of hydrolysis reactions	Kulishova et al. 2010, Cordero-Soto et al. 2020, García-Martínez et al. 2022
Separation of the gaseous mixture (substrates and products) can be accomplished via a single step of fractional condensation	Bárzana et al. 1989a, Kulishova and Zharkov 2017
(B) Environmental and economic advantages	
The absence of solvents reduces contamination wastes	Goubet et al. 2002, Kulishova and Zharkov 2017
High production rates for minimal plant sizes reducing costs	Lamare et al. 2004
Eco-friendly product recovery using simplified downstream processes	Lamare et al. 2004
Development of white biotechnology processes	Cordero-Soto et al. 2020

when the first device used live immobilized microorganisms on the surface of an ammonia sensitive electrode. This device was used to detect the amino acid arginine and it was called a "bio-selective sensor" and later on "biosensor" and this term has remained since then to designate the union between biological material and a physical transducer. From that moment on, the design and applications of biosensors in different fields of analytical chemistry have continued to grow.

Biosensors are devices that provide qualitative, quantitative or semi-quantitative information on the surrounding environment based on specific biochemical reactions, the biosensors belong to a subgroup of chemical sensors that incorporates a biological or biomimetic sensing element and are defined as a compact analysis device that incorporates a biological recognition element (nucleic acid, ribosome, enzyme, antibody, receptor, tissue and cell) or biomimetic (polymers of intrinsic microporosity (PIMs), aptamers, nucleic acid probes (PNAs)) combined with a transduction system processing the signal caused by the interaction between the recognition element and the analyte.

Some of the advantages of the biosensors are that they generate the physicochemical, enzymatical, biological, microbiological, compositional, and clinical/health signals, with high sensitivity, selectivity, and high reproducibility. Low cost and long-life of devices, easy usability (no specialized personnel needed), ease of transport and a quick response analysis are other useful features of biosensors. They are automatable and can be miniaturized, allow multi-tasking , and require few operative elements; can be incorporated into microscopic systems, and also allow to application of to real-time monitoring (Patel 2002, Mello and Kubota 2002, Velasco-García and Mottram 2003).

Biosensors classification

To build a biosensor, each component must be identified separately and then, an assembly of the different components leads to an integrated device. In the case of enzymatic biosensors, the recognition element is an enzyme or the biochemical change that the enzyme is catalyzing which may be accompanied by the change of the emitted signal (light, pH, heat, mass change or gas). Biosensors are classified according to their type of interactions between the recognition element and transduction system (Fig. 2).

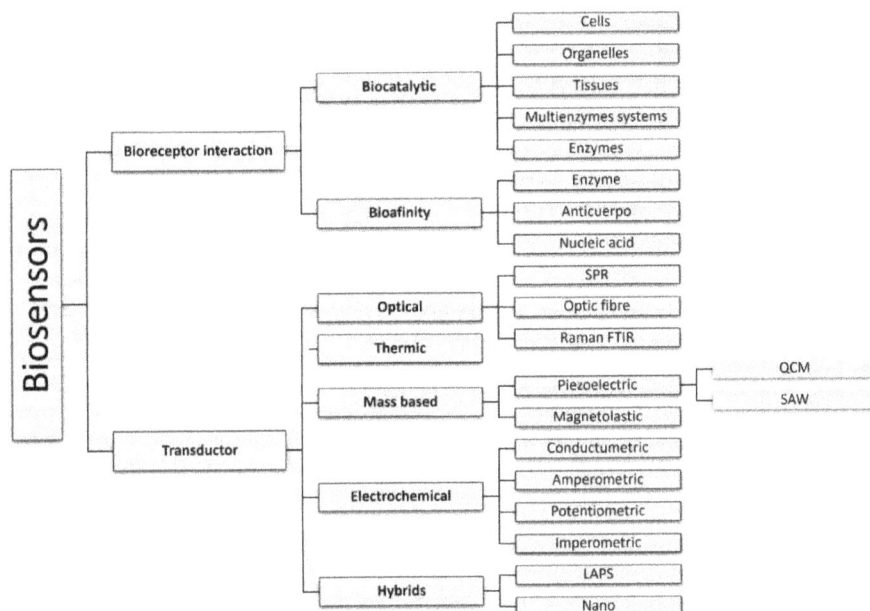

Figure 2. Biosensors classification. SPR (Surface Plasmon Resonance); FTIR (Fourier Transformed InfraRed); QCM (Quartz Crystal Microbalance); SAW (Surface Acoustic Wave); LAPS (Light-Addressable Potentiometric Transducer).

Type of interaction: biocatalytic or bio-affinity

Biocatalytic sensors

They are based on the use of biocatalysts, which are elements that facilitate a chemical reaction in which one or more known products are formed from one or more substrates without consumption of the biocatalyst, which is regenerated and can be used again. They can be used to perceive the presence of the substrates of the reaction, by detecting their disappearance of some co-substrate other than of the one to be detected or by the appearance of some known product. They are made up by subcellular organelles, whole cells, tissues, enzymes, multienzyme systems, or by the means of the use of reversible enzyme inhibitors; and are characterized by their regenerative capacity. However, among the typical recognition elements of this type of interaction, the enzymes are more frequently used. They are immobilized molecules of protein or ribozyme nature, highly specific and very efficient. The response signal can be measured by combining the signal response to a calorimetric, potentiometric, amperometric, piezoelectric or optoelectric transducer (Halámek et al. 2005). Some enzymes are not stable when isolated, or their purification is difficult or very expensive, in these circumstances, cell organelles, whole cells, or tissues containing the enzymes in a natural environment are used.

Bio-affinity sensors

They are based on the interaction of the analyte with the recognition element, without a catalytic transformation, but rather produce an equilibrium reaction in which an analyte-receptor complex is formed (Wang 2005). Such interactions generate signals that require to be detected by high sensitivity systems. To measure the interaction (analyte-receptor), since there is no consumption of substrates or product generation, a competing element that competes for the binding site with the analyte for the receptor, such as an enzyme that gives a complementary biocatalytic reaction which is detected by the transduction system. A system for direct detection of the interaction between the receptor and analyte is based on changes that occur on the surface, or by optical changes that occur as a consequence of this interaction. This type of interaction sometimes has an operating range of narrow concentrations, because the saturation of the receptor can occur and often does not allow continuous monitoring of analyte concentration. The recognition element must be in direct contact with the sample and may not incorporate an outer membrane to separate the recipient element from the matrix of the sample, so some of these biosensors have difficulties operating in complex biological matrices. There are different types of bio-affinity receptors (other than enzymes) such as antibodies, lectins, receptors, whole cells, nucleic acids, molecular imprinting polymers (MIPs), aptamers and peptide nucleic acids (PNAs) (Wang 2005).

Recognition element

A recognition element may be selected, and it can be either biological or biomimetic, the material depends on the characteristics of the compound to analyze. For example, when trying to detect an allergenic substance, a biological recognition element is selected, then antibodies are used. Among the biological ones, the recognition elements frequently used are antibodies, enzymes, organelles, whole cells, tissues, and nucleic acids; while the most used biomimetic elements are MIPs, PNAs, and aptamers (Syazana and Minhaz 2017).

Transduction system

The nature of the interaction between the element of recognition and the analyte constitutes a determining factor for the selection of the transduction system since it determines the variation in the physicochemical properties that will occur as a consequence of the interaction. The most referenced transduction systems are: Optical transducers (Patel 2002, Mello and Kubota 2002, Velasco-García and Mottram 2003), fiber optic sensors, surface plasmon resonance-based systems, electrochemical transducers, mass based or nanomechanical (Tamayo et al. 2003), magnetoelastic transducers (Skinner et al. 2022), thermometric transducers, and hybrid transducers.

Industrial applications of the SG biosensors

The introduction of enzyme reactions or whole cells in a SG system has important advantages in terms of economics, technical features, time/sample-analysis, qualitative/quantitative diagnosis, non-invasive methods; becoming a real tool for the detection of several molecules that, in turn, provide relevant information about issues related to the monitoring of the presence of biomarkers in several processes (Fig. 3) related to the food industry, clinical/health monitoring systems, cosmetic

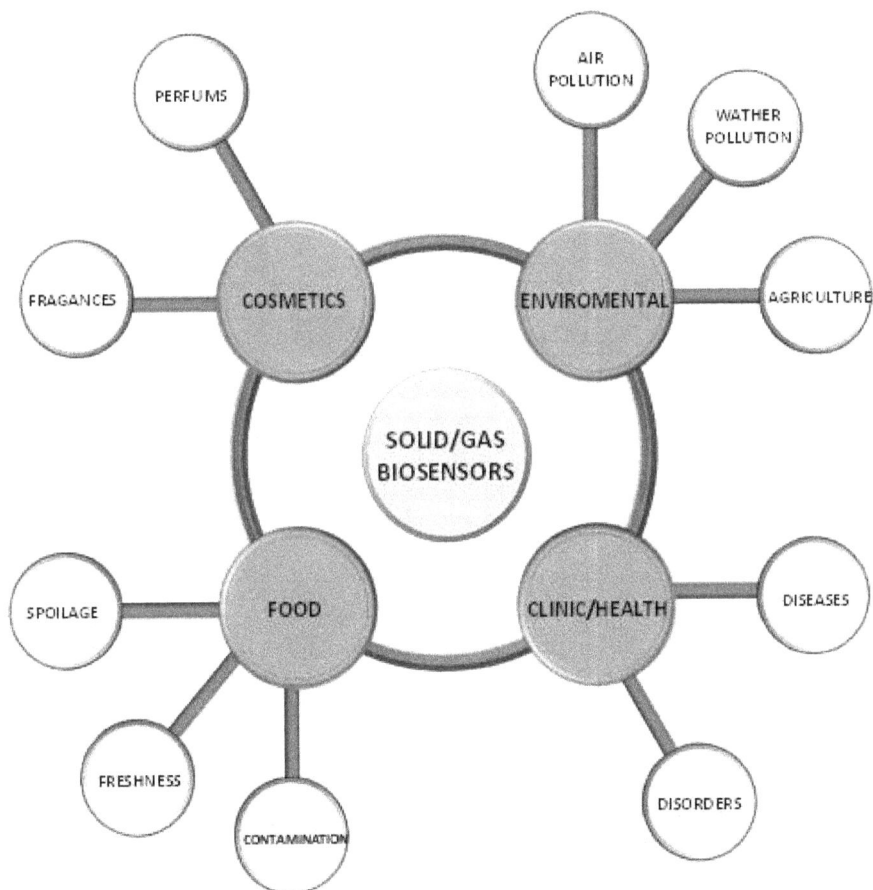

Figure 3. Industrial applications of the SG biosensors.

and pharmaceutical sectors, widening the scope for the potential applications of biocatalysts. In food analysis, it has been encouraged by an increased demand for quality from food safety authorities, sellers, and consumers, and a reduction in the analytical times linked with electrochemical detection of many quality factors and the need for very effective detection in-process operations, and for the effective identification of contaminants. The limitations of conventional food analysis methods, which require expensive and time-consuming devices, have guided the development of biosensors. Biosensors cater to the researchers who need inexpensive and small sensors that can be operated online.

Food industry

In food matrices, xenobiotic compounds such as pesticides, dioxins, drugs, additives, and polyaromatic hydrocarbons stand out in the chemical composition analysis. Also, pathogens and toxins of bacterial origin, traceability processes and genetically modified organisms, allergens, antinutrients, control process and stability (Table 2).

Table 2. Applications of the SG biosensors in the food industry.

Biomarker	Disorder/indicator	Reference
Acetic acid	Meat, fish, or fruit spoilage	Tampio et al. 2019, Fortin et al. 2009, Cova et al. 2022
Acetaldehyde	Freshness indication	Tampio et al. 2019, Fortin et al. 2009, Cova et al. 2022
Alcohols	Freshness indication	Tampio et al. 2019, Fortin et al. 2009, Cova et al. 2022
1-Butanol	Freshness indication	Preis et al. 2013
Amylbutyrate	Food safety diagnostics	Jin et al. 2012
Butyric acid	Food spoilage	Tampio et al. 2019, Fortin et al. 2009, Cova et al. 2022
Caproic acid	Food spoilage	Tampio et al. 2019, Fortin et al. 2009, Cova et al. 2022
2-Ethyl-hexanol	Freshness indication	Preis et al. 2013
1-Hexanol	Freshness indication, salmonella contamination	Preis et al. 2013, Alessandroni et al. 2022
3-Methyl-1-butanol	Salmonella contamination	Alessandroni et al. 2022
Trimethylamine	Fish spoiling	Odeyemi et al. 2018
Valeric acid	Food decomposition	Tampio et al. 2019, Fortin et al. 2009, Cova et al. 2022

Clinical/Health industry

A diseased organ or tissue may release gaseous metabolites (biomarkers) that inhibit or activate the expression of certain specific enzymes and depending on the time-line of the disease and the severity of the condition, their concentrations in the breath increase. Therefore, the use of enzymes as a biological recognition element for the detection and diagnosis of these compounds is quite promising. In spite of some difficulties (sample collection and concentration, oral hygienic habits) that have to be overcome during the breath monitoring, analysis by biosensors is much easier than traditional tests based upon blood, urine, sputum and feces (Das et al. 2020) for clinical analysis (Table 3).

Table 3. Applications of the SG biosensors in Clinic/Health industry.

Biomarker	Disorder/indicator	Reference
Acetone	Diabetes, Diabetic ketoacidosis, Starvation	Toyooka et al. 2013, Yang et al. 2016
Aldehydes	Alcoholic liver diseases, Alzheimer's, atherosclerosis, childhood cancer, diabetes, hemochromatosis, lung cancer, metabolic/genetic disorders in synthesis or metabolism of aldehydes, oxidative stress, Parkinson's, amyotrophic lateral sclerosis, smoking Wernickle's encephalopathy Wilson's diseases	O'Brien et al. 2005, Ivanova 2002, Li et al. 2003, Shinpo et al. 2000
Ammonia	*Excess*: Alzheimer's type II, brain swelling, hepatic encephalopathy, kidney failure, liver dysfunction, peptide ulcer *Reduced*: Asthma, inflammation in airways	Berg et al. 2002, Kim et al. 2018, Essiet 2013, Butterworth 2003, Kearney et al. 2002, Amano 2002, Van den Broek et al. 2007
Ethane	Anorexia, irritable bowel syndrome, inflammatory bowel diseases, oxidative stress, ulcerative colitis, vitamin E deficiency	Mazzone et al. 2007, Knutson et al. 2000, Wendland et al. 2001, O'Brien et al. 2005
Hydrogen sulfide	Asthma, dental health, inflammation in airways	Suarez et al. 2000, Essiet 2013
Isoprene	*Excess*: Diabetes, lipids metabolism disorder, *Reduced*: Lung cancer	McGrath et al. 2001, Hibbard and Killard 2011, Bajtarevic et al. 2009
Nitric oxide	Asthma, cystic fibrosis, liver transplant rejection, pulmonary obstruction	Barnes and Kharitonov 1996, Maziak et al. 1998, Dotsch et al. 1996, Wang 2002
Pentane	Apnea, Arthritis, ischemic heart disease, myocardia infraction, liver diseases, oxidative stress, physical and mental stress, schizophrenia, sepsis	Mazzone et al. 2007, Knutson et al. 2000, Wendland et al. 2001, O'Brien et al. 2005

Cosmetics industry

In the cosmetics industry, SG biosensors are used during the production of perfumes and fragrances. Some of them are used as a component to be used in a perfume formula composition, while others are used to give a special aroma to face and body creams (Table 4).

Environmental industry

The development of SG biosensors for the air or water pollution caused by the different industries (pharmacy, food, chemical solvents, gas grills; gas heaters, plastic

Table 4. Applications of the SG biosensors in the cosmetics industry.

Biomarker	Indicator	Reference
α-Pinene	Perfumes and fragrances development (Oils of coniferous trees (α-pinene) and citrus fruit peels (limonene))	Weston-Green et al. 2021
Cinnamyl alcohol	Perfumes and fragrances development (Capable of stimulating smell of sweet hyacinth with balsamic and spicy notes)	Cova et al. 2019
Citronellol	Perfumes and fragrances capable of stimulating (Rosy, sweet and citrus aroma)	Zuliani et al. 2020
Eugenol	Perfumes and fragrances development	Goldsmith et al. 2011
Geraniol	Perfumes and fragrances development	Lee et al. 2018
Limonene	Perfumes and fragrances development (Oils of coniferous trees (α-pinene) and citrus fruit peels (limonene))	Weston-Green et al. 2021

and paints among others, as well as motor vehicles' fuels) has great importance as the global warming has terribly increased due to the polluting gases released without adequate environmental regulation. There is an urgency to improve the number and sensitivity of the portable GS analytical devices (Table 5).

SG bioreactors for novel bioprocesses

Catalytic conversion of gaseous substrates using enzymes or cells represents a novel concept in bioprocessing (Bárzana et al. 1989b). Several SG systems have been reported involving reactions such as esterification (Lamare et al. 2001, Csenádi et al. 2012), transesterification (Parvaresh et al. 1992), alcoholysis (Bousquet-Dubouch et al. 2001, Graber et al. 2003), enantioselective reduction (Nagayama et al. 2010, Kulishova et al. 2010, Nagayama et al. 2012). Recently Cordero-Soto et al. (2020) reviewed the potential application of SG biocatalysis to produce natural aroma compounds as food additives and for pharmaceutical applications (Fig. 4).

Food industry

Esterification reactions conducted by lipases are widely used for industrial bioprocesses such as the production of high-value organic esters with aromatic, flavoring, and emulsifying properties sought by the food industry (Vilas Bôas and de Castro 2022). The most frequently used commercial lipases are produced by *Candida antarctica* (Novozym® 435), and *Rhizomucor miehei* (Lipozyme®) (Faggiano et al. 2022). One of the main advantages of lipases is the retention of activity and selectivity in non-conventional organic media, allowing their application as biocatalysts in enantioselective synthetic reactions (Goubet et al. 2008).

Lamare et al. (2004) reported schematic diagrams of laboratory and pre-industrial scale continuous SG bioreactors developed to produce natural esters. Substrates were

Table 5. Applications of the SG biosensors in the environmental industry.

Biomarker	Pollution/presence indicator	Reference
Acetic acid	Cellulosic materials such as wood and paper	Cova et al. 2022
Acetone	Solvents; wallpaper and furniture polish	Cova et al. 2022, Jung et al. 2021
Acetaldehyde	Motor vehicles fuels, gas grills; gas heaters	Jung et al. 2021
Benzene	Motor vehicles fuels	Lerner et al. 2018
Butane	Gas grills; gas heaters; gas torches; end-life fridges, and freezers	Cova et al. 2022 Tong et al. 2018
Carbon disulfide	Volcanic eruptions; marshes	Cova et al. 2022
Carbon Tetrachloride	Fire extinguishers; cleaning products	Cova et al. 2022
Dichloromethane	Pain removers or adhesives	Jung et al. 2021
Formaldehyde	Plastic furniture items; fiberboards adhesives, insulating materials, polyacetal plastics-based materials	Jung et al. 2021
Heptane	Motor vehicles fuels	Chin and Batterman 2012
Isopropyl alcohol	Solvents; sanitizes solutions	Cova et al. 2022, Jia et al. 2008
Isoprenoids	Polystyrene objects, rigid panels, and furnishings	Cova et al. 2022
Methyl chloride	Solvents; fire extinguishers	Cova et al. 2022
Nitrogen oxides	Air pollution from pharmaceutical, food and brick production industries	Hu et al. 2019
Propane	Gas grills; gas heater	Cova et al. 2022
Sulphur dioxide	Air pollution from pharmaceutical, food and brick production industries	Hu et al. 2019
Styrene	Polystyrene, acrylonitrile-butadiene-styrene (ABS), styrene-butadiene, latex	Lerner et al. 2018
Toluene	Motor vehicles fuels, gas grills; gas heaters	Chin and Batterman 2012, Cova et al. 2022, Jung et al. 2021
Tetrachloroethylene	Solvent used in the textile and metal industries	Lerner et al. 2018
Vinyl chloride	PVC pipes, wire, cable coatings, and textiles; burnt tobacco	Cova et al. 2022

vaporized in a liquid/gas flashing unit and warmed up in a gas/gas heat exchanger to the reaction temperature before the gas mixture was injected into the packed-bed bioreactor. The thermodynamic activities in the gaseous flow were maintained at this step by controlling each partial pressure.

García-Martínez et al. (2022) recently reported the modelling of the esterification reaction of propionic acid and isobutyl alcohol by immobilized *Candida antarctica* lipases in an SG bioreactor for obtaining isobutyl propionate, an ester with a fruity flavor. This study showed that the SG system is a novel strategy to produce flavor and aroma compounds, since the separation of the final product is easier than the

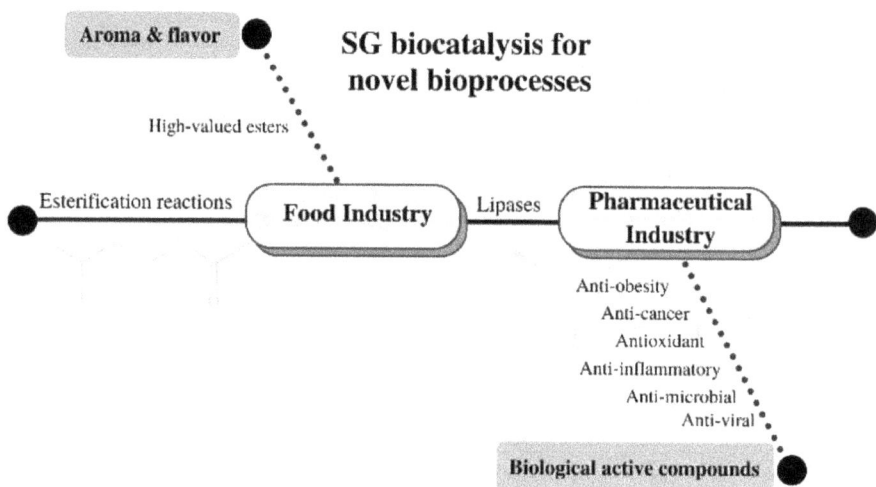

Figure 4. SG bioreactors for novel industrial bioprocesses.

conventional production in a liquid bioreactor, where several extraction techniques have been evaluated to extract the final compound, increasing the cost of operation (Cordero-Soto et al. 2020).

The synthesis of isoamyl acetate, an ester with banana flavor has been studied using a lipase immobilized on functionalized activated carbon inserted with metallic nanoparticles as novel support that showed an increase in enzymatic activity, reaching a bioconversion above 90% (de Oliveira et al. 2022). Immobilized lipases (*Candida rugosa* and porcine pancreatic lipase) in calcium alginate gel were evaluated to produce the flavor esters isoamyl acetate, ethyl valerate (green apple flavor), and butyl acetate (pineapple flavor) in a solvent-free system and hexane medium (Ozyilmaz and Gezer 2010). Other valuable esters have been obtained using *Yarrowia lipolytica* lipase through esterification reactions with hexane such as ethyl octanoate, a fruity flowery flavor ester produced after 4 h (87.3%), cetyl stearate (92.9%), and stearyl palmitate (90.2%) produced after 6 h (de Souza et al. 2019).

Pharmaceutical industry

The preparation of high-value compounds to improve human health has been recently related to the holistic consumption of natural compounds with biological activity. Bioprocesses that involve the use of enzymes have been extensively studied to produce compounds with antioxidant, anti-inflammatory, anti-obesity, anti-microbial, anti-viral, and anti-cancer activities (Cordero-Soto et al. 2020).

Faggiano et al. (2022) mentioned that the enzymatic acylation of flavonoids involves operation parameters such as molar ratio, temperature, and a_w, besides the nature and size of the acyl donor that control regioselectivity and conversions. The glycosylated flavonoids have a wide application in pharmaceutical and wellness industries.

Lipases have been studied for the esterification by erythorbic acid with lauric acid to produce erythorbyl laurate, a compound with surface-active, foam stabilization, antioxidant, anti-microbial and properties (Yu et al. 2022). A mixture of lauric acid esters was obtained in a solvent-free system at a molar ratio of 1.0 of lauric acid to glycerol, using immobilized lipase B from *Candida antarctica*, Novozym® 435, where erythorbyl laurate and glyceryl laurate were obtained with conversion of 99.52% of lauric acid (Yu et al. 2022). As an example of a SG biocatalysis process, Fig. 5 presents a simplified scheme for the isobutyl propionate synthesis.

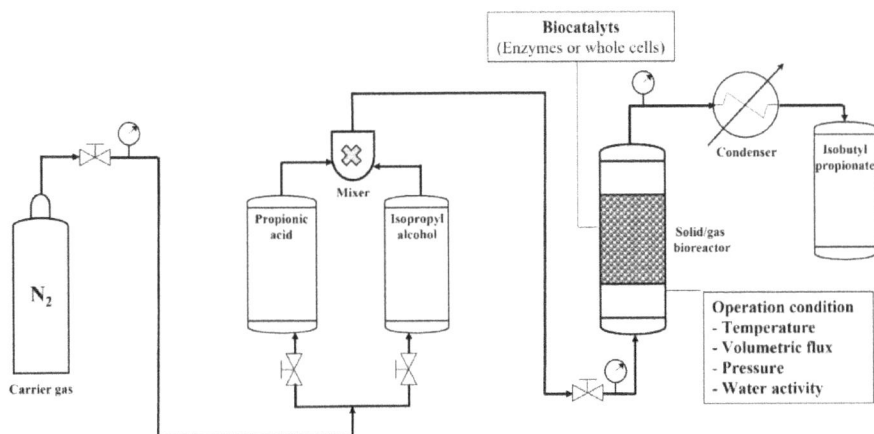

Figure 5. A simplified diagram of SG system for isobutyl propionate synthesis.

Remediation of flue gases

SG system applied in environmental remediation is a promising technology for polluted air treatment, particularly, for bioremediation of volatile organic compounds (VOCs), the most common VOCs being halogenated compounds, aldehydes, alcohols, ketones and aromatic compounds. These compounds are used in industry and agriculture as herbicides, insecticides, solvents, or intermediates for chemical synthesis, and have been selected as pollutants of primary concern by the United States Environmental Protection Agency (USEPA) (Kulishova and Zharkov 2017). The known health effects in humans may include irritation of respiratory tracts, eye and nose, allergic skin reaction, and affecting kidney and central nervous system because of their recalcitrance, toxicity, carcinogenicity, and potential teratogenicity (Goubet et al. 2008).

The traditional treatment of waste air involves technologies such as catalytic oxidation or adsorption onto activated carbon with biofilters (solid/liquid/gas bioreactor). Direct treatment of gaseous effluents by a SG biocatalyst exploits the ability of cells, lyophilized enzymes, or enzymes immobilized on a solid support to catalyze transformations of substrates and products which are in gaseous/vapor phase (Lamare et al. 2004).

Bioremediation of halogen hydrocarbons by SG system

Haloalkane dehalogenases (EC.3.8.1.3) are the only enzyme known to be capable of direct hydrolytic dehalogenation of the halogenated compound. This enzyme can cleave the bond between a primary carbon atom and a halogen atom of an aliphatic alkane (Ang et al. 2018).

Dravis et al. (2000) evaluated the pure enzyme haloalkane dehalogenase from the bacteria *Rhodococcus rhodochrous* to convert vapor-phase halogenated aliphatic hydrocarbons 1, chlorobutane, 1,3 dichloropropane to their corresponding alcohols 1-butanol and 3-chloro-1-propanol. This enzyme was also immobilized on a commercial support, λ-alumina impregnated with polyethylenamine (PEI-Alumina), and the enzyme was found to be more stable in the immobilized and lyophilized form than in the dissolved state (Dravis et al. 2000). This work with purified haloalkane dehalogenase from *Rhodococcus rhdochrous* showed that $a_w \sim 1$ was required for optimal enzyme functioning, and the enzymatic activity decreased by 80% at $a_w = 0.5$ and the optimum is 0.9. However, the purification step of the enzyme increases the cost of the treatment process. Erable et al. (2005) evaluated the use of *R. erythropolis* whole-dehydrated cells to convert halogenated alkanes to their analogous alcohols and HCl in a SG biofilter. In this study, optimal dehalogenase activity for the lyophilized cells was obtained at a_w of 0.9, while a critical a_w of 0.4 was necessary for the enzyme to become active. Dehalogenase activity increases with temperature (60°C); however, the increase in activity was associated with the decrease in stability. This could be related to the combined action of temperature and the relatively high-a_w (0.6 to 0.8) used with the whole cells (Erable et al. 2005, Goubet et al. 2008, Marchand et al. 2009). Additionally, the loss of the dehalogenase activity of whole cells in an SG bioreactor using recombinant *E. coli* BL21 (DE3) expressing the DhaA halidohydrolase was associated with HCl accumulation. To overcome this, a buffer salt was used to neutralize the acid produced during the reaction (Marchand et al. 2009). As an example of an SG flue gases bioremediation, Fig. 6 presents a

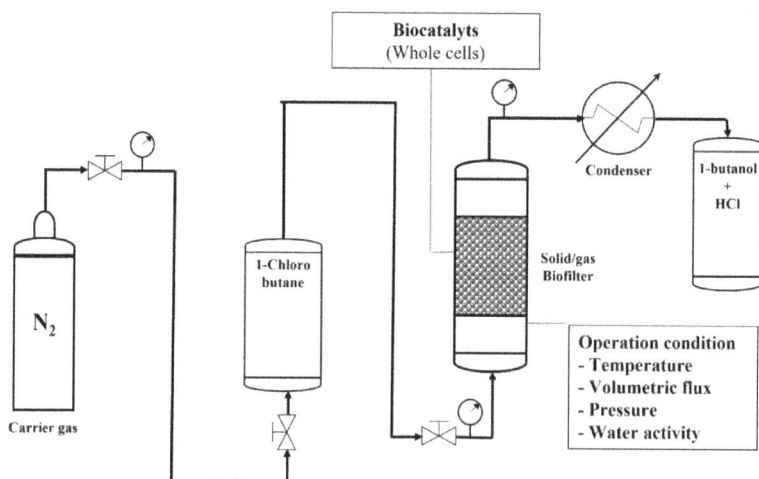

Figure 6. A simplified diagram of a continuous SG biofilter biotransformation of halogenated compounds.

simplified scheme for the degradation of 1-chlrobutane to its corresponding alcohols (1-butanol) and HCl in an SG biofilter.

Engineering aspects involved in SG biocatalysis

The conceptual design or scaling-up of technology for applications of SG biocatalysis needs the establishment of engineering fundamentals focused on understanding the effect of thermodynamics, kinetics, and transport phenomena on the performance of the biocatalyst and, hence, the bioreactor. In this regard, different SG applications are analyzed in terms of the reactor engineering essentials involved during their operation. Figures 7 and 8 display engineering diagrams of the applications where SG biocatalysis becomes the heart of the processes: the synthesis of high-value compounds (Cordero-Soto et al. 2020, García-Martínez et al. 2022); the remediation

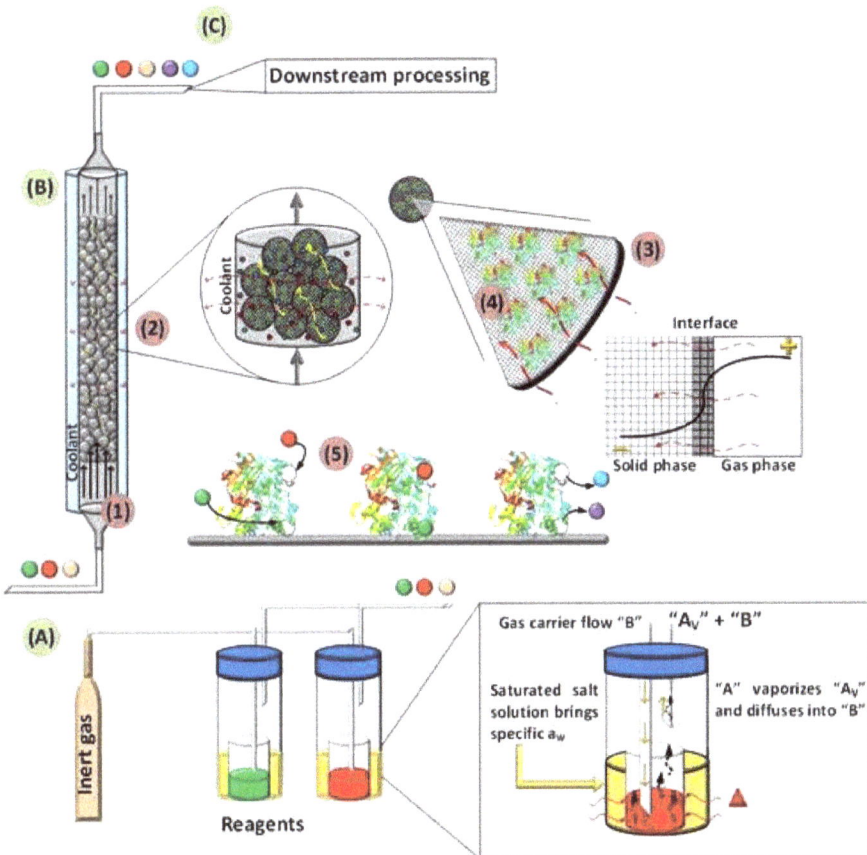

Figure 7. General scheme of SG system: (A) saturator, (B) SG bioreactor and (C) product recovery (downstream processing). Transport mechanisms involved in a tubular SG bioreactor: (1) convection; (2) dispersion; (3) interparticle; (4) intraparticle and (5) kinetics.

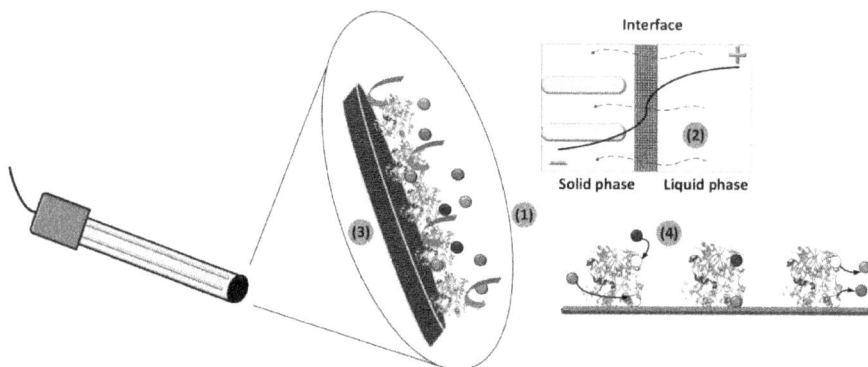

Figure 8. General scheme of a SG biosensor: Transport mechanisms involved: (1) convection; (2) interparticle; (3) intraparticle; and (4) kinetics.

of flue gases for the degradation of toxic volatile organic compounds (Hwee 2022, Erable et al. 2005, 2009); and biosensors for different industrial applications, as mentioned earlier (Bankole et al. 2022). In the SG bioreactors, three sections are recognized, i.e., the saturator, the bioreactor, and the downstream processing, while in the biosensor, the reaction section is the only one component. Depending on the application, different mechanisms are involved in every bioprocess step such as thermodynamics and transport phenomena in both the saturator and the downstream processing; and thermodynamics, kinetics, and transport phenomena in the bioreactor or biosensor, the zone where the SG biocatalysis reaction occurs. In the following sections, the reactor engineering fundamentals involved in the saturator, bioreactors and biosensor are presented and thoroughly discussed, the engineering examination of the downstream processing is, however not covered here.

Saturators

The saturator aims at controlling the a_w and species concentration in the bioreactor. Particularly, the feedstock contains an inert gas saturated with substrates or the target compound to be treated at a specific a_w, as mentioned in Section 2. Regarding the effect of the a_w on the performance of the SG bioreactor, it is essential for improving the catalytic activity, the thermal and catalytic stability, and the catalytic enantioselectivity (Bárzana et al. 1989a, Halling 2000, Bellissent-Funel et al. 2016). An inert gas, normally nitrogen or helium to improve heat transfer processes, is bubbled through the saturator to transport species into the bioreactor. Depending on the operational residence time of the inert gas in the saturator, temperature, total pressure, concentration of the substrates and water in the feedstock can be strategically set. It is worth mentioning that the concentration of these compounds in the feedstock can be controlled at the required values by varying the operating conditions fixed in the saturator. The governing mass transport equation involved in the saturator reads:

$$V_s \frac{dC_{is}}{dt} = -Q_s C_{is} + V_s k_{gl} a_{gl} (C_{is}* - C_{is})$$

(1)

where V_s is the volume of fluid in the saturator, Q_s is the saturator volumetric flow rate, t is time, and C_{is} and $C_{is}*$ are the concentration of species i in the saturator and the saturation vapor concentration under identical conditions at equilibrium, respectively. The concentration of species i relates to the partial pressure by means of the ideal gases' law, such that:

$$C_{is} = \frac{p_{is}}{R_g T_s} = \frac{P_s y_{is}}{R_g T_s}$$

(2)

$$C_{is}* = \frac{p_{is}^o}{R_g T_s}$$

(3)

where p_{is} is the partial pressure of species i in the saturator, P_s and T_s are the operational pressure and temperature in the saturator, R_g is the gas ideal constant. The maximum partial pressure of species i is, thus, determined by applying Raoult's equation:

$$p_{is} = x_i p_{is}^o$$

(4)

where p_{is}^o is the vapor pressure of the component i in the saturator and x_i is the mole fraction of species i in the mixture. The vapor pressure of every compound is, then, determined by applying Antoine's equation:

$$\log_{10} p_{is}^o = A_{is} - \frac{B_{is}}{T_s + C_{is}}$$

(5)

here, T_s is the operational temperature in the saturator and A_{is}, B_{is} and C_{is} are specific empirical constants obtained from the literature for every species i contained in the saturation section. Concerning the partial pressure of water at the bioreactor inlet, it is normally examined using the definition of the a_w, an operating variable defined as the ratio between the partial pressure of water and the saturation vapor pressure under identical conditions but at equilibrium:

$$a_w = \frac{p_{vw}}{p_{vw}^o}$$

(6)

As identified from Eq. (1), the outlet concentration of species i at the saturator becomes the inlet concentration at the bioreactor inlet. Thus, to operate in a thermodynamic control regime, the residence time and interfacial transport rate need to be large enough to saturate the inert gas with the substrates at the required a_w. Residence time is increased by decreasing the flow rate, while the interfacial mass transfer is increased by improving the contact area between gas and substrates such that the bubbling of the inert gas through the saturator is the key aspect of the saturator design. Once a thermodynamic regime control is accomplished, the concentration of species and water in the inert gas is controlled by manipulating the operational pressure and temperature, i.e., the larger the temperature and the lower the operational pressure, the larger the concentration of species at the outlet of the saturator.

SG bioreactors

Once the engineering bases for the design and operation of the saturator have been established, the phenomenological and operational analysis of the SG bioreactor becomes pertinent. Once substrates and water carried by the inert gas are fed into the bioreactor and operational conditions are fixed, different mechanisms are involved: fluid dynamics, heat and mass transport, and kinetics. What follows in each of these is discussed. These mechanisms are also schematized in Fig. 7.

Fluid dynamics: Velocity profiles arise because of both the void fraction profiles created by the packing arrangement in the bioreactor and tangential shear stresses originating due to the bioreactor's wall surface. This interstitial velocity impacts heat and mass transport phenomena at different levels (Hernández-Aguirre et al. 2022). Predictions of the interstitial velocity rely on the adequacy for quantifying viscous and kinetic energy losses due to the presence of the biocatalyst packed in the SG bioreactor. Thus, an adequate quantification of the fluid dynamic field depends on the determination of the transport descriptors such as void fraction and the apparent permeability contained in the terms accounting for solid-fluid interactions (Gómez-Ramos et al. 2019, Hernández-Aguirre et al. 2022). The fluid dynamic model for an SG bioreactor, including synthesis reactions or remediation, is the well-known Navier-Stokes-Darcy-Forchheimer equations. The analysis of fluid dynamic mechanisms and the methodology for quantifying them has been broadly reviewed (Castillo-Araiza et al. 2008, Cordero-Soto et al. 2020, García-Martínez et al. 2022, Hernández-Aguirre et al. 2022).

Mass and heat transfer phenomena: Five important processes are involved during the operation of SG bioreactors: (i) Heat and mass convective mechanisms (García-Martínez et al. 2022), (ii) dispersive mass and conductive heat transfer mechanisms (Castillo-Araiza 2021), (iii) interparticle heat transfer and mass transfer mechanisms (Castillo-Araiza 2021, Schlichting and Gersten 2000), (iv) intraparticle heat transfer and mass transfer mechanisms (Cordero-Soto et al. 2020), and (v) heat transfer from the core of the bioreactor to the cooling system (Castillo-Araiza 2021, García-Martínez et al. 2022, Cordero-Soto et al. 2020, Castillo-Araiza et al. 2007).

Kinetics: A proper determination of bioreaction rates offers the framework for connecting the microscopic kinetic performance of the biocatalyst with the macroscopical functioning of the SG bioreactor. The characterization of kinetics is based on the development of an intrinsic or extrinsic model, in which the reaction mechanism at the molecular level needs to be transferred to mathematical equations. From a macroscopic point of view, this reaction mechanism accounts for different

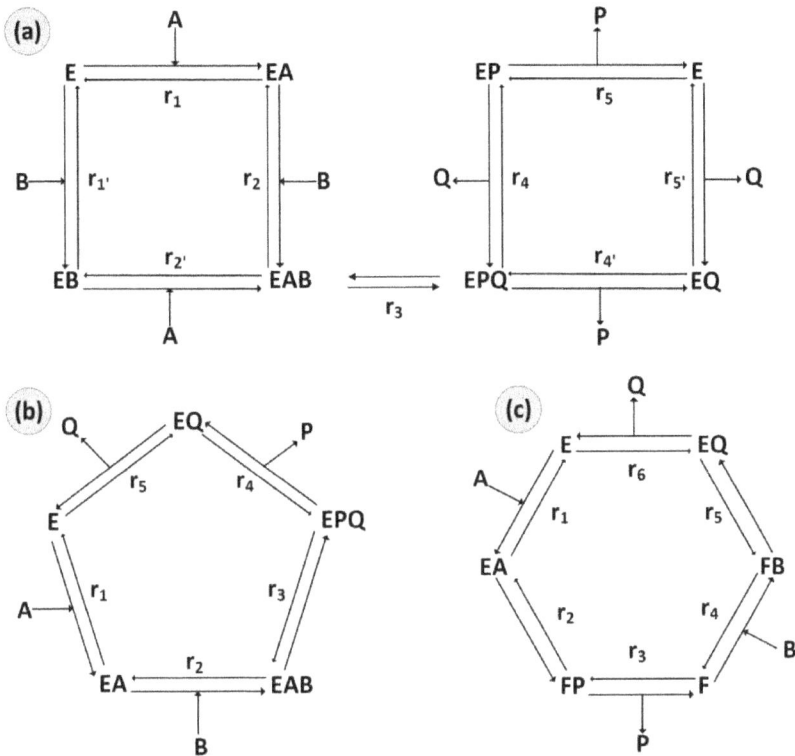

Figure 9. Macroscopic reaction mechanism involved in SG Bi-Bi catalysis: (a) random Bi-Bi; (b) ordered Bi-Bi; and (c) Ping Pong Bi-Bi. A and B denote reactants; P and Q associate with products; E is the active enzyme and F a chemically modified form of E. F, FB, FP are intermediate complexes.

scenarios related to adsorption reaction, equilibrium information, inhibition and desorption. Figure 9 displays different mechanism scenarios that capture microscopic elemental reaction steps at the macroscopical level. Nevertheless, their adequacy when transferred to a mathematical model must be tested by following well-known reaction engineering basics as broadly stated in the literature for chemical and biological reactions (Alvarado-Camacho et al. 2022, García-Martínez et al. 2022, Cordero-Soto et al. 2020, Castillo-Araiza et al. 2017).

Note that the mathematical formulation follows the law of kinetic mass action and the mean-field approximation such that reaction rates are written in terms of bioreactor variables such as the molar flow of species i, total pressure and temperature. Thus, reaction thermodynamics and kinetics are transferred to the bioreactor model, which accounts for the transport phenomena involved in the SG bioreactor. When quantifying kinetics, the role of thermodynamics must be accounted for during the analysis. To this end, the first step of characterizing kinetics for a global reaction $(A + B = P + W)$ rests on the macroscopic molar balance, which should be stated as a function of the reaction extent as presented in Table 6.

Table 6. Molar balance in terms of the extent of the bioreaction.

Species	V	Initial number of moles	Final number of moles
A	1	n_A	$n_A - \xi$
B	1	n_B	$n_B - \xi$
P	1	n_P	$n_E + \xi$
W	1	n_W	$n_W + \xi$
I	0	n_{i2}	n_{i2}
Sum	0	$n_i = n_A + n_B + n_E + n_W + n_{i2}$	$n_f = n_i = n_t$

Thus, the equilibrium constant, in terms of the extent of the reaction is expressed as follows:

$$K(T) = \exp^{\frac{-\Delta G^\circ}{RT}} = \frac{Y_P Y_W}{Y_A Y_B} = \frac{\left(\dfrac{n_P + \xi}{n_t}\right)\left(\dfrac{n_W + \xi}{n_t}\right)}{\left(\dfrac{n_A - \xi}{n_t}\right)\left(\dfrac{n_B - \xi}{n_t}\right)} = \frac{(n_E + \xi)(n_W + \xi)}{(n_A - \xi)(n_B - \xi)} \tag{7}$$

where the extent of the reaction can be defined as follows:

$$\xi = \frac{n_i^{eq} - n_i^i}{v_i} \tag{8}$$

where n_i^{eq} is the moles of species i at thermodynamic equilibrium, n_i^i denotes the initial moles of species i, and v_i is the stoichiometric number of species i. The solution of Eqs. (7) and (8) leads to determine both the equilibrium constant involved in the kinetic model and the equilibrium conversion.

SG biosensors

Considering that a biosensor is a device capable of measuring signals related to a species i involved during the biological reaction, the main mechanisms involved during its operation are convective bulk and interfacial transport phenomena as elucidated in Fig. 8. The biocatalyst, as mentioned earlier, can be enzyme-based, tissue-based, or DNA-based catalytic materials (Mehrotra 2016). Because of the phenomenological and operational similarities taking place around the biocatalyst, mechanisms involved during the operation of the biosensor are related to those for SG bioreactors used for synthesis and remediation, such as intraparticle and interparticle transport mechanisms, kinetics, and convective transport phenomena which are now driven by natural or macroscopical convective forces. To this end, reaction mechanisms presented in Fig. 9 and boundary layer theory extensively discussed in Schlichting and Gersten (2000) can be applied to either understand the performance of biosensors or design them.

Challenges

For all SG biocatalysis applications, the main challenges are the development of active and stable biocatalysts using isolated enzymes or enzymes contained in lyophilized whole cells. Related to SG biosensors, the main challenges are associated with the identification of the largest number of VOCs related to biomarkers that help to detect *on-line* products related to deterioration or contamination in industrial processes, chemical and microbial changes, maturity and freshness of fruit and vegetable products, presence of toxic compounds. The production of specific molecules such as volatile analytes can also denote a specific human disease. Another challenge is the enhancement of the sensitivity and accuracy of the transducers and the output of the signal processors. Regarding the development of novel bioprocesses to produce high-value volatile compounds, this technology is limited to the volatility of the substrates and products, therefore, it is necessary to find technological alternatives to expand the range of substrates. In the remediation of flue gases, the main challenges are the development of immobilized enzymes or whole cells capable to direct transform gaseous toxic compounds from an industrial effluent. A major understanding of the microscopic interaction between thermodynamics, kinetics and transport phenomena is required for the construction of a better reactor design and increase the production of high-value compounds.

Conclusions

The technical characteristics of the SG system offer technological, environmental, and economic advantages. It is an attractive technology with applications in biosensors, biocatalysis and flue gases bioremediation. SG technology has a broad impact on several industries such as food, pharmaceutical, clinical, health, environmental, etc. SG enzymatic biosensors, due to their high sensitivity and electrocatalytic activity, represent a good alternative for the detection of compounds of analytical interest, mainly in the areas of environmental monitoring, clinical and food analysis. Its ease of application, short response times, selectivity and the possibility of miniaturization make enzyme-based sensor devices an intense and extensive field of research, with mature and developing technologies. SG biocatalysis in a packed-bed bioreactor generates interesting applications for the catalytic conversion of volatile compounds allowing enantioselective synthetic reactions of high-value bioproducts such as aromas as food additives and pharmaceutical applications. Nowadays, remediation of flue gases to eliminate gaseous toxic compounds from industrial effluents is a promising technology for the environmental industry.

Authors' contribution

The manuscript was written through the contribution of all authors. Sergio Huerta-Ochoa: Conceptualized the chapter organization, integration, and revision of the whole chapter, also authored sections related to the introduction, definition, and description of the SG biocatalysis, as well as challenges and conclusions. Carlos

Omar Castillo-Araiza: Revision of the whole chapter, and authored Section 6 related to the engineering aspects involved in SG biocatalysis. Itza Nallely Cordero-Soto: Authored Section 4 related to SG bioreactors for novel bioprocesses. Yahir Alejandro Cruz-Martínez: Conceptualized and prepared figures accounted for in Section 6. Lilia Arely Prado-Barragán: Collaborated in the chapter organization, integration, and revision of the whole chapter, also authored Section 3 related to SG biosensors. Olga Miriam Rutiaga-Quiñones: Authored Section 5 related to remediation of flue gases.

Acknowledgments

We are deeply grateful to Dr. Deepak Kumar Verma (Former Researcher, Agricultural and Food Engineering Department, Indian Institute of Technology Kharagpur, India) for review, correction, and technical support on the final versions of this chapter.

References

Alessandroni, L., G. Caprioli, F. Faiella, D. Fiorini, R. Galli, X.H. Huang et al. 2022. A shelf-life study for the evaluation of a new biopackaging to preserve the quality of organic chicken meat. Food Chem. 371: 131134.

Alvarado-Camacho, C., J. Poissonnier, J.W. Thybaut and C.O. Castillo. 2022. Unravelling the redox mechanism and kinetics of a highly active and selective Ni-based material for the oxidative dehydrogenation of ethane. React. Chem. Eng. 7: 619–640.

Amano, A. 2002. Monitoring ammonia to assess halitosis. Oral Surg. Oral Med. Oral Pathol. 94: 692–696.

Andrade, C.A.S., M.D.L. Oliveira, V.R. Hering, T.E.S. Faulin, S.P. Dulcineia and D.S.P. Abdalla. 2011. Biosensors for detection of low-density lipoprotein and its modified forms. Book Chapter 9. In Biosensors for Health, Environment and Biosecurity. Edited by Pier Andrea Serra. IntechOpen.

Ang, T.F., J. Maiangwa, A.B. Salleh, Y.M. Normi and T.C. Leow. 2018. Dehalogenases: From improved performance to potential microbial dehalogenation applications. Molecules 23(5): 1100.

Bajtarevic, A., C. Ager, M. Pienz, M. Klieber, K. Schwarz, M. Ligor, T. Ligor, W. Filipiak, H. Denz and M. Fiegl. 2009. Noninvasive detection of lung cancer by analysis of exhaled breath. BMC Cancer 9: 348.

Bankole, O.E., D.K. Verma, M.L.C. González, J.G. Ceferino, J. Sandoval-Cortés and C.N. Aguilar. 2022. Recent trends and technical advancements in biosensors and their emerging applications in food and bioscience. Food Biosci. 47: 101695.

Barnes, P.J. and S.A. Kharitonov. 1996. Exhaled nitric oxide: A new lung function test. Thorax. 51: 233.

Bárzana, E., A.M. Klibanov and M. Karel. 1987. Enzyme-catalyzed, gas phase reaction. Appl. Biochem. Biotechnol. 15: 25–34.

Bárzana, E., A.M. Klibanov and M. Karel. 1989a. Enzymatic oxidation of ethanol in the gaseous phase. Biotech. Bioeng. 34: 1178–1185.

Bárzana, E., A.M. Klibanov and M. Karel. 1989b. Calorimetric method for the enzymatic analysis of gases: The determination of ethanol and formaldehyde vapors using solid alcohol oxidase. Anal. Biochem. 182: 109–115.

Bellissent-Funel, M.C., A. Hassanali, M. Havenith, R. Henchman, P. Peter Pohl, F. Sterpone et al. 2016. Water determines the structure and dynamics of proteins. Chem. Rev. 116(13): 7673–7697.

Berg, J.M., J.L. Tymoczko and L. Stryer. 2002. Protein turnover and amino acid catabolism. pp. 633. In: W.H. Freeman (ed.). Biochemistry. 5th Ed. New York, NY. USA.

Bousquet-Dubouch, M.P., M. Graber, N. Sousa, S. Lamare and M.D. Legoy. 2001. Alcoholysis catalyzed by Candida antarctica lipase B in a gas/solid system obeys a Ping Pong Bi Bi mechanism with competitive inhibition by the alcohol substrate and water. Biochim. Biophys. Acta 1550: 90–99.

Butterworth, R.F. 2003. Hepatic encephalopathy. Alcohol Res. Health 27: 240.

Castillo-Araiza, C.O. 2021. Decoding complexity during the modeling of the microscopic energetic path in a wall-cooled packed-bed reactor. Energy: Science, Technology, and Management 1(1): 1–18.

Castillo-Araiza, C.O., H. Jiménez-Islas and F. López-Isunza. 2007. Heat-transfer studies in packed-bed catalytic reactors of low tube/particle diameter ratio. Ind. Eng. Chem. Res. 46(23): 7426–7435.

Castillo-Araiza, C.O. and F. Lopez-Isunza. 2008. Hydrodynamic models for packed beds with low tube-to-particle diameter ratio. Int. J. Chem. React. 6(1).

Castillo-Araiza, C.O., D. Palmerín-Carreño, A. Prado-Barragán and S. Huerta-Ochoa. 2017. On the conceptual design of a partitioning technology for the bioconversion of (+)-valencene to (+)-nootkatone on whole cells: Experimentation and modelling. Chem. Eng. Process. 122: 493–507.

Chin, J.Y. and S.A. Batterman. 2012. VOC composition of current motor vehicle fuels and vapors, and collinearity analyses for receptor modeling. Chemosphere 86: 951–958.

Clark, L.C. and C. Lyons. 1962. Electrode systems for continuous monitoring in cardiovascular surgery. Ann. NY Acad. Sci. 102: 29–45.

Cordero-Soto, I.N., C.O. Castillo-Araiza, L.E. García Martínez, A. Prado-Barragán and S. Huerta-Ochoa. 2020. Solid/Gas biocatalysis for aroma production: An alternative process of white biotechnology. Biochem. Eng. J. 164: 107767.

Cova, C.M., A. Zuliani, M.J. Muñoz-Batista and R. Luque. 2019. Efficient Ru-based scrap waste automotive converter catalysts for the continuous-flow selective hydrogenation of cinnamaldehyde. Green Chem. 21: 4712–4722.

Cova, C.M., E. Rincón, E. Espinosa, L. Serrano and A. Zuliani. 2022. Paving the way for a green transition in the design of sensors and biosensors for the detection of volatile organic compounds (VOCs). Biosensors 12(2): 51.

Csanádi, Z., R. Kurdi and K. Bélafi-Bakó. 2012. Ethyl-acetate synthesis in gas phase by immobilised lipase. Hung. J. Ind. Chem. 40(1): 39–44.

Das, T., S. Das, M. Karmakar, S. Chakraborty, D. Saha and M. Pal. 2020. Novel barium hexaferrite based highly selective and stable trace ammonia sensor for detection of renal disease by exhaled breath analysis. Sens. Actuators B Chem. 325: 128765.

Debeche, T., C. Marmet, L. Kiwi-Minsker, A. Renken and M.A. Juillerat. 2005. Structured fiber supports for gas phase biocatalysis. Enzyme Microb. Technol. 36: 911–916.

de Oliveira, T.P., M.P.F. Santos, M.J.P. Brito and C.M. Veloso. 2022. Incorporation of metallic particles in activated carbon used in lipase immobilization for production of isoamyl acetate. J. Chem. Technol. Biotechnol. 7043.

de Souza, C.E.C., B.D. Ribeiro and M.A.Z. Coelho. 2019. Characterization and application of *Yarrowia lipolytica* lipase obtained by solid-state fermentation in the synthesis of different esters used in the food industry. Appl. Biochem. Biotechnol. 189(3): 933–959.

Dimoula, K., M. Pohl Büchs and A.C. Spiess. 2009. Substrate and water adsorption phenomena in a gas/solid enzymatic reactor. Biotechnol. J. 4: 712–721.

Dotsch, J., S. Demirakca, H.G. Terbrack, G. Huls, W. Rascher and P.G. Kuhl. 1996. Irway nitric oxide in asthmatic children and patients with cystic fibrosis. Eur. Respir. J. 9: 2537.

Dravis, B.C., K.E. LeJeune, A.D. Hetro and A.J. Russell. 2000. Enzymatic dehalogenation of gas phase substrates with haloalkane dehalogenase. Biotechnol. Bioeng. 69(3): 235–241.

Dunn, R.V. and R.-M. Daniel. 2004. The use of gas-phase substrates to study enzyme catalysis at low hydration. Phil. Trans. R. Soc. Lond. 359: 1309–1320.

Erable, B., I. Goubet, A. Seltana and T. Maugard. 2009. Non-conventional gas phase remediation of volatile halogenated compounds by dehydrated bacteria. J. Environ. Manage. 90(8): 2841–2844.

Erable, B., T. Maugard, I. Goubet, S. Lamare and M.D. Legoy. 2005. Biotransformation of halogenated compounds by lyophilized cells of *Rhodococcus erythropolis* in a continuous solid–gas biofilter. Process Biochem. 40(1): 45–51.

Essiet, I.O. 2013. Diagnosis of kidney failure by analysis of the concentration of ammonia in exhaled human breath. J. Emerg. Trends Eng. Appl. Sci. 4: 859.

Faggiano, A., M. Ricciardi and A. Proto. 2022. Catalytic routes to produce polyphenolic esters (PEs) from biomass feedstocks. Catalysts 12(4): 447.

Fortin, C., H.L. Goodwin and M.R. Thomsen. 2009. Consumer attitudes toward freshness indicators on perishable food products. J. Food Distrib. Res. 40: 1–15.

García-Martínez, L.E., C.O. Castillo-Araiza, G. Quijano and S. Huerta-Ochoa. 2022. On the modelling and surface response analysis of a non-conventional wall-cooled solid/gas bioreactor with application in esterification. Chem. Eng. J. 437: 135063.

Graber, M., M.P. Bousquet-Dubouch, S. Lamare and M.D. Legoy. 2003. Alcoholysis catalyzed by *Candida antarctica* lipase B in a gas/solid system: effects of water on kinetic parameters. Biochim. Biophys. Acta 1648: 24–32.

Goldsmith, B.R., J.J. Mitala, J. Josue, A. Castro, M.B. Lerner, T.H. Bayburt et al. 2011. Biomimetic chemical sensors using nanoelectronic readout of olfactory receptor proteins. ACS Nano 5: 5408–5416.

Gómez-Ramos, G.A., C.O. Castillo-Araiza, S. Huerta-Ochoa, M. Couder-García and A. Prado-Barragán. 2019. Assessment of hydrodynamics in a novel bench-scale wall-cooled packed bioreactor under abiotic conditions. Chem. Eng. J. 375: 121945.

Goubet, I., M. Graber, S. Lamare, T. Maugard and M.D. Legoy. 2008. Solid/gas biocatalysis. pp. 255–278. *In*: G. Carrea and S. Riva (eds.). Organic Synthesis with Enzymes in Non-Aqueous Media. WILEY-VCH Verlag GmbH & Co. KGaA, Weinheim.

Goubet, I., T. Maugard, S. Lamare and M.D. Legoy. 2002. Role of water activity and temperature on activity and stability of dried whole cells of *Saccharomyces cerevisiae* in a continuous solid–gas bioreactor. Enzyme Microb. Technol. 31: 425–430.

Halámek, J., J. Pribyl, A. Makower, P. Skládal and F.W. Scheller. 2005. Sensitive detection of organophosphates in river water by means of a piezoelectric biosensor. Anal. Bioanal. Chem. 382(8): 1904–11.

Halling, P.J. 2000. Biocatalysis in low-water media: Understanding effects of reaction conditions. Curr. Opin. Chem. Biol. 4: 74–80.

Hernández-Aguirre, A., E. Hernández-Martínez, F. López-Isunza and C.O. Castillo. 2022. Framing a novel approach for pseudo continuous modeling using Direct Numerical Simulations (DNS): Fluid dynamics in a packed bed reactor. Chem. Eng. J. 429: 132061.

Hibbard, T. and A.J. Killard. 2011. Breath ammonia analysis: Clinical application and measurement. Breath Ammon. Clin. App. Meas. 41: 21.

Hu, Y., Z.Y. Li, Y.T. Wang, L. Wang, H.T. Zhu, L. Chen et al. 2019. Emission factors of NOx, SO2, PM and VOCs in pharmaceuticals, brick and food industries in Shanxi, China. Aerosol Air Qual. Res. 19: 1785–1797.

Hwee, N. 2022. Methane Remediation Using Biocatalysts in Batch Gas-Solid Mass Transfer System Challenges and Prospects. Technical Report. Lawrence Livermore National Laboratory.

Ivanova, S., F. Batliwalla, J. Mocco, S. Kiss, J. Huang, W. Mack et al. 2002. Neuroprotection in cerebral ischemia by neutralization of 3-aminopropanal. Proc. Natl. Acad. Sci. USA 99(8): 5579–5584.

Jia, C., S. Batterman and Ch. Godwin. 2008. VOCs in industrial, urban and suburban neighborhoods-Part 2: Factors affecting indoor and outdoor concentrations. Atmos. Environ. 42(9): 2101–2116.

Jin, H.J., S.H. Lee, T.H. Kim, J. Park, H.S. Song, T.H. Park et al. 2012. Nanovesicle-based bioelectronic nose platform mimicking human olfactory signal transduction. Biosens. Bioelectron. 35: 335–341.

Jung, C.R., Y. Nishihama, S.F. Nakayama, K. Tamura, T. Isobe, T. Michikawa et al. 2021. Indoor air quality of 5000 households and its determinants. Part B: Volatile organic compounds and inorganic gaseous pollutants in the Japan Environment and Children's study. Environ. Res. 197: 111135.

Kearney, D.J., T. Hubbard and D. Putnam. 2002. Breath ammonia measurement in *Helicobacter pylori* infection. Dig. Dis. Sci. 47: 25232.

Kim, D., J.M. Ko, Y.M. Kim, G.H. Seo, G.H. Kim, B.H. Lee et al. 2018. Low prevalence of argininosuccinate lyase deficiency among inherited urea cycle disorders in Korea. J. Hum. Genet. 63(8): 911–917.

Knutson, M.D., G.J. Handelman and F.E. Viteri. 2000. Methods for measuring ethane and pentane in expired air from rats and humans. Free Radic. Biol. Med. 28: 514.

Kulishova, L.M. and D.O. Zharkov. 2017. Solid/Gas biocatalysis. Biokhimiya 82(2): 196–207.

Kulishova, L.M., K. Dimoula, M. Jordan, A. Wirtz, D. Hofmann, B. Santiago-Schübel et al. 2010. Factors influencing the operational stability of NADPH-dependent alcohol dehydrogenase and an NADH-dependent variant thereof in gas/solid reactors. J. Mol. Catal. B Enzym. 67: 271–283.

Kurkal, V., R.M. Daniel, J.L. Finney, M. Tehei, R.V. Dunn and J.C. Smith. 2005. Enzyme activity and flexibility at very low hydration. Biophys. J. 89: 1282–1287.

Lamare, S., M.D. Legoy and M. Graber. 2004. Solid/gas bioreactors: Powerful tools for fundamental research and efficient technology for industrial applications. Green Chem. 6: 445–458.

Lee, M., H. Yang, D. Kim, M. Yang, T.H. Park and S. Hong. 2018. Human-like smelling of a rose scent using an olfactory receptor nanodisc-based bioelectronic nose. Sci. Rep. 8: 13945.

Lerner, J.E.C., M.D. Gutiérrez, D. Mellado, D. Giuliani, L. Massolo, E.Y. Sanchez et al. 2018. Characterization and cancer risk assessment of VOCs in home and school environments in gran La Plata, Argentina. Environ. Sci. Pollut. Res. 25: 10039–10048.

Li, W., X.M. Yuan, S. Ivanova, K.J. Tracey, J.W. Eaton and U.T. Brunk. 2003. 3-Aminopropanal, formed during cerebral ischaemia, is a potent lysosomotropic neurotoxin. Biochem. J. 371: 429.

Marchand, P., S. Lamare, M.D. Legoy and I. Goubet. 2009. Dehalogenation of gaseous 1-chlorobutane by dehydrated whole cells: Influence of the microenvironment of the halidohydrolase on the stability of the biocatalyst. Biotechnol. Bioeng. 103: 687–695.

Maziak, W., S. Loukides, S. Culpitt, P. Sullivan, S.A. Kharitonov and P.J. Barnes. 1998. Exhaled nitric oxide in chronic obstructive pulmonary disease. Am. J. Respir. Crit. Care Med. 157: 998.

Mazzone, P.J., J. Hammel, R. Dweik, J. Na, C. Czich, D. Laskowski et al. 2007. Diagnosis of lung cancer by the analysis of exhaled breath with a colorimetric sensor array. Thorax. 62: 565.

McGrath, L.T., R. Patrick and B. Silke. 2001. Breath isoprene in patients with heart failure. Eur. J. Heart Fail. 3: 423.

Mehrotra, P. 2016. Biosensors and their applications—A review. J. Oral Biol. Craniofacial Res. 6: 153–159.

Mello, L.D. and L.T. Kubota. 2002. Review of the use of biosensors as analytical tools in the food and drink industries. Food Chemistry 77: 237–256.

Mizobuchi, M. and K. Nagayama. 2015. Reduced pressure gas phase bioreactor as a tool for stereoselective reduction catalyzed by alcohol dehydrogenase from *Parvibaculum lavamentivorans*. Biochem. Eng. J. 93: 11–16.

Nagayama, K., A.C. Spiess and J. Büchs. 2010. Immobilization conditions of ketoreductase on enantioselective reduction in a gas-solid bioreactor. Biotechnol. J. 5: 520–525.

Nagayama, K., A.C. Spiess and J. Büchs 2012. Enhanced catalytic performance of immobilized *Parvibaculum lavamentivorans* alcohol dehydrogenase in a gas phase bioreactor using glycerol as an additive. Biochem. Eng. J. 52: 301–303.

O'Brien, J., A.G. Siraki and N. Shangari. 2005. Aldehyde sources, metabolism, molecular toxicity mechanisms, and possible effects on human health. Crit. Rev. Toxicol. 35: 609.

Odeyemi, O.A., C.M. Burke, C.C.J. Bolch and R. Stanley. 2018. Seafood spoilage microbiota and associated volatile organic compounds at different storage temperatures and packaging conditions. Int. J. Food Microbiol. 280: 87–99.

Ozyilmaz, G. and E. Gezer. 2010. Production of aroma esters by immobilized *Candida rugosa* and porcine pancreatic lipase into calcium alginate gel. J. Mol. Catal. B Enzym. 64(3-4): 140–145.

Parvaresh, F., H. Robert, D. Thomas and M.D. Legoy. 1992. Gas phase transesterification reactions catalyzed by lipolytic enzymes. Biotechnol. Bioeng. 39: 467–473.

Patel, P.D. 2002. (Bio)sensors for measurement of analytes implicated in food safety: A review. Trends Anal. Chem. 21(2): 96–115.

Preis, S., D. Klauson and A. Gregor. 2013. Potential of electric discharge plasma methods in abatement of volatile organic compounds originating from the food industry. J. Environ. Manag. 114: 125–138.

Pulvin, S., M.D. Legoy, R. Lortie, M. Pensa and D. Thomas. 1986. Enzyme technology and gase catalysis: Alcohol dehydrogenase example. Biotechnol. Lett. 8(11): 783–784.

Schlichting, H. and K. Gersten. 2000. Fundamentals of boundary-layer theory. pp. 29–49. *In*: H. Schlichting and K. Gersten (eds.). Boundary-Layer Theory. Springer, Berlin, Heidelberg.

Shinpo, K., S. Kikuchi, H. Sasaki, A. Ogata, F. Moriwaka and K. Tashiro. 2000. Selective vulnerability of spinal motor neurons to reactive dicarbonyl compounds, intermediate products of glycation, *in vitro*: Implication of inefficient glutathione system in spinal motorneurons. Brain Res. 861: 151.

Skinner, W.S., S. Zhang, R.E. Guldberg and K.G. Ong. 2022. Magnetoelastic sensor optimization for improving mass monitoring. Sensors 22: 827–839.

Suarez, F.L., K.J. Furne, J. Springfield and D.M. Levitt. 2000. Morning breath odor: Influence of treatments on sulfur gases. J. Dent. Res. 79: 1773.

Syazana, A.L. and U.A. Minhaz. 2017. Introduction to food biosensors. pp. 1–21. *In*: U.A. Minhaz, Z. Mohammed and T. Eiichi (eds.). Food Chemistry, Function and Analysis No. 1 Food Biosensors. The Royal Society of Chemistry.

Tamayo, J., M. Alvarez and L.M. Lechuga. 2003. Digital tuning of the quality factor of micromechanical resonant biological detectors. Sens. Actuators B: Chem. 89: 33–39.

Tampio, E.A., L. Blasco, M.M. Vainio, M.M. Kahala and S.E. Rasi. 2019. Volatile fatty acids (VFAs) and methane from food waste and cow slurry: Comparison of biogas and VFA fermentation processes. Glob. Chang. Biol. Bioenergy 11: 72–84.

Tong, R., Z. Lei, Y. Xiaoyi, L. Jiefeng, Z. Peining and L. Jianfeng. 2018. Emission characteristics and probabilistic health risk of volatile organic compounds from solvents in wooden furniture manufacturing. J. Clean. Prod. 208: 1096–1108.

Toyooka, T., S. Hiyama and Y. Yamada. 2013. A prototype portable breath acetone analyzer for monitoring fat loss. J. Breath Res. 7: 036005.

Van den Broek, A.M., L. Feenstra and C. de Baat. 2007. A review of the current literature on aetiology and measurement methods of halitosis. J. Dent. 35: 627.

Velasco-García, M. and T. Mottram. 2003. Biosensor technology addressing agricultural problems. Biosyst. Eng. 84(1): 1–12.

Vilas Bôas, R.N. and H.F. de Castro. 2022. A review of synthesis of esters with aromatic, emulsifying, and lubricant properties by biotransformation using lipases. Biotechnol. Bioeng. 119(3): 725–742.

Wang, J. 2005. Nanomaterial-based amplified transduction of biomolecular interactions. Small Small: Nano Micro. 1(11): 1036–1043.

Wang, R. 2002. Two's company, three's a crowd: Can H2S be the third endogenous gaseous transmitter? FASEB J. 16: 1792.

Wendland, B.E., E. Aghdassi, C. Tam, J. Carrrier, A.H. Steinhart, S.L. Wolman et al. 2001. Lipid peroxidation and plasma antioxidant micro-nutrients in Crohn's disease. Am. J. Clin. Nutr. 74(2): 259–264.

Weston-Green, K., H. Clunas and C.J. Naranjo. 2021. A review of the potential use of pinene and linalool as terpene-based medicines for brain health: Discovering novel therapeutics in the flavours and fragrances of cannabis. Front. Psychiatry 12: 1309.

Yang, F. and A.J. Russell. 1996a. The role of hydration in enzyme activity and stability: 1. Water adsorption by alcohol dehydrogenase in a continuous gas phase reactor. Biotechnol. Bioeng. 49: 700–708.

Yang, F. and A.J. Russell. 1996b. The role of hydration in enzyme activity and stability: 2. Alcohol dehydrogenase activity and stability in a continuous gas phase reactor. Biotechnol. Bioeng. 49: 709–716.

Yang, J., Q. Nie, H. Liu, M. Xian and H. Liu. 2016. A novel MVA-mediated pathway for isoprene production in engineered *E. coli*. BMC Biotech. 16: 5.

Yu, H., Y. Byun and P.S. Chang. 2022. Lipase-catalyzed two-step esterification for solvent-free production of mixed lauric acid esters with antibacterial and antioxidative activities. Food Chem. 366: 130650.

Zuliani, A., C.M. Cova, R. Manno, V. Sebastian, A.A. Romero and R. Luque. 2020. Continuous flow synthesis of menthol via tandem cyclisation-hydrogenation of citronellal catalysed by scrap catalytic converters. Green Chem. 22: 379–387.

Chapter 5

Biochemistry and Biomolecules of Halophiles

Recent Trends and Prospects

Sumit Kumar, Bibhuti Bhusan Das, Nitin Srivastava
and *Sunil Kumar Khare**

Introduction

The word halophile is derived from Greek, meaning "salt-loving." Halophiles are distinct from other life forms in their requirement of salt for growth and physiological functions. Halophiles, one of the largest classes among extremophiles, have the distinctive capability to flourish in surroundings rich in elevated salt concentrations (Oren 2002a, Ma et al. 2010, Martínez et al. 2022). They possess the inherent ability to function and carry out metabolic activities in extreme conditions. They are the natural inhabitants of the saline habitat, which comprises 70% of the earth's surface. Therefore, their metabolic and physiological activities are adapted to function under high salt conditions. Kushner (1993) provided the most widely accepted classification related to halophiles. He divided them into five categories, namely, non-halophiles (< 0.2 M, salt); slight halophiles (0.2–0.5 M salt); moderate halophiles (0.5–2.5 M, salt); borderline extreme halophiles (1.5–4.0 M salt) and extreme halophiles (2.5–5.2 M, salt).

Their characteristic adjustment to extreme variations in the exterior salt concentration makes them prospective contenders for usefulness in bioprocesses. The exceptional metabolism of halophiles and the capability to survive and grow under a multitude of polyextreme conditions, i.e., high salinity, extreme pH

Enzyme and Microbial Biochemistry Laboratory, Department of Chemistry, Indian Institute of Technology Delhi, Hauz Khas, New Delhi-110016, India.
* Corresponding author: skkhare@chemistry.iitd.ac.in; skhare@rocketmail.com

and temperatures, hydrophobic conditions, UV radiation, high pressure, toxic compounds, low oxygen concentration, and heavy metals, makes them an ideal reservoir for diverse bioactive compounds (Lach et al. 2021). Research on halophiles has gained considerable attention in recent years because of their potential usefulness in food industries, enhanced oil recovery, biodegradation of toxic pollutants, and as a source of organic osmotic stabilizers, bioplastics, enzymes, exopolysaccharides, and bacteriorhodopsin (Ventosa and Nieto 1995, Margesin and Schinner 2001, Oren 2010, Litchfield 2011, Karan et al. 2012, Hemamalini and Khare 2018, Corral et al. 2019, Uma et al. 2020, Khalil et al. 2021, Jimoh et al. 2022). Novel halophilic biomolecules can also be utilized for specific purposes, e.g., bacteriorhodopsin for biocomputing, pigments for food coloring, osmolytes in cosmetics, etc. They also offer an excellent solution for the environmental bioremediation of saline industrial wastes. Enzymes, exopolysaccharides, metabolites, pigments, and solutes made by halophiles have high marketable importance. Nevertheless, they have been less studied, with not many articles available on devising the production processes based upon them and their products.

The potential of halophiles to produce industrially relevant biomolecules has been especially paid less attention. These microbes can be used to produce bioactives, metabolites, and other industrially important chemicals. Discovering the significant purposes of bioactives from halophiles could be an attractive proposition in the direction of commercialization. The vast marine/saline ecosystems are mostly less explored for their microbial diversity and the biotechnological potential. The research on their basic biology and bioactive molecules offers excellent biotechnological potential. Despite having the versatility to produce a range of industrially important biomolecules, halophiles have not been developed as cell factories for economical production.

Diversity among halophilic microorganisms

The majority of halophiles have been isolated and characterized from the environments, such as saline soil, saline water, salt lakes, soda lakes, saltern crystallizer ponds, the Dead Sea, commercial salts, salted foods and salterns (Ventosa et al. 1998, Oren 2002a, Oren 2008, Karan et al. 2012, Gibtan et al. 2017, Ghosh et al. 2019). In the universal phylogenetic tree of life based on 16S rRNA gene sequence, they are well represented class in Archaea, Bacteria, and Eukarya domain (Fig. 1).

Halophilic archaea are the primitive aerobic halophiles thought to have evolved from anaerobic methanogens. According to a study, the closest evolutionary sibling of haloarchaea, Marine Group IV archaea (Hikarchaeia), has lost practically all genes involved in methanogenesis and the Wood-Ljungdahl pathway. On the other hand, it obtained a huge number of genes for aerobic respiration and salt/UV resistance (Martijn et al. 2020). Halophilic Archaea is largely represented by the *Halobacteriaceae* family of order *Halobacteriales*. Lately, two new orders, *Haloferacales* and *Natrialbales*, have been proposed, encompassing the unique families *Haloferacaceae* and *Natrialbaceae* (Gupta et al. 2015). Predominant genera among halophilic Archaea are *Halobacterium, Haloarcula, Halococcus, Haloferax, Halorubrum, Natrialba,*

Bacteria   **Archaea**   **Eukarya**

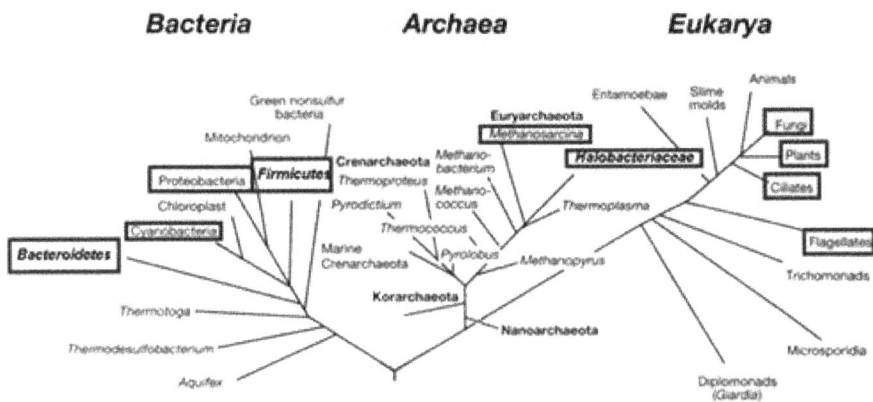

Figure 1. Distribution of halophilic microorganisms in the universal phylogenetic tree. Groups marked with the boxes contain at least one halophilic representative (Reprinted under Creative Commons Attribution License from Oren 2008).

and *Natronococcus*. Some of the described genera *Halogranum*, *Halolamina*, *Haloplanus*, *Halosarcina*, and *Halorientalis* are also predominant and ubiquitous in saline habitats (Youssef et al. 2012). The culture-independent molecular techniques have revealed that square archaeon *Haloquadratum waslbyi* and nanohaloarchaea are abundant at elevated salt concentrations, but more varied archaeal species are found in environments with intermediate salinities (Ventosa et al. 2015). Extremely halophilic methanogenic Archaea are classified in the *Methanosarcinaceae* family (Kamekura 1998). The salt requirement of haloarchaea is exceptionally high, ranging from 100 to 300 g/L with optima in the range of 150–200 g/L (Grant et al. 1998). The Dead Sea, Great Salt Lake, Lake Magadii, and Deep Lake have been major spots for their studies (Grant et al. 1999, Baliga et al. 2004, DasSarma and DasSarma 2012). Recent studies have established that Nanohaloarchaea are one of the dominant microorganisms inhabiting the hypersaline habitats (La Cono et al. 2020, Zhao et al. 2020, Najjari et al. 2021). Microbiome analysis using metagenomics and single-cell genomics has revealed their presence in diverse global habitats (Rinke et al. 2013). Their characteristics have mainly been revealed by metagenomic studies showing small genome sizes (Narasingarao et al. 2012, Rinke et al. 2013). It has been revealed that they range from 0.1–0.8 μm in size and live in symbiotic association with other haloarchaea. Small genome and cell size are unifying characteristics of this class of halophilic microorganisms classified in phylum Nanohaloarchaeota under DPANN superphylum (named after the first members discovered: Diapherotrites, Parvarchaeota, Aenigmarchaeota, Nanoarchaeota, and Nanohaloarchaeota) (Rinke et al. 2013).

Eukaryotic halophiles have received less attention as compared to Archaea and Bacteria. *Dunaliella salina*, a unicellular algae, is the most widely studied halophilic eukarya. Some halophilic fungi, black yeast, and protozoans are also reported in the high salt environment (Oren 2008). Bacteria constitute the most extensive collection of halophiles. A large number among them has been isolated and characterized across the globe. They are mostly moderately halophilic and the most versatile group in the sense that they can carry out metabolic activities over a wide range of salinities. *Halomicronema, Halothermothrix, Marinococcus, Salinicoccus, Halobacillus, Oceanobacillus, Virgibacillus, Geomicrobium, Actinopolyspora, Halomonas, Marinobacter, Chromohalobacter,* and *Pseudoalteromonas* are some predominant genera of halophilic bacteria that have been isolated and characterized in detail across the various saline habitats of the world. Halophilic cyanobacteria and actinomycetes are also reported from saline ecosystems. Viruses termed halovirus have also been reported from saline environments. They control the microbial population in these ecosystems. Among 100 halovirus identified, 90 infect haloarchaea, while others are predators of bacteria and eukaryotes (Luk et al. 2014).

Culture-dependent and -independent diversity studies of halophiles have been done for various saline habitats to understand the microbial communities. However, some of the isolation and diversity studies have been done from the viewpoint of harnessing their biotechnological potential. Hypersaline environments in south Spain (Sánchez-Porro et al. 2003), Tuzkoy salt mine, Turkey (Birbir et al. 2004), Howz Soltan Lake, Iran (Rohban et al. 2009), Tunisian Solar Saltern (Baati et al. 2010), different saline habitats of India (Kumar et al. 2012), salt lakes, Algeria (Menasria et al. 2018) and Saline and Hypersaline Lakes of Romania (Ruginescu et al. 2020) have been investigated with the perspective of potential enzyme producing isolates. Hydrocarbon degraders have been isolated from the hypersaline coastal area of the Arabian Gulf (Al-Mailem et al. 2010), hypersaline pond in France (Tapilatu et al. 2010), salt marshes in Bolivia, crystallizer ponds in Chile, and Cabo Rojo (Puerto Rico), sabkhas (salt flats) in the Persian Gulf (Saudi Arabia) and the Dead Sea (Israel and Jordan) (Bonfá et al. 2011), and saltern near Da Gang Oilfield, China (Zhao et al. 2017). Halophilic bacteria isolated from the South African saltpan have been characterized for three potential applications. The diverse range of bacteria isolated could produce hydrolases (cellulase and lipase), bioremediate hydrocarbons, and produce secondary metabolites for prospective industrial and pharmaceutical applications (Selvarajan et al. 2017). Bioprospecting of sabkhas in Kuwait has been done to isolate squalene producing moderately halophilic bacteria (Alagarsamy et al. 2021).

Morphology of halophiles

Halophiles are typically rods or cocci that do not produce spores. Cocci can be found solitary or in groups. They are generally non-motile (without flagella), but in some cases, motility is observed with polar tufts of flagella. The strength of the salt present can influence the size and morphology of halophilic bacteria. Rod-like structures become elongated or bulbous or even exist as cocci when the salt concentration varies. Replacement of sodium chloride with other salts is reported to cause extreme pleomorphism. Interestingly the morphology of the cells is also influenced by varying culture media (Ventosa et al. 1998). The presence of C-50 carotenoids (α- bacterioruberin and derivatives) in cell vesicles causes halophile colonies to appear red or orange (Oren 2008).

Adaptive features of halophilic microorganisms

The endurance of halophiles in saline water has raised much inquisitiveness about their adaptive mechanism. Obviously, the high salt concentration in environs should lead to exosmosis, pulling out cytoplasmic water and literally dehydrating the cell. To prevent such osmotic water migration out of their cytoplasm, halophiles adopt two key strategies. Both approaches function by boosting the cell's internal osmolarity.

Compatible-solute strategy: This is typical among halophiles who are halotolerant or moderate halophiles. Despite the low internal salt concentration, the osmotic balance between the cytoplasm and the external medium is maintained by (i) active pumping of the ions out of the cells and (ii) by low molecular-weight organic osmotic solutes, known as osmolytes (Oren 2002b). Compatible solutes are said to be made up of four different sorts of molecules, viz., polyols (usually glycerol), sugars (sucrose, trehalose), amino acids (proline, glutamic acid), and quaternary amines (glycine betaine, ectoine, hydroxyectoine). Among these, glycine betaine and ectoine occur most commonly. Amino acids (glutamate, proline, etc.), and some sugars (sucrose, fructose, glucose, or trehalose) are generally accumulated by slight halophiles. Glycine betaine or glutamate betaine; glycosyl glycerol, ectoine, glutamate, K^+; trehalose, or sucrose are all accumulated by moderately and halotolerant bacteria. The halophiles usually accumulate more than one type of osmolyte depending on the growth conditions (Detkova and Boltyanskaya 2007). The concentration of osmolytes typically ranges from 1 mM to 2 M, depending on the extracellular osmolarity. Since the macromolecular machinery of the cell tolerates and functions in the presence of osmolytes over a wide range of concentrations, these compounds are termed as compatible solutes (Lentzen and Schwarz 2006).

Osmotic balance by compatible solutes gives a high degree of adaptability to halophilic microorganisms (Oren 2002b). They also serve as protectants against UV radiation (Kunte 2006) and low water conditions such as freezing and drying (Roberts 2005).

"Salt-in" strategy: This approach is mainly adapted by the halophilic Archaea of the family *Halobacteriaceae* and the anaerobic halophilic *Haloanaerobiales*. The adaptation entails selective inflow of K^+ ions into the cytoplasm and their accumulations (typically stabilized with Cl^-) up to isotonic level, matching the external environment (Oren 2000). Potassium ions are favoured over sodium ions because of their low water binding (Dennis and Shimmin 1997). The chloride ions are in the same concentration as the surroundings, while sodium is lower, and potassium is several times higher (Detkova and Boltyanskaya 2007). These organisms' membranes include a large number of Na^+/H^+ antiporters, which provide a proton gradient for the removal of Na^+ from the cytoplasm (Padan et al. 2001).

Salinibacter ruber, a halophilic bacterium, employs a salt-in strategy and is believed to be a close relative of haloarchaea. They share many phenotypic characteristics with haloarchaea, which is believed to result from horizontal gene transfer because of co-evolution in the same extreme halophilic environment (Anton et al. 2013). Some halophiles, e.g., *Halobacillus halophilus*, apply both, viz., compatible-solute and "salt-in" strategies (Oren 2008, Saum and Müller 2008).

Adaptive features of halophilic proteins

A high concentration of salt accumulated in the cytoplasm critically affects proteins and other macromolecules. The presence of salt results in protein aggregation due to greater hydrophobic interactions, increased hydration of ions, reduced accessibility of free water, and prevention of intra- and intermolecular electrostatic interactions. Halophilic proteins adapt to overcome these effects and maintain native conformation and functionality (Dennis and Shimmin 1997, Detkova and Boltyanskaya 2007, Sinha and Khare 2014, Kumar et al. 2016a).

Halophilic proteins are negatively charged and require cations for stabilization. Because potassium is present in high concentrations inside cells, these proteins have evolved to require potassium as a counter ion rather than sodium. Potassium ions are preferred over sodium ions because of potassium ions' significantly low water binding nature (Dennis and Shimmin 1997). Ions in solution interact with nearby water molecules to form hydration spheres. As a result, any water attached to the ion is unavailable to proteins. The amount of accessible water decreases as the concentration of ions inside the cell rises. Proteins also require hydration spheres to avoid aggregation (Paul et al. 2008). As a result, they have to compete with the K^+ for the water. Halophilic proteins adapt for this competition by decreasing the number of hydrophobic amino acids and increasing hydrophilic amino acids on their surface (Bolhuis et al. 2008). They contain a large excess of acidic amino acids, aspartic and glutamic acid. This also results in an acidic proteome of halophiles with lower pI values of proteins. They contain fewer hydrophobic amino acids than non-halophilic species (Madern et al. 2000, Paul et al. 2008).

Another adaptive trait is that there are considerably fewer large hydrophobic amino acid residues than small hydrophobic amino acid residues (glycine, alanine, and valine) and borderline (serine and threonine) hydrophobic amino acids (Madern et al. 2000, Kastritis et al. 2007, Tadeo et al. 2009). Malate dehydrogenase enzymes have been studied in detail for protein adaptation in halophiles. Its evolutionary history and changes in biochemical properties have also been delineated in halophiles (Blanquart et al. 2021). The 3D structure modeling of *Geomicrobium* sp. protease showed a comparatively higher percentage of small (glycine, alanine, and valine) and borderline (serine and threonine) hydrophobic residues on its surface. The structural study also revealed that acidic residues were enriched at the expense of basic residues (Fig. 2). Salt played a vital role in maintaining these enzymes' secondary and tertiary structures, as evidenced by CD and fluorescence spectroscopy. The presence of salt modulates the activity and stability of halophilic enzymes (Karan and Khare 2011).

It can finally be concluded that the halophilic proteins have (i) high acidic amino acid content on the surface, (ii) low lysine content, (iii) comparatively low hydrophobicity at the core of the protein, and (iv) more salt bridges. As a typical example, high content of acidic residues play significant roles in (i) binding of essential water molecules, (ii) binding of salt ions, (iii) preventing protein aggregation, and (iv) providing flexibility to protein structure through electrostatic repulsion in *Halobacterium salinarum* dodecin and *Haloarcula marismortui* malate dehydrogenase (HmMDH) (Mevarech et al. 2000, Siddiqui and Thomas 2008).

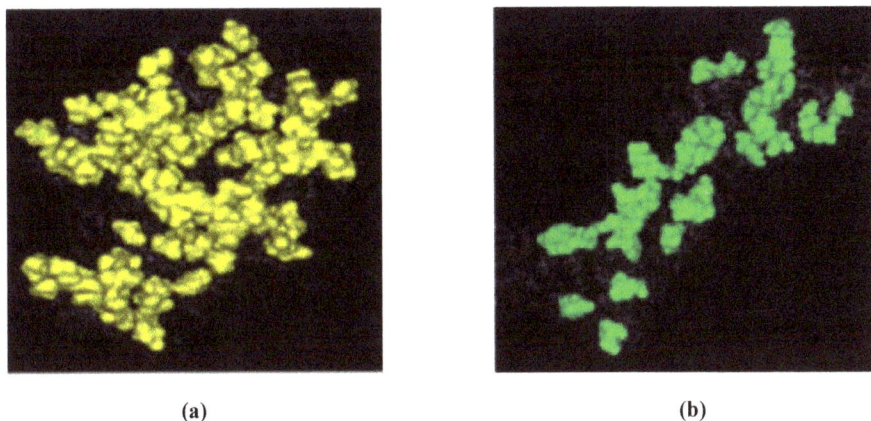

(a)	**(b)**

Figure 2. 3D structure of the *Geomicrobium* sp. protease. (a) The hydrophobic patches present on the protein surface have been given a yellow surface. (b) The negatively charged residues (aspartate and glutamate) present on the protein surface have been given a green surface (Reprinted under Creative Commons Attribution License from Karan et al. 2011).

Some of the unique biochemistry of halophiles results from their adaptation mechanism in salty environments. This also results in the production of some signature bioactive biomolecules of industrial importance by them. Research and significance of biomolecules from halophiles of biotechnological importance are discussed in the following sections.

Industrial molecules from halophiles

Bacteriorhodopsin

Bacteriorhodopsin (BR) is a light-activated protein that is found in haloarchaea's purple membrane. The photoactive BR generates a proton gradient across the membrane by acting as a light-driven proton pump. This is utilized to generate chemical energy in the form of Adenosine Triphosphate (ATP). It comprises seven transmembrane helices containing integral membrane protein with a broad light-absorption maximum of 570 nm. It is composed of bacterioopsin protein moiety and covalently bound chromophore retinal. The Retinal is linked through a protonated Schiff Base to the Lysine 216 side chain. BR is a remarkably stable protein and maintains photoactivity under a wide range of different environmental conditions, viz., active from 0–45°C (stability up to 80°C in water and 140°C in dry conditions); pH range of 1–11; absence of salt. It is resistant to protease hydrolysis and remains active and stable in the presence of sunlight over an extended period. Its stability is attributed to two-dimensional crystalline structure and the source organism Archaea (high salt and temperature stable) (Ashwini et al. 2017, Singh et al. 2021, Pfeifer et al. 2021). BR photoactivity has led to the development of various bioelectronic technologies based on it. Few fields of such potential uses are solar cells, fuel cells, optical devices, biosensors, artificial retinas, nanosensors, storage devices, biodefense, camouflage, holographic memory, and security ink (Singh et al. 2021).

Commercially BR is produced by Merck AG, BOCSCI Inc., and Halotek companies and is available currently in the market at about 553 €/mg. Its demand is ever-increasing but can be obtained in a small amount due to difficulty in cultivating haloarchaea and poor yield. It became vital to improve the production yield and cut costs. In recent times many approaches have been applied to improve its yield by media composition optimization, variation of agitation speed, light intensity, oxygen saturation, and inhibitory metabolites removal (Pfeifer et al. 2021).

Bacteriorhodopsin is in demand for both research and commercial purposes. Various haloarchaea can produce BR, but *H. salinarum* is the model organism for production and economic development (Shakuri et al. 2016, Pfeifer et al. 2021). The highest yield to date has been attained by a mutant strain KSK03307 produced through UV-mediated directed evolution for the highest BR yield and low carotenoid content. Productivity of 0.6 g L^{-1} h^{-1} was achieved in a 3L fed-batch cultivation (Kalenov et al. 2016). The problem in cultivation and poor yield results in availability of a small amount of BR after the fermentation process. This results in the high cost of commercially available BR. It has become imperative to enhance production capability and reduce the production cost to get economically viable BR (Ghasemi et al. 2008, Kalenov et al. 2016). As highlighted above, reducing the production cost and increasing the novel applications of BR are the primary objectives of researchers. In past years, there have been several approaches to increase BR yield from *H. salinarum*. Media composition, nitrogenous sources, oxygen availability, light intensity, agitation speed, and elimination of inhibitory substances have been optimized to get a higher yield (Ghasemi et al. 2008, Kahaki et al. 2014, Pfeifer et al. 2021).

Until now, the tedious step of sucrose density gradient was in practice during downstream process of BR. An aqueous two-phase separation (ATPS) process was developed for BR purification, which reduced the handling period by 10-fold to 2.5 h (Shiu et al. 2013). *Haloarcula marismortui* BR was also purified by the aqueous two-phase separation method and showed significant photoactivity under high salt concentration. Its photoelectric activity was assessed by coating Indium Tin Oxide (ITO) glass and measuring generated DC-voltage upon light illumination (Alsafadi et al. 2018). Later the process was improved by developing a one-step-three-phase extraction system (A3PS). It resulted in a recovery of 89.7% BR with better purity. The BR extracted by A3PS was of better purity and showed 60% higher photocurrent generation than the one obtained by using ATPS (Shiu et al. 2014).

In one study, 12 haloarchaeal strains were isolated from hypersaline solar saltern of Kottakuppam, Tamilnadu, and out of these, 11 strains showed BR gene presence based on PCR screening. Using the gene of one isolate, functional recombinant BR was cloned, expressed, and purified from *Escherichia coli* (Verma et al. 2020). In another study, a bio-sensitized solar cell was developed utilizing BR containing cell lysate of *Halostagnicola larsenii* RG2.14 (MCC 2809) (Kanekar et al. 2020). Some reviews have extensively discussed the future devices, energy harvesters, and sensors based on BR. BR use in biosensors, fuel cells, water-splitting devices, and solar cells have been highlighted in these reviews (Ashwini et al. 2017, Singh et al. 2021). Halotek company commercially produces BR from *H. salinarum*, and this archaeal production is considered a B-TRL 3C stage (Pfeifer et al. 2021). Research on BR-based bioelectronic devices has been done extensively and is mainly based on photochemical and photoelectric applications. At present, the focus is on developing biohybrid electronic devices of high efficiency (Li et al. 2018).

Carotenoids

Carotenoids are secondary metabolites universally synthesized by plants, algae, yeast, bacteria, and archaea. Haloarchaea and moderately halophilic bacteria also synthesize these isoprenoid pigments. They perform the function of photosynthesis, photoprotection, and shield from oxidative stress. Carotenoids have health benefits and are used as food additives, colorants, and in cosmetics. Lycopene, β-carotene, and bacterioruberin are some principal carotenoids synthesized by halophiles. Lycopene is the precursor for most carotenoids and has anti-carcinogenic and antioxidative properties. It has found applications as food additives and in pharmaceuticals and cosmetics (Maoka 2020). β-Carotene is derived from lycopene and acts as provitamin A. Because of its function as vitamin A, it is used as a food supplement and has anticancer and antioxidative properties. Bacterioruberin C50 carotenoids are the main carotenoids produced by haloarchaea. It has shown photoprotection from gamma and UV rays and is believed to be involved in photoprotection. It also has 2.8 times higher antioxidant potential than β-carotene and finds application in cosmetics as a skin protection agent. It could be a suitable substitute for butyl-hydroxytoluol (BHT), a synthetic antioxidant used in personal care products (Pfeifer et al. 2021). At present, majority of carotenoids in industrial applications are synthetic. Its synthesis is complex and costly, and some can have side effects also. Thus, the naturally sourced carotenoids are much sought-after, and halophiles are an ideal chassis for its production. Carotenoids hold great biotechnological potential in biomedicine; in particular, it's antioxidants. It can be used as chemotherapeutic and chemopreventive agents in breast cancer. Its production potential and application from halophiles need special attention (Giani et al. 2021, Pfeifer et al. 2021).

Dunaliella salina, halophilic algae, is a major producer of carotenoids and is a success story in halophile biotechnology (Vachali et al. 2012). Carotenoid extracted from halotolerant *Kocuria* sp. QWT-12 exhibited anticancerous activity on human breast cancer cell lines. The inhibitory concentrations did not affect the viability of the normal fibroblast cell line (Rezaeeyan et al. 2017). Carotenoids of haloarchaea *Halobacterium halobium* isolated from Tunisian solar saltern substantially decreased the viability of the HepG2 cancer cell line (Abbes et al. 2013). *Haloferax mediterranei* genetic engineering has shown an increased yield of lycopene and the potential for commercialization. Metabolic engineering of *H. mediterranei* resulted in a high accumulation of lycopene (Zuo et al. 2018). Bacterioruberin has been identified in various haloarchaea, i.e., *Haloferax volcanii*, *H. mediterranei*, *H. salinarum*, and *Haloarcula vallismortis* (Ronnekleiv 1995, Mandelli et al. 2012, Jehlička et al. 2013). Osmotic stress by providing low salt concentration has been shown to increase the production of bacterioruberin in haloarchaea. *H. mediterranei* exhibited an increase in yield at lower salt concentrations, and it was the highest reported production by a wild-type strain (Chen et al. 2015).

In one research, a metagenome and culture-dependent analysis showed the widespread presence of carotenoid genes in the coastal area of the Arabian sea. The culture-based study showed that *Haloferax* is the major genus present, and they are the predominant producers of carotenoids (Moopantakath et al. 2021). Carotenoids were isolated from the 12 haloarchaeal strains isolated from hypersaline solar saltern of Kottakuppam, Tamilnadu [India]. The isolated carotenoids were lycopene and bacterioruberin (Verma et al. 2020). The most significant advantage of using halophiles for carotenoid production is in their downstream process. They can be harvested by cell lysis in deionized water and arecheaper as compared to tedious processes in other organisms such as *E. coli*. Halotek company produces commercial bacterioruberin with the trade name "Halorubin" and may be considered a B-TRL 3C stage (Pfeifer et al. 2021).

Halophilic enzymes

A primary research goal in enzyme biotechnology is to find new sources of innovative and industrially valuable enzymes. Enzymes used in bioprocesses must be stable at high temperatures, pH, and in the presence of solvents, salts, and toxicants, among other factors. In this regard, halophiles have emerged as a significant repository of new enzymes in recent years. Halophile-derived enzymes have unique structural features and catalytic ability, allowing them to maintain metabolic and physiological functions under high salt environments. In the presence of high salt concentrations, they perform the same function as their non-halophilic counterparts. Some of these enzymes have been shown to be active and stable in multiple severe conditions (Sellek and Chaudhuri 1999, Gomes and Steiner 2004, Karan and Khare 2010, Kumar et al. 2016b). Some of their interesting properties include: (i) optimum activity and stability at high salt concentrations, (ii) the protective effect of salt in maintaining the structure, (iii) increased resistance towards denaturation, and (iv) the capacity to catalyze in low water or non-aqueous media (Madern et al. 2000, Antranikian et al. 2005, Tokunaga et al. 2008, Karan et al. 2012).

The potential of halophilic enzymes has been extensively reviewed (Hough and Danson 1999, Sellek and Chaudhuri 1999, Madern et al. 2000, Eichler 2001, Gomes and Steiner 2004, Ghosh et al. 2005, Oren 2010, Qiu et al. 2021). Some intriguing properties have been reported for xylanases, amylases, proteases, asparaginase, nitrilase, cellulases, chitinases, xylanases, esterases, and lipases from halophiles. These have mainly been investigated from the genera *Haloarcula*, *Haloferax*, *Halobacterium*, *Marinobacter*, *Halomonas*, *Bacillus*, *Halococcus*, and *Chromohalobacter*. Their industrial properties are summarized in Table 1.

One of the novel characteristics found in halophilic enzymes is their solvent stability (Karan et al. 2012, Sinha and Khare 2013, Kumar et al. 2016b). This is due to the fact that they work in surroundings with minimal water activity due to high salt levels. Solvent stable enzymes are potentially beneficial in synthetic applications (Gupta and Roy 2004, Gaur et al. 2008). In the presence of 66% benzene, toluene, and chloroform, amylase from *Haloarcula* sp. strain S-1 remained active and stable (Fukushima et al. 2005). Solvent stable proteases have been earlier reported from

Table 1. The enzymes from halophiles and their biotechnological potential.

Halophilic microorganism(s)	Enzymes	Potential applications	Reference(s)
Haloquadratum walsbyi	Alcohol dehydrogenases	Solvent- and thermos-stable	Cassidy and Paradisi 2018
Haloferax volcanii		Synthesis of enantiopure aromatic alcohols	Alsafadi et al. 2017
Alkalibacterium sp. SL3	α-Amylases	Solvent stable and surfactant-resistant	Wang et al. 2019
Haloferax sp. HA10		Detergent stable	Bajpai et al. 2015
Exiguobacterium sp.		Solvent stable, Bakery industry	Chang et al. 2013
Marinobacter sp. EMB8		Salt and solvent stable; MOS synthesis	Kumar and Khare 2012
Saccharopolyspora sp.		Detergents formulation	Chakraborty et al. 2011
Nesterenkonia sp.		Starch hydrolysis	Shafiei et al. 2010
Bacillus sp.		Detergent formulations	Kiran and Chandra 2008
Halomonas elongata	L-Asparaginase	Anticancer agent	Ghasemi et al. 2017
Clonostachys rosea	Chitinases	Halophilic	Pasqualetti et al. 2019
Planococcus rifitoensis strain M2-26		Salt and heat stable	Essghaier et al. 2010
Aspergillus flavus	Cellulases	Thermo- and salt-tolerant; Bioethanol production	Bano et al. 2019
Thalassobacillus sp. LY18		Salt and solvent stable	Li et al. 2012
Marinobacter sp. MSI032		Stable at alkaline pH	Shanmughapriya et al. 2010
Chromohalobacter canadensis	Esterases	Thermotolerant	Wang et al. 2020
Haloarcula marismortui		Alkaline and salt stable	Camacho et al. 2009
Paracoccus marcusii	β-Galactosidases	Haloalkaline; production of prebiotic oligosaccharides	Kalathinathan et al. 2021
Halorubrum lacusprofundi		Salt and solvent stable	Karan et al. 2013
Haloferax alicantei		Salt stable	Holmes et al. 1997
Halolactibacillus sp. SK71	Glucoamylase	Solvent stable; bioethanol production	Yu and Li 2014
Aquisalibacillus elongatus	Laccase	Polyextremotolerant; delignification of sugar beet pulp	Rezaei et al. 2017
Marinobacter litoralis SW-45	Lipases	Haloalkaine; fatty acid esters synthesis	Musa et al. 2019

Table 1 contd. ...

...Table 1 contd.

Halophilic microorganism(s)	Enzymes	Potential applications	Reference(s)
Alkalispirillum sp. NM-ROO2		Solvent stable; alkyl levulinates production	Mesbah 2019
Marinobacter lipolyticus SM19		Eicosapentaenoic acid (EPA) production; solvent stable	Pérez et al. 2011
Halomonas sp. IIIMB2797	Nitrilase	Solvent stable; useful in biotransformations	Singh et al. 2018
Halococcus salifodinae	Proteases	Solvent- and surfactant-resistant	Hou et al. 2021
Bacillus iranensis		Alkaline and salt stable	Ghafoori et al. 2016
Bacillus sp. EMB9		Solvent stable	Sinha and Khare 2013
Geomicrobium sp. EMB2		Solvent stable, detergent formulations	Karan and Khare 2010
Halobacterium halobium		Solvent stable, used for peptide synthesis	Ryu et al. 1994
Halomonas sp. CSM-2	ω-Transaminase	Furfural conversion in sea water medium	Kelly et al. 2019
Flammeovirga pacifica strain WPAGA1	Xylanases	Low temperature active and salt stable	Cai et al. 2018
Thermoanaerobacterium saccharolyticum NTOU1		Salt stable	Hung et al. 2011
Halophilic bacterium CL8		pH, salt and heat stable	Wejse et al. 2003

H. salinarum (Kim and Dordick 1997). *Geomicrobium* sp. EMB2 (Karan and Khare 2010) and *Bacillus* EMB9 (Sinha and Khare 2013) proteases showed activity and stability in a wide range of hydrophobic solvents. In the study on hydrolases from different saline habitats of India, solvent stability emerged as a generic feature. All the halophilic hydrolases investigated were stable in hexane, cyclohexane, decane, dodecane, and toluene. This unique feature can make them potentially useful for application in non-aqueous enzymology (Kumar et al. 2012).

Halophilic enzymes seem quite capable for industrial applications involving high salt or hypersaline conditions. Furthermore, they have the potential to work under polyextreme conditions of high pH, temperature, and presence of hydrophobic solvents. In this regard, various studies have highlighted their diverse biotechnological applications. Halophilic proteases find applications in detergent, leather tanning, food, and pharmaceutical industries. Solvent-stable protease of *Halobacterium halobium* has been used for peptide synthesis (Ryu et al. 1994). *Halobacterium* sp. strain LBU50301 isolated from salt-fermented fish samples (*budu*) served as a source of proteases (Chuprom et al. 2016). Proteases have also been reported from haloarchaeal strains such as *Natronolimnobius innermongolicus*, *Natrialba magadii*, and *Halogranum rubrum* (Mellado et al. 2005, Selim et al. 2014, Gao et al. 2017).

Recently a co-culture technology was developed for enhanced production of protease from halophilic *Marinirhabdus* sp. and *Marinobacter hydrocarbonoclasticus* (Anh et al. 2021). Solvent-stable halophilic lipase from *Marinobacter lipolyticus* SM19 has been used for Eicosapentaenoic acid (EPA) Production (Pérez et al. 2011). *Salicola* strain IC10 lipase has been reported to be alkali and salt stable (de Lourdes Moreno et al. 2009). A solvent stable lipase from *Haloarcula* sp. G41 showed usage for biodiesel production (Li and Yu 2014). It has been highlighted for use in marine algae-based biofuel production (Schreck and Grunden 2014). Recently, a haloalkaline lipase from *Bacillus flexus* PU2 showed effective biofilm removal of pathogenic *Vibrio parahaemolyticus* (Palanichamy et al. 2022). A recent review on halophilic lipase has highlighted its use in biodiesel production, food flavor modification, and waste treatment (Qiu et al. 2021). α-Amylase from halophiles finds applications in starch hydrolysis and maltooligosaccharide synthesis (Kumar and Khare 2015). Solvent-stable amylase from *Exiguobacterium* sp. has been noted for application in the bakery industry (Chang et al. 2013). *Halomonas meridiana* amylase has been found to be alkali and salt stable (Coronado et al. 2000). *Nesterenkonia* sp. strain F amylase was used for starch hydrolysis (Shafiei et al. 2010). Polyextremophilic amylase of *Aspergillus gracilis* has been found suitable for bioremediation of wastes containing starch and solvents (Ali et al. 2014). Cellulases and chitinases from halophiles have not been studied much. Salt, solvent, and alkaline stable cellulase from *Thalassobacillus* sp. LY18 has been purified and characterized (Li et al. 2012). The potential of halophilic cellulase has also been highlighted for saccharification of lignocelluloses pretreated with ionic liquids (Gunny et al. 2014). Halophilic heat and salt stable chitinase production has been reported from *Planococcus rifitoensis* strain M2-26 (Essghaier et al. 2010). It is clear that halophilic enzymes are stable across a wide range of harsh situations. As a result, they are suggested as the catalyst of choice for applications in (i) hypersaline waste treatment, (ii) peptide synthesis, (iii) detergents, (iv) textile industry, (v) pharmaceuticals, and (vi) food processing (Oren 2010, Delgado-García et al. 2012, Reed et al. 2013, Qiu et al. 2021).

Halocin

Microbes are naturally endowed with the capacity to produce antimicrobial molecules that have physiological significance in the natural ecosystem, possibly due to interspecies competition. Proteinaceous antibiotics secreted by Bacteria and Eukarya domains are called bacteriocin and eukaryocin, respectively. Similarly, one secreted by Archaea are called archaeocins (O'Connor and Shand 2002). Among archaeocins, one secreted by extreme halophiles of domain Archaea are called halocins (Meseguer and Rodriguez-Valera 1986), and those from the hyperthermophilic crenarchaea *Sulfolobus* are called sulfolobicin (Prangishvili et al. 2000). Haloarchaea are known producers of peptide antibiotics halocins. From the application point of view, halocins can also be helpful in the leather industry. Salts are applied to leather hides during processing. This results in infestation by halophilic Archaea, and further lipolytic and proteolytic activity of haloarchaea causes damage to leather hides. The halocin has a specific inhibitory range as it inhibits the growth of particular haloarchaea and not all. Furthermore, the organisms secreting it are immune to it and don't get affected. When leather gets infested by a particular haloarchaea, we need to treat it with specific halocin (from other haloarchaea) to inhibit its growth (Birbir and Eryilmaz 2007, Kumar et al. 2021). Treatment with halocins will benefit and help to minimize spoilage in the leather industry.

Halocins were first reported by Francisco Rodriguez-Valera in 1982 (Rodriguez-Valera et al. 1982), and their production is universal among haloarchaeal rods of the family *Halobacteriaceae* (Torreblanca et al. 1994, Shand and Leyva 2007). Despite their ubiquitous nature, very few have been characterized and researched in detail. Halocins vary in size ranging from 35–3.6 kDa. Protein halocins are sensitive to environmental stress, such as the absence of salt and high temperatures (Ghanmi et al. 2016). Peptide halocins of a size smaller than 10 kDa are called microhalocins. They are very robust, showing activity under low salt conditions, high temperature, and in the presence of an acid, alkali, and organic solvents (Shand and Leyva 2007, Kumar and Tiwari 2017). Microhalocins can be a model system to understand structure/ function relationships under harsh conditions because of their exceptional stability under such conditions. Some recent reviews have focused on the antimicrobial properties of halocins (de Castro et al. 2020, Kumar et al. 2021, Rani et al. 2021).

Some diverse types of halocins studied are S8 from Haloarchaeon S8a (Great Salt Lake, Utah) (Price and Shand 2000), SechA from *H. mediterranei* SechA (Sečovlje solar salterns crystallizers, Slovenia) (Pasic et al. 2008), KPS1 from *H. volcanii* KPS1 (Kovalam solar saltern) (Kavitha et al. 2011), SH10 from *Natrinema* sp. BTSH10 (saltpan of Kanyakumari, Tamilnadu) (Karthikeyan et al. 2013), HA3 from *Haloferax larsenii* HA3 (Pachpadra salt lake, Rajasthan) (Kumar and Tiwari 2017), and HA4 from *Haloferax larsenii* HA4 (Kaur and Tiwari 2021). Genetic characteristics of only halocin H4, S8, and C8 have been investigated (Besse et al. 2015). The maturation, secretion, and immunity property of halocin C8 have been studied in detail, and it has been established that a single gene encodes for antimicrobial peptide and immunity (Sun et al. 2005). It is apparent that halocins have been less explored at the gene level. Halocins share some common features. Most of the halocins are induced, and their production level increases during the transition from stationary to exponential phase (O'Connor and Shand 2002). Consequently, they are an excellent system for studying phase-specific gene regulation in haloarchaea (Cheung et al. 1997). It is observed that halocins have twin arginine translocation (Tat) signal sequences at the amino terminus and are secreted by the Tat pathway. The proteinaceous nature of halocins is also confirmed by the fact that they lose activity after treatment with proteases.

Regarding their mode of action, they alter the cell permeability or inhibit the Na+/H+ antiporter and proton flux. Halocins eradicate sensitive organisms by swelling and finally resulting in cell lysis (O'Connor and Shand 2002, Sun et al. 2005). The activity spectrum of some of the halocins is broad, showing antagonism towards a wide range of Archaea, while others have a narrow range. Halocins have also shown antimicrobial activity against Gram-positive and Gram-negative human pathogens and thus can be a futuristic molecule to treat diseases related to antibiotic resistant microbes (Kavitha et al. 2011). *Haloferax gibbonsii*'s halocin H6 has been found to inhibit Na+/H+ exchanger (NHE) in halobacterial cells (Meseguer et al. 1995). This halocin was also found to inhibit NHE in mammalian cells. Halocin H6 treated dog ischemia-reperfusion model showed a considerable reduction of premature ventricular ectopic beats and infarct size, while blood pressure and heart rate stayed unaffected. This is the first report of biological molecule showing inhibitory action in NHE of eukaryotic cells and this effect can be harnessed for pharmaceutical applications (Lequerica et al. 2006). Such studies are scanty considering the importance of halocins from the perspective of biotechnological applications, the study of the basic biology of haloarchaea, and ecological significance.

Other than the biomolecules discussed above, many more molecules from halophiles have been researched and suggested for future applications. Bioplastics are one among them; biodegradable Poly-3-hydroxybutyrate (PHB) production from *Halomonas halophila* (Kucera et al. 2018) and *Haloferax* sp. MA10 (Tekin et al. 2012) and Poly (3-hydroxybutyrate-co-3-hydroxyvalerate) (PHBHV) production from *Natrinema ajinwuensis* (Mahansaria et al. 2018) and *H. mediterranei* ES1 (Han et al. 2017) have been researched. Halophilic actinomycetes have been marked as a source of antibiotics and other drug molecules. Halophilic actinomycetes are scantly explored as a source of active biomolecules. Though approximately 10000 drug molecules have been derived from normal actinomycetes, those from underexplored hypersaline environments have been studied to a lesser extent. Actinomycetes are prolific makers of secondary metabolites with various biological functions, including antibacterial, antifungal, antiviral, antioxidant, anti-inflammatory, antitumor, and immunomodulatory activities. Due to the high rediscovery rate of existing compounds from actinomycetes, there has been a growing interest in the development of new antimicrobial drugs from the novel and halophilic actinomycetes (Subramani and Aalbersberg 2013, Goel et al. 2021, Santhaseelan et al. 2022). The saline environment is mainly underutilized as a source of actinomycetes, with the potential to yield novel, bioactive natural compounds. Furthermore, some bioactive molecules of halophiles have been commercialized and are success stories. Bitop AG, Dortmund, Germany, uses ectoine produced by *Halomonas elongata* as a skincare product in cosmetics. This company uses extremolytes produced by halophiles in a range of products (https://www.bitop.de/en). HALOTEK Biotechnologie GmbH, Leipzig, Germany, uses the biotechnological potential of haloarchaea for products of personal care, dietary supplements, functional feeds, and life sciences. These products are mainly based on bacteriorhodopsin, halorubrin, and lipids primarily derived from *H. salinarum* (https://halotek.de/).

Recent trends: non-sterile fermentation using halophiles

In recent years halophiles have been identified as a natural choice for non-sterile fermentation because of their essential high salt requirement for growth. Non-sterile fermentation has an advantage over sterile one due to the exclusion of sterilization steps, easier maintenance, simple bioreactor design, and effortless operation. The cost-effective non-sterile fermentation turns out to be the preference for the production of biofuels and platform chemicals. It has been utilized for hydrogen, bioplastics, lactic acid, lipids, and 1,3-propanediol production. Various strategies employed for non-sterile fermentations are starvation, antimicrobial agents, extreme pH, temperature, immobilization, etc. (Chen and Wan 2017). The presence of salt prevents the growth of non-halophilic/mesophilic bacteria. Thus, the metabolites of halophiles can be produced under open and continuous fermentation processes without any contamination (Li et al. 2014). Production of metabolites under non-sterile conditions turns out to be economical and a significant development toward its biotechnological application.

Cultivating halophiles to obtain high biomass is difficult and causes constraints in the successful commercialization of compounds obtained from halophiles. Furthermore, sterilization of media components before fermentation is an energy-intensive process. Added to this, maintenance of sterile conditions throughout the fermentation process adds much cost to the overall production process. The non-sterile fermentation process, in which the sterilization process is not done and the fermentation process is carried out without maintaining sterile conditions, provides a valuable alternative to cut this cost.

Halophiles have three unique advantages over other microbes for fermentative production of different metabolites. First, the downstream process is easier as the cells lyse in the absence of salts, and the recovery process of the product becomes effortless and green. Secondly, sea water instead of fresh water can be used and this significantly improves the economy of the process, adding to environmental benefits. In this direction, halophiles and their enzymes hold tremendous potential for developing sea water-based biorefineries. This will have an enormous environmental and economic benefit (Scapini et al. 2022). Thirdly, non-sterile fermentation is possible with halophiles, which makes the whole process very economical. In this direction, efforts have been made to develop halophiles as the foundation for inexpensive bioprocesses (Liu et al. 2019). The use of contamination-resistant microorganisms, such as halophiles, thermophiles, alkaliphiles, acidophiles, etc., for bioproduction, is termed as "next-generation industrial biotechnology (NGIB)". For NGIB, halophiles can be the most successful chassis for bioproduction (Zhang et al. 2018).

Halophiles are a promising cell factory for economic fermentation through the non-sterile fermentation approach and have been successfully used for bioplastic production (Mitra et al. 2020). *Halomonas* TD01 was used to develop an unsterile and continuous fermentation process for polyhydroxybutyrate (PHB) production. The process was run for 14 days to achieve a 40 g/L cell dry weight containing 60% PHB (Tan et al. 2011). Engineered *Halomonas bluephagenesis* TD01 was successfully used for Poly(3-hydroxybutyrate-co-4-hydroxybutyrate) production in open non-sterile fermentation conditions. The process was further scaled up to a 1000 L pilot fermenter to achieve 83 g/L CDW comprising 61% P(3HB-co-4HB)

(Chen et al. 2017). A marine *Bacillus subtilis* was used for fibrinolytic enzyme production employing non-sterile fermentation (Pan et al. 2019). The use of halophiles for the economical production of bio-commodities by employing non-sterile fermentation is largely untapped and needs further focus and research.

Conclusions

Halophiles are a diverse group of organisms found in all three domains of life. Renewed focus for research on these microorganisms is required to study their basic biology and harness their biotechnological potential. Because of their unique adaptability in harsh and saline environments, they produce diverse novel biomolecules. The stability of these molecules under polyextreme conditions make them suitable for industrial applications. These bioactive molecules ranging from antibiotics, bacteriorhodopsin, bioplastics, carotenoids, enzymes, extremolytes, halocin, etc., find applications in industrial bioprocesses and healthcare. Furthermore, halophiles can serve as chassis for the economical production of these biomolecules by employing non-sterile fermentation.

Acknowledgments

The financial support by the Department of Biotechnology (Government of India), Indian Institute of Technology Delhi, and MHRD, Government of India is gratefully acknowledged. Authors SK and BBD are grateful to the Council of Scientific and Industrial Research (CSIR, Govt. of India) for the Senior Research Associateship (Scientists' Pool Scheme) and Research Fellowship, respectively.

References

Abbes, M., H. Baati, S. Guermazi, C. Messina, A. Santulli, N. Gharsallah et al. 2013. Biological properties of carotenoids extracted from *Halobacterium halobium* isolated from a Tunisian solar saltern. BMC Complement. Altern. Med. 13: 255.

Alagarsamy, S., S.F.K. Habeebullah and F. Al-Yamani. 2021. Bioprospecting potentials of moderately halophilic bacteria and the isolation of squalene producers from Kuwait sabkha. Int. Microbiol. 24: 373–384.

Ali, I., A. Akbar, B. Yanwisetpakdee, S. Prasongsuk, P. Lotrakul and H. Punnapayak. 2014. Purification, characterization, and potential of saline waste water remediation of a polyextremophilic α-amylase from an obligate halophilic *Aspergillus gracilis*. Biomed. Res. Int. 2014: 1–7.

Al-Mailem, D.M., N.A. Sorkhoh, H. Al-Awadhi, M. Eliyas and S.S. Radwan. 2010. Biodegradation of crude oil and pure hydrocarbons by extreme halophilic archaea from hypersaline coasts of the Arabian Gulf. Extremophiles 14: 321–328.

Alsafadi, D., S. Alsalman and F. Paradisi. 2017. Extreme halophilic alcohol dehydrogenase mediated highly efficient syntheses of enantiopure aromatic alcohols. Org. Biomol. Chem. 15: 9169–9175.

Alsafadi, D., F.I. Khalili, H. Juwhari and B. Lahlouh. 2018. Purification and biochemical characterization of photo-active membrane protein bacteriorhodopsin from *Haloarcula marismortui*, an extreme halophile from the Dead Sea. Int. J. Biol. Macromol. 118: 1942–1947.

Anh, H.T.H., E. Shahsavari, N.J. Bott and A.S. Ball. 2021. Application of co-culture technology to enhance protease production by two halophilic bacteria, *Marinirhabdus* sp. and *Marinobacter hydrocarbonoclasticus*. Molecules 26: 3141.

Antranikian, G.V., E. Constantinos and C. Bertoldo. 2005. Extreme environments as a resource for microorganisms and novel biocatalysts. Adv. Biochem. Eng. Biotechnol. 96: 219–262.

Antón, J., M. Lucio, A. Peña, A. Cifuentes, J. Brito-Echeverría, F. Moritz et al. 2013. High metabolomic microdiversity within co-occurring isolates of the extremely halophilic bacterium *Salinibacter ruber*. PLoS One 8: e64701.

Ashwini, R., S. Vijayanand and J. Hemapriya. 2017. Photonic potential of haloarchaeal pigment bacteriorhodopsin for future electronics: A review. Curr. Microbiol. 74: 996–1002.

Baati, H., R. Amdouni, N. Gharsallah, A. Sghir and E. Ammar. 2010. Isolation and characterization of moderately halophilic bacteria from Tunisian solar saltern. Curr. Microbiol. 60: 157–161.

Bajpai, B., M. Chaudhary and J. Saxena. 2015. Production and characterization of α-amylase from an extremely halophilic archaeon, *Haloferax* sp. HA10. Food Technol. Biotechnol. 53: 11–17.

Baliga, N.S., R. Bonneau, M.T. Facciotti, M. Pan, G. Glusman, E.W. Deutsch et al. 2004. Genome sequence of *Haloarcula marismortui*: A halophilic archaeon from the Dead Sea. Genome Res. 14: 2221–2234.

Bano, A., X. Chen, S. Prasongsuk, A. Akbar, P. Lotrakul, H. Punnapayak et al. 2019. Purification and characterization of cellulase from obligate halophilic *Aspergillus flavus* (TISTR 3637) and its prospects for bioethanol production. Appl. Biochem. Biotechnol. 189: 1327–1337.

Besse, A., J. Peduzzi, S. Rebuffat and A. Carre-Mlouka. 2015. Antimicrobial peptides and proteins in the face of extremes: Lessons from archaeocins. Biochimie 118: 344–355.

Birbir, M. and S. Eryilmaz. 2007. Inhibiting lipolytic haloarchaeal damage on brine cured hides with halocin producer strains. J. Soc. Leather. Technol. Chem. 91: 69–72.

Birbir, M., A. Ogan, B. Calli and B. Mertoglu. 2004. Enzyme characteristics of extremely halophilic archaeal community in Tuzkoy salt mine, Turkey. World J. Microbiol. Biotechnol. 20: 613–621.

Blanquart, S., M. Groussin, A. Le Roy, G.J. Szöllosi, E. Girard, B. Franzetti et al. 2021. Resurrection of ancestral malate dehydrogenases reveals the evolutionary history of halobacterial proteins: Deciphering gene trajectories and changes in biochemical properties. Mol. Biol. Evol. 38: 3754–3774.

Bolhuis, A., D. Kwan and J.R. Thomas. 2008. Halophilic adaptations of proteins. pp. 71–104. *In*: K.S. Siddiqui and T. Thomas (eds.). Protein Adaptation in Extremophiles. Nova Science Publishers, Inc. New York, USA.

Bonfá, M.R., M.J. Grossman, E. Mellado and L.R. Durrant. 2011. Biodegradation of aromatic hydrocarbons by Haloarchaea and their use for the reduction of the chemical oxygen demand of hypersaline petroleum produced water. Chemosphere 84: 1671–1676.

Cai, Z.W., H.H. Ge, Z.W. Yi, R.Y. Zeng and G.Y. Zhang. 2018. Characterization of a novel psychrophilic and halophilic β-1, 3-xylanase from deep-sea bacterium, *Flammeovirga pacifica* strain WPAGA1. Int. J. Biol. Macromol. 118: 2176–2184.

Camacho, R.M., J.C. Mateos, O. González-Reynoso, L.A. Prado and J. Córdova. 2009. Production and characterization of esterase and lipase from *Haloarcula marismortui*. J. Ind. Microbiol. Biotechnol. 36: 901–909.

Cassidy, J. and F. Paradisi. 2018. *Haloquadratum walsbyi* yields a versatile, NAD+/NADP+ dual affinity, thermostable, alcohol dehydrogenase (HwADH). Mol. Biotechnol. 60: 420–426.

Chakraborty, S., A. Khopade, R. Biao, W. Jian, X.Y. Liu, K. Mahadik et al. 2011. Characterization and stability studies on surfactant, detergent and oxidant stable α-amylase from marine haloalkaliphilic *Saccharopolyspora* sp. A9. J. Mol. Catal. B Enzym. 68: 52–58.

Chang, J., Y.S. Lee, S.J. Fang, I.H. Park and Y.L. Choi. 2013. Recombinant expression and characterization of an organic-solvent-tolerant α-amylase from *Exiguobacterium* sp. DAU5. Appl. Biochem. Biotechnol. 169: 1870–1883.

Chen, C.W., S. Hsu, M.-T. Lin and Y. Hsu. 2015. Mass production of C50 carotenoids by *Haloferax mediterranei* in using extruded rice bran and starch under optimal conductivity of brined medium. Bioprocess Biosyst. Eng. 38: 2361–2367.

Chen, X., J. Yin, J. Ye, H. Zhang, X. Che, Y. Ma et al. 2017. Engineering *Halomonas bluephagenesis* TD01 for non-sterile production of poly (3-hydroxybutyrate-co-4-hydroxybutyrate). Bioresour. Technol. 244: 534–541.

Chen, Z. and C. Wan. 2017. Non-sterile fermentations for the economical biochemical conversion of renewable feedstocks. Biotech. Lett. 39: 1765–1777.

Cheung, J., K.J. Danna, E.M. O'Connor, L.B. Price and R.F. Shand. 1997. Isolation, sequence, and expression of the gene encoding halocin H4, a bacteriocin from the halophilic archaeon *Haloferax mediterranei* R4. J. Bacteriol. 179: 548–551.

Chuprom, J., P. Bovornreungroj, M. Ahmad, D. Kantachote and S. Dueramae. 2016. Approach toward enhancement of halophilic protease production by *Halobacterium* sp. strain LBU50301 using statistical design response surface methodology. Biotechnol. Rep. 10: 17–28.

Coronado, M.J., C. Vargas, J. Hofemeister, A. Ventosa and J.J. Nieto. 2000. Production and biochemical characterization of an α-amylase from the moderate halophile *Halomonas meridiana*. FEMS Microbiol. Lett. 183: 67–71.

Corral, P., M.A. Amoozegar and A. Ventosa. 2019. Halophiles and their biomolecules: Recent advances and future applications in biomedicine. Mar. Drugs 18: 33.

DasSarma, S. and P. DasSarma. 2012. Halophiles, Encyclopedia of Life Sciences. Wiley, London.

Delgado-García, M., B. Valdivia-Urdiales, C.N. Aguilar-González, J.C. Contreras-Esquivel and R. Rodríguez-Herrera. 2012. Halophilic hydrolases as a new tool for the biotechnological industries. J. Sci. Food Agric. 92: 2575–2580.

Dennis, P.P. and L.C. Shimmin. 1997. Evolutionary divergence and salinity-mediated selection in halophilic Archaea. Microbiol. Mol. Biol. Rev. 61: 90–104.

Detkova, E.N. and Y.V. Boltyanskaya. 2007. Osmoadaptation of haloalkaliphilic bacteria: Role of osmoregulators and their possible practical application. Microbiology 76: 511–522.

de Castro, I., S. Mendo and T. Caetano. 2020. Antibiotics from Haloarchaea: What can we learn from comparative genomics? Mar. Biotechnol. 22: 308–316.

de Lourdes Moreno, M., M.T. García, A. Ventosa and E. Mellado. 2009. Characterization of *Salicola* sp. IC10, a lipase-and protease-producing extreme halophile. FEMS Microbiol. Ecol. 68: 59–71.

Eichler, J. 2001. Biotechnological uses of archaeal extremozymes. Biotechnol. Adv. 19: 261–278.

Essghaier, B., M. Rouaissi, A. Boudabous, H. Jijakli and N. Sadfi-Zouaoui. 2010. Production and partial characterization of chitinase from a halotolerant *Planococcus rifitoensis* strain M2-26. World J. Microbiol. Biotechnol. 26: 977–984.

Fukushima, T., T. Mizuki, A. Echigo, A. Inoue and R. Usami. 2005. Organic solvent tolerance of halophilic α-amylase from a haloarchaeon, *Haloarcula* sp. strain S-1. Extremophiles 9: 85–89.

Gao, R., T. Shi, X. Liu, M. Zhao, H. Cui and L. Yuan. 2017. Purification and characterisation of a saltstable protease from the halophilic archaeon *Halogranum rubrum*. J. Sci. Food. Agr. 97: 1412–1419.

Gaur, R., A. Gupta and S.K. Khare. 2008. Purification and characterization of lipase from solvent tolerant *Pseudomonas aeruginosa* PseA. Process Biochem. 43: 1040–1046.

Ghafoori, H., M. Askari and S. Sarikhan. 2016. Purification and characterization of an extracellular haloalkaline serine protease from the moderately halophilic bacterium, *Bacillus iranensis* (X5B). Extremophiles 20: 115–123.

Ghanmi, F., A. Carré-Mlouka, M. Vandervennet, I. Boujelben, D. Frikha, H. Ayadi et al. 2016. Antagonistic interactions and production of halocin antimicrobial peptides among extremely halophilic prokaryotes isolated from the solar saltern of Sfax, Tunisia. Extremophiles 20: 363–374.

Ghasemi, A., S. Asad, M. Kabiri and B. Dabirmanesh. 2017. Cloning and characterization of *Halomonas elongata* L-asparaginase, a promising chemotherapeutic agent. Appl. Microbiol. Biotechnol. 101: 7227–7238.

Ghasemi, M.F., A. Shodjai-Arani and N. Moazami. 2008. Optimization of bacteriorhodopsin production by *Halobacterium salinarium* PTCC 1685. Process Biochem. 43: 1077–1082.

Ghosh, D., M. Saha, B. Sana and J. Mukherjee. 2005. Marine enzymes. Adv. Biochem. Eng. Biotechnol. 96: 189–218.

Ghosh, S., S. Kumar and S.K. Khare. 2019. Microbial diversity of saline habitats: An overview of biotechnological applications. pp. 65–92. *In*: B. Giri and A. Varma (eds.). Microorganisms in Saline Environments: Strategies and Functions. Springer Nature, Switzerland.

Giani, M., Z. Montero-Lobato, I. Garbayo, C. Vílchez, J.M. Vega and R.M. Martínez-Espinosa. 2021. *Haloferax mediterranei* cells as C50 carotenoid factories. Mar. Drugs 19: 100.

Gibtan, A., K. Park, M. Woo, J.K. Shin, D.W. Lee, J.H. Sohn et al. 2017. Diversity of extremely halophilic archaeal and bacterial communities from commercial salts. Front. Microbiol. 8: 799.

Goel, N., S.W. Fatima, S. Kumar, R. Sinha and S.K. Khare. 2021. Antimicrobial resistance in biofilms: Exploring marine actinobacteria as a potential source of antibiotics and biofilm inhibitors. Biotechnol. Rep. 30: e00613.

Gomes, J. and W. Steiner. 2004. The biocatalytic potential of extremophiles and extremozymes. Food Technol. Biotechnol. 42: 223–235.

Grant, S., W.D. Grant, B.E. Jones, C. Kato and L. Li. 1999. Novel archaeal phylotypes from an East African alkaline saltern. Extremophiles 3: 139–145.

Grant, W.D., R.T. Gemmell and T.J. Mcgenity. 1998. Halophiles. pp. 93–132. *In*: K. Horikoshi and W.D. Grant (eds.). Extremophiles: Microbial Life in Extreme Environments. Wiley-Liss, New York, USA.

Gunny, A.A.N., D. Arbain, R.E. Gumba, B.C. Jong and P. Jamal. 2014. Potential halophilic cellulases for *in situ* enzymatic saccharification of ionic liquids pretreated lignocelluloses. Bioresour. Technol. 155: 177–181.

Gupta, M.N. and I. Roy. 2004. Enzymes in organic media: Forms, functions and applications. Eur. J. Biochem. 271: 2575–2583.

Gupta, R.S., S. Naushad and S. Baker. 2015. Phylogenomic analyses and molecular signatures for the class *Halobacteria* and its two major clades: A proposal for division of the class *Halobacteria* into an emended order *Halobacteriales* and two new orders, *Haloferacales* ord. nov. and *Natrialbales* ord. nov., containing the novel families *Haloferacaceae* fam. nov. and *Natrialbaceae* fam. nov. Int. J. Syst. Evol. Microbiol. 65: 1050–1069.

Han, J., L.P. Wu, X.B. Liu, J. Hou, L.L. Zhao, J.Y. Chen et al. 2017. Biodegradation and biocompatibility of haloarchaea-produced poly (3-hydroxybutyrate-co-3-hydroxyvalerate) copolymers. Biomaterials 139: 172–186.

Hemamalini, R. and S.K. Khare. 2018. Halophilic lipase does forms catalytically active aggregates: Evidence from *Marinobacter* sp. EMB5 lipase (LipEMB5). Int. J. Biol. Macromol. 119: 172–179.

Holmes, M.L., R.K. Scopes, R.L. Moritz, R.J. Simpson, C. Englert, F. Pfeifer et al. 1997. Purification and analysis of an extremely halophilic β-galactosidase from *Haloferax alicantei*. Biochim. Biophys. Acta Protein Struct. Mol. Enzymol. 1337: 276–286.

Hou, J., X.M. Yin, Y. Li, D. Han, B. Lü, J.Y. Zhang et al. 2021. Biochemical characterization of a low salt-adapted extracellular protease from the extremely halophilic archaeon *Halococcus salifodinae*. Int. J. Biol. Macromol. 176: 253–259.

Hough, D.W. and M.J. Danson. 1999. Extremozymes. Curr. Opin. Chem. Biol. 3: 39–46.

Hung, K.S., S.M. Liu, W.S. Tzou, F.P. Lin, C.L. Pan, T.Y. Fang et al. 2011. Characterization of a novel GH10 thermostable, halophilic xylanase from the marine bacterium *Thermoanaerobacterium saccharolyticum* NTOU1. Process Biochem. 46: 1257–1263.

Jehlička, J., H.G.M. Edwards and A. Oren. 2013. Bacterioruberin and salinixanthin carotenoids of extremely halophilic Archaea and Bacteria: A Raman spectroscopic study. Spectrochim. Acta A Mol. Biomol. Spectrosc. 106: 99–103.

Jimoh, A.A., O.O. Ikhimiukor and R. Adeleke. 2022. Prospects in the bioremediation of petroleum hydrocarbon contaminants from hypersaline environments: A review. Environ. Sci. Pollut. Res. Int. 29: 35615–35642.

Kahaki, F.A., V. Babaeipour, H.R. Memari and M.R. Mofid. 2014. High overexpression and purification of optimized bacterio-opsin from *Halobacterium salinarum* R1 in *E. coli*. Appl. Biochem. Biotechnol. 174: 1558–1571.

Kalathinathan, P., K. Pulicherla, A. Sain, S. Gomathinayagam, R. Jayaraj, S. Thangaraj et al. 2021. New alkali tolerant β-galactosidase from *Paracoccus marcusii* KGP–A promising biocatalyst for the synthesis of oligosaccharides derived from lactulose (OsLu), the new generation prebiotics. Bioorg. Chem. 115: 105207.

Kalenov, S.V., M.M. Baurina, D.A. Skladnev and A.Y. Kuznetsov. 2016. High-effective cultivation of *Halobacterium salinarum* providing with bacteriorhodopsin production under controlled stress. J. Biotechnol. 233: 211–218.

Kamekura, M. 1998. Diversity of extremely halophilic bacteria. Extremophiles 2: 289–295.

Kanekar, P.P., S.O. Kulkarni, C.V. Jagtap, V.S. Kadam and H.M. Pathan. 2020. A novel approach for the development of bio-sensitized solar cell using cell lysate of a haloarchaeon *Halostagnicola larsenii* RG2. 14 (MCC 2809) containing bacteriorhodopsin. Sol. Energy 212: 326–331.

Karan, R. and S.K. Khare. 2010. Purification and characterization of a solvent-stable protease from *Geomicrobium* sp. EMB2. Environ. Technol. 31: 1061–1072.

Karan, R. and S.K. Khare. 2011. Stability of haloalkaliphilic *Geomicrobium* sp. protease modulated by salt. Biochemistry (Moscow) 76: 686–693.

Karan, R., M.D. Capes, P. DasSarma and S. DasSarma. 2013. Cloning, overexpression, purification, and characterization of a polyextremophilic β-galactosidase from the Antarctic haloarchaeon *Halorubrum lacusprofundi*. BMC Biotechnol. 13: 3.

Karan, R., R.K.M. Singh, S. Kapoor and S.K. Khare. 2011. Gene identification and molecular characterization of solvent stable protease from a moderately haloalkaliphilic bacterium *Geomicrobium* sp. EMB2. J. Microbiol. Biotechnol. 21: 129–135.

Karan, R., S. Kumar, R. Sinha and S.K. Khare. 2012. Halophilic microorganisms as sources of novel enzymes. pp. 555–579. *In*: T. Satyanarayana and B.N. Johri (eds.). Microorganisms in Sustainable Agriculture and Biotechnology. Springer, Netherlands.

Karthikeyan, P., S.G. Bhat and M. Chandrasekaran. 2013. Halocin SH10 production by an extreme haloarcheon *Natrinema* sp. BTSH10 isolated from salt pans of South India. Saudi J. Biol. Sci. 20: 205–212.

Kastritis, P.L., N.C. Papandreou and S.J. Hamodrakas. 2007. Haloadaptation: Insights from comparative modeling studies of halophilic archaeal DHFRs. Int. J. Biol. Macromol. 41: 447–453.

Kaur, R. and S.K. Tiwari. 2021. Purification and characterization of a new halocin HA4 from *Haloferax larsenii* HA4 isolated from a salt lake. Probiotics Antimicrob. Proteins 13: 1458–1466.

Kavitha, P., A.P. Lipton, A.R. Sarika and M.S. Aishwarya. 2011. Growth characteristics and halocin production by a new isolate, *H. volcanii* KPS1 from Kovalam solar saltern (India). Res. J. Biol. Sci. 6: 257–262.

Kelly, S.A., T.S. Moody and B.F. Gilmore. 2019. Biocatalysis in seawater: Investigating a halotolerant ω-transaminase capable of converting furfural in a seawater reaction medium. Eng. Life Sci. 19: 721–725.

Khalil, A.C., V.L. Prince, R.C. Prince, C.W. Greer, K. Lee, B. Zhang et al. 2021. Occurrence and biodegradation of hydrocarbons at high salinities. Sci. Total Environ. 762: 143165.

Kim, J. and J.S. Dordick. 1997. Unusual salt and solvent dependence of a protease from an extreme halophile. Biotechnol. Bioeng. 55: 471–479.

Kiran, K.K. and T.S. Chandra. 2008. Production of surfactant and detergent-stable, halophilic, and alkalitolerant alpha-amylase by a moderately halophilic *Bacillus* sp. strain TSCVKK. Appl. Microbiol. Biotechnol. 77: 1023–1031.

Kucera, D., I. Pernicová, A. Kovalcik, M. Koller, L. Mullerova, P. Sedlacek et al. 2018. Characterization of the promising poly (3-hydroxybutyrate) producing halophilic bacterium *Halomonas halophile*. Bioresour. Technol. 256: 552–556.

Kumar, S. and S.K. Khare. 2012. Purification and characterization of maltooligosaccharide-forming alpha-amylase from moderately halophilic *Marinobacter* sp. EMB8. Bioresour. Technol. 116: 247–251.

Kumar, S. and S.K. Khare. 2015. Chloride activated halophilic alpha-amylase from *Marinobacter* sp. EMB8: Production optimization and nanoimmobilization for efficient starch hydrolysis. Enzyme Res. 2015: 859485.

Kumar, S., J. Grewal, A. Sadaf, R. Hemamalini and S.K. Khare. 2016b. Halophiles as a source of polyextremophilic α-amylase for industrial applications. AIMS Microbiol. 2: 1–26.

Kumar, S., R. Karan, S. Kapoor, S.P. Singh and S.K. Khare. 2012. Screening and isolation of halophilic bacteria producing industrially important enzymes. Braz. J. Microbiol. 43: 1595–1603.

Kumar, S., R.H. Khan and S.K. Khare. 2016a. Structural elucidation and molecular characterization of *Marinobacter* sp. alpha-amylase. Prep. Biochem. Biotechnol. 46: 238–246.

Kumar, V. and S.K. Tiwari. 2017. Activity-guided separation and characterization of new halocin HA3 from fermented broth of *Haloferax larsenii* HA3. Extremophiles 21: 609–621.

Kumar, V., B. Singh, M.J. van Belkum, D.B. Diep, M.L. Chikindas, A.M. Ermakov et al. 2021. Halocins, natural antimicrobials of Archaea: Exotic or special or both? Biotechnol. Adv. 53: 107834.

Kunte, H.J. 2006. Osmoregulation in bacteria: Compatible solute accumulation and osmosensing. Environ. Chem. 3: 94–99.

Kushner, D.J. 1993. Growth and nutrition of halophilic bacteria. pp. 87–103. *In*: R.H. Vreeland and L.I. Hochstein (eds.). The Biology of Halophilic Bacteria. CRC Press, Florida, USA.

La Cono, V., E. Messina, M. Rohde, E. Arcadi, S. Ciordia, F. Crisafi et al. 2020. Symbiosis between nanohaloarchaeon and haloarchaeon is based on utilization of different polysaccharides. Proc. Natl. Acad. Sci. 117: 20223–20234.

Lach, J., P. Jęcz, D. Strapagiel, A. Matera-Witkiewicz and P. Stączek. 2021. The methods of digging for "Gold" within the salt: Characterization of halophilic prokaryotes and identification of their valuable biological products using sequencing and genome mining tools. Genes 12: 1756.

Lentzen, G. and T. Schwarz. 2006. Extremolytes: Natural compounds from extremophiles for versatile applications. Appl. Microbiol. Biotechnol. 72: 623–634.

Lequerica, J.L., J.E. O'Connor, L. Such, A. Alberola, I. Meseguer, M. Dolz et al. 2006. A halocin acting on Na+/H+ exchanger of Haloarchaea as a new type of inhibitor in NHE of mammals. J. Physiol. Biochem. 62: 253–262.

Li, T., X. Chen, Q. Qu, J. Chen and G.Q. Chen. 2014. Open and continuous fermentation: Products, conditions and bioprocess economy. Biotechnol. J. 9: 1503–1511.

Li, X. and H.Y. Yu. 2014. Characterization of an organic solvent-tolerant lipase from *Haloarcula* sp. G41 and its application for biodiesel production. Folia Microbiol. 59: 455–463.

Li, X., H.L. Wang, T. Li and H.Y. Yu. 2012. Purification and characterization of an organic solvent-tolerant alkaline cellulase from a halophilic isolate of *Thalassobacillus*. Biotechnol. Lett. 34: 1531–1536.

Li, Y.T., Y. Tian, H. Tian, T. Tu, G.Y. Gou, Q. Wang et al. 2018. A review on bacteriorhodopsin-based bioelectronic devices. Sensors 18: 1368.

Litchfield, C.D. 2011. Potential for industrial products from the halophilic archaea. J. Ind. Microbiol. Biotechnol. 38: 1635–1647.

Liu, C., D.K. Baffoe, Y. Zhan, M. Zhang, Y. Li and G. Zhang. 2019. Halophile, an essential platform for bioproduction. J. Microbiol. Methods 166: 105704.

Luk, A.W., T.J. Williams, S. Erdmann, R.T. Papke and R. Cavicchioli. 2014. Viruses of haloarchaea. Life 4: 681–715.

Ma, Y., E.A. Galinski, W.D. Grant, A. Oren and A. Ventosa. 2010. Halophiles 2010: Life in saline environments. Appl. Environ. Microbiol. 76: 6971–6981.

Madern, D., C. Ebel and G. Zaccai. 2000. Halophilic adaptation of enzymes. Extremophiles 4: 91–98.

Mahansaria, R., A. Dhara, A. Saha, S. Haldar and J. Mukherjee. 2018. Production enhancement and characterization of the polyhydroxyalkanoate produced by *Natrinema ajinwuensis* (as synonym) ≡ *Natrinema altunense* strain RM-G10. Int. J. Biol. Macromol. 107: 1480–1490.

Mandelli, F., V.S. Miranda, E. Rodrigues and A.Z. Mercadante. 2012. Identification of carotenoids with high antioxidant capacity produced by extremophile microorganisms. World J. Microbiol. Biotechnol. 28: 1781–1790.

Maoka, T. 2020. Carotenoids as natural functional pigments. J. Nat. Med. 74: 1–16.

Margesin, R. and F. Schinner. 2001. Biodegradation and bioremediation of hydrocarbons in extreme environments. Appl. Microbiol. Biotechnol. 56: 650–663.

Martijn, J., M.E. Schön, A.E. Lind, J. Vosseberg, T.A. Williams, A. Spang et al. 2020. Hikarchaeia demonstrate an intermediate stage in the methanogen-to-halophile transition. Nat. Commun. 11: 1–14.

Martínez, G.M., C. Pire and R.M. Martínez-Espinosa. 2022. Hypersaline environments as natural sources of microbes with potential applications in biotechnology: The case of solar evaporation systems to produce salt in Alicante County (Spain). Curr. Res. Microb. Sci. 3: 100136.

Mellado, E., C. Sánchez-Porro and A. Ventosa. 2005. Proteases produced by halophilic bacteria and archaea. pp. 181–190. *In*: J.L. Barredo (ed.). Microbial Enzymes and Biotransformations. Humana Press, New Jersey, USA.

Menasria, T., M. Aguilera, H. Hocine, L. Benammar, A. Ayachi, A.S. Bachir et al. 2018. Diversity and bioprospecting of extremely halophilic archaea isolated from Algerian arid and semi-arid wetland ecosystems for halophilic-active hydrolytic enzymes. Microbiol. Res. 207: 289–298.

Mesbah, N.M. 2019. Covalent immobilization of a halophilic, alkalithermostable lipase LipR2 on Florisil® nanoparticles for production of alkyl levulinates. Arch. Biochem. Biophys. 667: 22–29.

Meseguer, I. and F. Rodriguez-Valera. 1986. Effect of halocin H4 on cells of *Halobacterium halobium*. J. Gen. Microbiol. 132: 3061–3068.

Meseguer, I., M. Torreblanca and T. Konishi. 1995. Specific inhibition of the halobacterial Na+/H+ antiporter by halocin H6. J. Biol. Chem. 270: 6450–6455.

Mevarech, M., F. Frolow and L.M. Gloss. 2000. Halophilic enzymes: Proteins with a grain of salt. Biophys. Chem. 86: 155–164.

Mitra, R., T. Xu, H. Xiang and J. Han. 2020. Current developments on polyhydroxyalkanoates synthesis by using halophiles as a promising cell factory. Microb. Cell Factories 19: 1–30.

Moopantakath, J., M. Imchen, R. Kumavath and R.M. Martínez-Espinosa. 2021. Ubiquitousness of *Haloferax* and carotenoid producing genes in Arabian Sea coastal biosystems of India. Mar. Drugs 19: 442.

Musa, H., F. Hafiz Kasim, A.A. Nagoor Gunny, S.C.B. Gopinath and M. Azmier Ahmad. 2019. Enhanced halophilic lipase secretion by *Marinobacter litoralis* SW-45 and its potential fatty acid esters release. J. Basic Microbiol. 59: 87–100.

Najjari, A., P. Stathopoulou, K. Elmnasri, F. Hasnaoui, I. Zidi, H. Sghaier et al. 2021. Assessment of 16S rRNA gene-based phylogenetic diversity of Archaeal communities in halite-crystal salts processed from natural Saharan saline systems of Southern Tunisia. Biology 10: 397.

Narasingarao, P., S. Podell, J.A. Ugalde, C. Brochier-Armanet, J.B. Emerson, J.J. Brocks et al. 2012. *De novo* metagenomic assembly reveals abundant novel major lineage of Archaea in hypersaline microbial communities. ISME J. 6: 81–93.

O'Connor, E.M. and R.F. Shand. 2002. Halocins and sulfolobicins: The emerging story of archaeal protein and peptide antibiotics. J. Ind. Microbiol. Biotechnol. 28: 23–31.

Oren, A. 2000. Life at high salt concentrations. pp. 263–282. *In:* M. Dworkin, S. Falkow, E. Rosenberg, K.-H. Scheifer and E. Stackebrandt (eds.). The Prokaryotes. A Handbook on the Biology of Bacteria: Ecophysiology, Isolation, Identification, Applications. Springer, New York.

Oren, A. 2002a. Molecular ecology of extremely halophilic archaea and Bacteria. FEMS Microbiol. Ecol. 39: 1–7.

Oren, A. 2002b. Adaptation of halophilic archaea to life at high salt concentrations. pp. 81–96. *In:* A. Läuchli and U. Lüttge (eds.). Salinity: Environments-Plants—Molecules. Kluwer Academic Publishers, Dordrecht, Netherlands.

Oren, A. 2008. Microbial life at high salt concentrations: Phylogenetic and metabolic diversity. Saline Syst. 4: 2–15.

Oren, A. 2010. Industrial and environmental applications of halophilic microorganisms. Environ. Technol. 31: 825–834.

Padan, E., M. Venturi, Y. Gerchman and N. Dover. 2001. Na$^{+/H+}$ antiporters. Biochim. Biophys. Acta Bioenerg. 1505: 144–157.

Palanichamy, E., A. Repally, N. Jha and A. Venkatesan. 2022. Haloalkaline lipase from *Bacillus flexus* PU2 efficiently inhibits biofilm formation of aquatic pathogen *Vibrio parahaemolyticus*. Probiotics Antimicrob. Proteins 1–11.

Pan, S., G. Chen, R. Wu, X. Cao and Z. Liang. 2019. Non-sterile submerged fermentation of fibrinolytic enzyme by marine *Bacillus subtilis* harboring antibacterial activity with starvation strategy. Front. Microbiol. 10: 1025.

Pasic, L., B.H. Velikonja and N.P. Ulrih. 2008. Optimization of the culture conditions for the production of a bacteriocin from halophilic archaeon Sech7a. Prep. Biochem. Biotechnol. 38: 229–245.

Pasqualetti, M., P. Barghini, V. Giovannini and M. Fenice. 2019. High production of chitinolytic activity in halophilic conditions by a new marine strain of *Clonostachys rosea*. Molecules 24: 1880.

Paul, S., S.K. Bag, S. Das, E.T. Harvill and C. Dutta. 2008. Molecular signature of hypersaline adaptation: Insights from genome and proteome composition of halophilic prokaryotes. Genome Biol. 9: R70.

Pérez, D., S. Martín, G. Fernández-Lorente, M. Filice, J.M. Guisán, A. Ventosa et al. 2011. A novel halophilic lipase, LipBL, showing high efficiency in the production of eicosapentaenoic acid (EPA). PLoS One 6: e23325.

Pfeifer, K., İ. Ergal, M. Koller, M. Basen, B. Schuster and K.M.R. Simon. 2021. Archaea biotechnology. Biotechnol. Adv. 47: 107668.

Prangishvili, D., I. Holz, E. Stieger, S. Nickell, J.K. Kristjansson and W. Zillig. 2000. Sulfolobicins, specific proteinaceous toxins produced by strains of the extremely thermophilic archaeal genus *Sulfolobus*. J. Bacteriol. 182: 2985–2988.

Price, L.B. and R.F. Shand. 2000. Halocin S8: A 36-amino-acid microhalocin from the haloarchaeal strain S8a. J. Bacteriol. 182: 4951–4958.

Qiu, J., R. Han and C. Wang. 2021. Microbial halophilic lipases: A review. J. Basic Microbiol. 61: 594–602.

Rani, A., K.C. Saini, F. Bast, S. Varjani, S. Mehariya, S.K. Bhatia et al. 2021. A review on microbial products and their perspective application as antimicrobial agents. Biomolecules 11: 1860.

Reed, C.J., H. Lewis, E. Trejo, V. Winston and C. Evilia. 2013. Protein adaptations in archaeal extremophiles. Archaea 2013: 373275.

Rezaeeyan, Z., A. Safarpour, M.A. Amoozegar, H. Babavalian, H. Tebyanian and F. Shakeri. 2017. High carotenoid production by a halotolerant bacterium, *Kocuria* sp. strain QWT-12 and anticancer activity of its carotenoid. EXCLI J. 16: 840.

Rezaei, S., A.R. Shahverdi and M.A. Faramarzi. 2017. Isolation, one-step affinity purification, and characterization of a polyextremotolerant laccase from the halophilic bacterium *Aquisalibacillus elongatus* and its application in the delignification of sugar beet pulp. Bioresour. Technol. 230: 67–75.

Rinke, C., P. Schwientek, A. Sczyrba, N.N. Ivanova, I.J. Anderson, J.F. Cheng et al. 2013. Insights into the phylogeny and coding potential of microbial dark matter. Nature 499: 431–437.

Roberts, M.F. 2005. Organic compatible solutes of halotolerant and halophilic microorganisms. Saline Syst. 1: 5–35.

Rodriguez-Valera, F., G. Juez and D.J. Kushner. 1982. Halocins: Salt-dependent bacteriocins produced by extremely halophilic rods. Can. J. Microbiol. 28: 151–154.

Rohban, R., M.A. Amoozegar and A. Ventosa. 2009. Screening and isolation of halophilic bacteria producing extracellular hydrolases from Howz Soltan Lake, Iran. J. Ind. Microbiol. Biotechnol. 36: 333–340.

Ronnekleiv, M. 1995. Bacterial carotenoids 53, C50-carotenoids 23; carotenoids of *Haloferax volcanii* versus other halophilic bacteria. Biochem. Syst. Ecol. 23: 627–634.

Ruginescu, R., I. Gomoiu, O. Popescu, R. Cojoc, S. Neagu, I. Lucaci et al. 2020. Bioprospecting for novel halophilic and halotolerant sources of hydrolytic enzymes in brackish, saline and hypersaline lakes of Romania. Microorganisms 8: 1903.

Ryu, K., J. Kim and J.S. Dordick. 1994. Catalytic properties and potential of an extracellular protease from an extreme halophile. Enzym. Microb. Technol. 16: 266–275.

Sánchez-Porro, C., S. Martín, E. Mellado and A. Ventosa. 2003. Diversity of moderately halophilic bacteria producing extracellular hydrolytic enzymes. J. Appl. Microbiol. 94: 295–300.

Santhaseelan, H., V.T. Dinakaran, H.U. Dahms, J.M. Ahamed, S.G. Murugaiah, M. Krishnan et al. 2022. Recent antimicrobial responses of halophilic microbes in clinical pathogens. Microorganisms 10: 417.

Saum, S.H. and V. Müller. 2008. Regulation of osmoadaptation in the moderate halophile *Halobacillus halophilus*: Chloride, glutamate and switching osmolyte strategies. Saline Syst. 4: 4–19.

Scapini, T., C. Dalastra, A.F. Camargo, S. Kubeneck, T.A. Modkovski, S.L.A. Júnior et al. 2022. Seawater-based biorefineries: A strategy to reduce the water footprint in the conversion of lignocellulosic biomass. Bioresour. Technol. 344: 126325.

Schreck, S.D. and A.M. Grunden. 2014. Biotechnological applications of halophilic lipases and thioesterases. Appl. Microbiol. Biotechnol. 98: 1011–1021.

Selim, S., N. Hagagy, M.A. Aziz, E.S. El-Meleigy and E. Pessione. 2014. Thermostable alkaline halophilic-protease production by *Natronolimnobius innermongolicus* WN18. Nat. Prod. Res. 28: 1476–1479.

Sellek, G.A. and J.B. Chaudhuri. 1999. Biocatalysis in organic media using enzymes from extremophiles. Enzyme Microb. Technol. 25: 471–482.

Selvarajan, R., T. Sibanda, M. Tekere, H. Nyoni and S. Meddows-Taylor. 2017. Diversity analysis and bioresource characterization of halophilic bacteria isolated from a South African saltpan. Molecules 22: 657.

Shafiei, M., A.A. Ziaee and M.A. Amoozegar. 2010. Purification and biochemical characterization of a novel SDS and surfactant stable, raw starch digesting, and halophilic α-amylase from a moderately halophilic bacterium, *Nesterenkonia* sp. strain F. Process Biochem. 45: 694–699.

Shakuri, S., A.M. Latifi, M. Mirzaei and S. Khodi. 2016. Isolating two native extreme halophilic bacterial strains producing bacteriorhodopsin protein from aran-bidgol lake. J. Appl. Biotechnol. Rep. 3: 447–452.

Shand, R.F. and K.J. Leyva. 2007. Peptide and protein antibiotics from the domain *Archaea*: Halocins and sulfolobicins. pp. 93–109. *In*: M.A. Riley and M.A. Chavan (eds.). Bacteriocins: Ecology and evolution. Springer, Berlin Heidelberg.

Shanmughapriya, S., G.S. Kiran, J. Selvin, T.A. Thomas and C. Rani. 2010. Optimization, purification, and characterization of extracellular mesophilic alkaline cellulase from sponge-associated *Marinobacter* sp. MSI032. Appl. Biochem. Biotechnol. 162: 625–640.

Shiu, P.-J., H.-M. Chen and C.-K. Lee. 2014. One-step purification of delipidated Bacteriorhodopsin by aqueous-three-phase system from purple membrane of *Halobacterium*. Food Bioprod. Process. 92: 113–119.

Shiu, P.-J., Y.-H. Ju, H.-M. Chen and C.-K. Lee. 2013. Facile isolation of purple membrane from *Halobacterium salinarum* via aqueous-two-phase system. Protein Expr. Purif. 89: 219–224.

Siddiqui, K.S. and T. Thomas. 2008. Protein Adaptation in Extremophiles. Nova Science Publishers, Inc., New York, USA.

Singh, P., S. Singh, N. Jaggi, K.H. Kim and P. Devi. 2021. Recent advances in bacteriorhodopsin-based energy harvesters and sensing devices. Nano Energy 79: 105482.

Singh, R.V., H. Sharma, A. Koul and V. Babu. 2018. Exploring a broad spectrum nitrilase from moderately halophilic bacterium *Halomonas* sp. IIIMB2797 isolated from saline lake. J. Basic Microbiol. 58: 867–874.

Sinha, R. and S.K. Khare. 2013. Characterization of detergent compatible protease of a halophilic *Bacillus* sp. EMB9: differential role of metal ions in stability and activity. Bioresour. Technol. 145: 357–361.

Sinha, R. and S.K. Khare. 2014. Protective role of salt in catalysis and maintaining structure of halophilic proteins against denaturation. Front. Microbiol. 5: 165.

Subramani, R. and W. Aalbersberg. 2013. Culturable rare Actinomycetes: Diversity, isolation and marine natural product discovery. Appl. Microbiol. Biotechnol. 97: 9291–9321.

Sun, C., Y. Li, S. Mei, L. Zhou and H. Xiang. 2005. A single gene directs both production and immunity of halocin C8 in a haloarchaeal strain AS7092. Mol. Microbiol. 57: 537–549.

Tadeo, X., B. López-Méndez, T. Trigueros, A. Laín, D. Castaño and O. Millet. 2009. Structural basis for the aminoacid composition of proteins from halophilic archea. PLoS Biol. 7: e1000257.

Tan, D., Y.S. Xue, G. Aibaidula and G.Q. Chen. 2011. Unsterile and continuous production of polyhydroxybutyrate by *Halomonas* TD01. Bioresour. Technol. 102: 8130–8136.

Tapilatu, Y.H., V. Grossi, M. Acquaviva, C. Militon, J.C. Bertrand and P. Cuny. 2010. Isolation of hydrocarbon-degrading extremely halophilic archaea from an uncontaminated hypersaline pond (Camargue, France). Extremophiles 14: 225–231.

Tekin, E., M. Ateş and Ö. Kahraman. 2012. Poly-3-hydroxybutyrate-producing extreme halophilic archaeon: *Haloferax* sp. MA10 isolated from Çamaltı Saltern, İzmir. Turk. J. Biol. 36: 303–312.

Tokunaga, H., T. Arakwa and M. Tokunaga. 2008. Engineering of halophilic enzymes: Two acidic amino acid residues at the carboxy-terminal region confer halophilic characteristics to *Halomonas* and *Pseudomonas* nucleoside diphosphate kinases. Protein Sci. 17: 1603–1610.

Torreblanca, M., I. Meseguer and A. Ventosa. 1994. Production of halocin is a practically universal feature of archaeal halophilic rods. Lett. Appl. Microbiol. 19: 201–205.

Uma, G., M.M. Babu, V.S.G. Prakash, S.J. Nisha and T. Citarasu. 2020. Nature and bioprospecting of haloalkaliphilics: A review. World J. Microbiol. Biotechnol. 36: 1–13.

Vachali, P., P. Bhosale and P.S. Bernstein. 2012. Microbial carotenoids. pp. 41–59. *In*: J.-L. Barredo (ed.). Microbial Carotenoids from Fungi. Humana, New Jersey.

Ventosa, A. and J.J. Nieto. 1995. Biotechnological applications and potentialities of halophilic microorganisms. World J. Microbiol. Biotechnol. 11: 85–94.

Ventosa, A., J.J. Nieto and A. Oren. 1998. Biology of moderately halophilic aerobic bacteria. Microbiol. Mol. Biol. Rev. 62: 504–544.

Ventosa, A., R.R. de la Haba, C. Sánchez-Porro and R.T. Papke. 2015. Microbial diversity of hypersaline environments: A metagenomic approach. Curr. Opin. Microbiol. 25: 80–87.

Verma, D.K., C. Chaudhary, L. Singh, C. Sidhu, B. Siddhardha, S.E. Prasad et al. 2020. Isolation and taxonomic characterization of novel Haloarchaeal isolates from Indian solar saltern: A brief review on distribution of bacteriorhodopsins and V-Type ATPases in Haloarchaea. Front. Microbiol. 11: 554927.

Wang, G., M. Luo, J. Lin, Y. Lin, R. Yan, W.R. Streit et al. 2019. A new extremely halophilic, calcium-independent and surfactant-resistant alpha-amylase from *Alkalibacterium* sp. SL3. J. Microbiol. Biotechnol. 29: 765–775.

Wang, M., L. Ai, M. Zhang, F. Wang and C. Wang. 2020. Characterization of a novel halotolerant esterase from *Chromohalobacter canadensis* isolated from salt well mine. 3 Biotech. 10: 1–13.

Wejse, P.L., K. Ingvorsen and K.K. Mortensen. 2003. Purification and characterisation of two extremely halotolerant xylanases from a novel halophilic bacterium. Extremophiles 7: 423–431.

Youssef, N.H., K.N. Ashlock-Savage and M.S. Elshahed. 2012. Phylogenetic diversities and community structure of members of the extremely halophilic Archaea (order Halobacteriales) in multiple saline sediment habitats. Appl. Environ. Microbiol. 78: 1332–1344.

Yu, H.Y. and X. Li. 2014. Characterization of an organic solvent-tolerant thermostable glucoamylase from a halophilic isolate, *Halolactibacillus* sp. SK 71 and its application in raw starch hydrolysis for bioethanol production. Biotechnol. Prog. 30: 1262–1268.

Zhang, X., Y. Lin and G.Q. Chen. 2018. Halophiles as chassis for bioproduction. Adv. Biosyst. 2: 1800088.

Zhao, D., S. Kumar, J. Zhou, R. Wang, M. Li and H. Xiang. 2017. Isolation and complete genome sequence of *Halorientalis hydrocarbonoclasticus* sp. nov., a hydrocarbon-degrading haloarchaeon. Extremophiles 21: 1081–1090.

Zhao, D., S. Zhang, Q. Xue, J. Chen, J. Zhou, F. Cheng et al. 2020. Abundant taxa and favorable pathways in the microbiome of soda-saline lakes in Inner Mongolia. Front. Microbiol. 11: 1740.

Zuo, Z.Q., Q. Xue, J. Zhou, D.H. Zhao, J. Han and H. Xiang. 2018. Engineering *Haloferax mediterranei* as an efficient platform for high level production of lycopene. Front. Microbiol. 9: 2893.

Chapter 6

Marine Enzymes

Exploiting Bacterial Resource for Blue Biotechnology

Yasmin Khambhaty

Introduction

Marine environments, including the subsurface are believed to contain about 3.67×10^{30} microorganisms (Whiteman et al. 1998) and with ~ 71% of the earth's surface covered by the ocean, this environment represents a vast pool of microbial diversity and exploitable biotechnology or "blue biotechnology". Accelerated research interest in this untapped bioresource, is not only to provide information on the key role these microbes play in marine biogeochemical cycles, but also in exploiting their ability to produce novel enzymes and metabolites/compounds with potential applications (Kennedy et al. 2008). Significant studies over the past few years has demonstrated that these organisms which are in "commensal" relationship with invertebrate hosts are from all the three domains; the bacteria, archaea and eukarya, many of which have not yet been successfully cultured. Nevertheless, organisms whose phylogeny can be quizzed by molecular approaches, including widely-used small subunit rRNA gene sequencing, and in few cases, whose secondary metabolite-producing clusters can be expressed in surrogate hosts may be cultured (Newman and Hill 2006). The progress in metagenomics have revolutionized the research in the field of blue biotechnology, enabling not only a glimpse into the uncultured microbes and mechanistic understanding of possible biogeochemical cycles but also led to the discovery of new enzymes for industrial bioconversions (Zhao et al. 2015). In this review, the exciting avenues that metagenomic based approaches offer

Microbiology Department, CSIR-Central Leather Research Institute, Adyar, Chennai 600 020, India.
Email: yasmink@clri.res.in

to uncultivable microorganisms; thereby allowing to exploit, the potential of this vast, and as yet untapped, marine microbial resource have been highlighted. The use of diverse bacteria as a possible resource to plethora of novel enzymes; and the importance of improving our understanding on how this diversity can be understood and sustainably exploited have also been discussed.

The metagenomic approach

It is commonly believed that more than 99% of bacteria in any given environment cannot be cultured using conventional methods (Sharma et al. 2005). Although about 200 genome projects are either completed or underway succeeding the publication of the first whole genome in 1995, speculations have been raised as to what is actually being sequenced, as these studies are limited only to culturable microbes (Newman and Hill 2006). Metagenomic based approaches have emerged as an attractive alternative allowing the assessment of the microbial genomes present within these environments (Cowan et al. 2005), and have potential advantages like; being culture independent, hypothesis independent and finally, a data resource which may be of enormous importance to the scientific community (Newman and Hill 2006). This approach was principally incepted to study the non-culturable microbiota and primarily focused on providing a better understanding of the comprehensive microbial community in varied environments. It involves the direct cloning of environmental DNA into large clone libraries to facilitate the analysis of the genes and the sequences within these libraries (Handelsman 2004) (Fig. 1). Enhanced DNA isolation techniques and progressive screening methodologies coupled with efficient cloning vectors such as bacterial artificial chromosomes (BACs) and cosmids, expression of large fragments of DNA and subsequent screening of huge cloned libraries for functional activities are possible (Lorenz and Eck 2005). This approach primarily divides into sequence-based and function-based screening; where in the case of former, PCR amplifications are used to identify target genes from conserved regions of known genes, while the latter is often carried out using robotic systems looking for a well-defined phenotype (Barone et al. 2014). The study of functional metagenomics is generally a challenge where several basic methods like gene construction, screening, expression, followed by bioinformatic analysis, protein product characterization such as optimum pH, temperature and activity analysis are carried out to discover novel enzymes (Nazir 2016). The coupling with additional innovative screening approaches such as Substrate-Induced Gene Expression screening (SIGEX) have also facilitated the cloning of catabolic operons (Uchiyama et al. 2005). A critical link from screening to the data analysis has been established by developing new systems, that relies on the sequencing of dense array of amplified DNA fragments through iterative cycles of enzymatic manipulation and imaging-based data collection (Shendure and Ji 2008). The so-called "third generation sequencing technologies", which include sequencing of individual molecules, the single-cell sequencing (Xu et al. 2009) involving fluorescence-based detection (Metzker 2010) is also gaining interest. Nevertheless, several challenges may be faced while using the metagenomic approach, viz., storage of the sample in a way that does not impact its viability, obtaining sufficient quantity

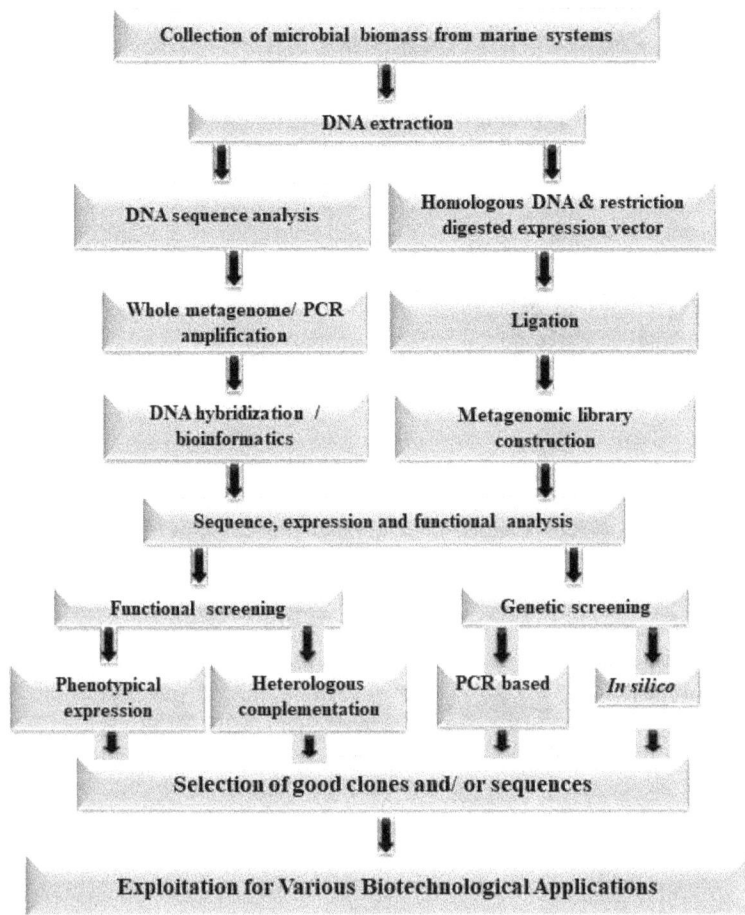

Figure 1. Metagenomics based approaches to identify novel enzymes from marine ecosystems.

of DNA to perform the analysis, the maintenance of gene reference catalogue for the significant amount of data generated, challenge in gene profiling with a risk of some reads mapping to two or more genes at the same time, etc (Kamble and Vavilala 2018). Hence, in order to achieve a holistic view, it requires an interdisciplinary effort and development of powerful bioinformatic tools that can help unravel the complexity of the biological world. An extensive *in silico* investigation on biological networks can prove effective in shedding light on connected components, their functionality and dynamics within the system.

Marine microbes: a potential and untapped source of novel enzymes

The world's oceans are colonized by microbes, which are both taxonomically diverse and metabolically complex and can function as both the primary producers and the primary recyclers. Their precise role in biogeochemical cycles is unknown; however, they are known to be involved in the cycling of nitrogen, carbon, oxygen, phosphorous, iron, sulphur and trace elements. Their colossal presence and biochemical versatility are believed to be responsible for the continuous enduring of these cycles, sustaining all living beings in these ecosystems (Karl 2007). The microbes that thrive under extreme and diverse niches of pressure, temperature and salinity are supposed to be an extraordinary natural source of stable and efficient enzymes, that could help improve the performance and sustainability of industrial processes making them more economical (Poli et al. 2017). Evolution has prompted the marine microbes to generate multifarious enzyme systems to adapt to the complicated marine environment ranging from nutritionally rich to nutritionally sparse locations where only a few organisms can thrive. Way back in 1985, Pace and colleagues introduced direct cloning from environmental samples (Pace et al. 1985) and the first study of screening based on functional genes was successfully conducted by Healy et al. (1995), reporting the isolation of functional genes that encode cellulase enzymes from the environment. Five years later, Rondon et al. (2000) used BAC as a vector for creation of a metagenome library from soil samples and found enzymes like lipases, amylases, and nucleases. Novel metagenomic methods such as DNA shuffling (Boubakri et al. 2006) and pre-amplification inverse-PCR (PAI-PCR) (Yamada et al. 2008) have also been employed to isolate new biocatalysts. PAI-PCR which was used to isolate glycosyl hydrolase genes from horse and termite guts, offered the potential to clone genes for which the copy number of target DNA sequences was low, while the shuffling approach, which was used to construct novel biocatalysts, simulated and accelerated the developments using molecular biological tools. Using a metagenomic library simplifies the discovery of genes and has facilitated the discovery of many novel enzymes from uncultivable bacteria (Sharpton 2014). The multifaceted marine environment consisting of thermal vents, cold sea region, high salinity, etc., may result in tremendous differences between enzymes synthesized by marine microbes as compared to terrestrial system (Imhoff et al. 2011). Organisms are required to endure extreme conditions such as high salinity, low temperatures which can occasionally alternate with high temperatures, increased depth meaning increased pressure and reduced light; nutritional rich regions that co-exist with locations where those resources are scarce or relatively uncommon, etc (Trincone 2013). And hence, enzymes from these organisms tend to exhibit distinct characteristics like stability and tolerance to high temperatures and acidic/alkaline environments. They also reveal a great deal of biological activities such as antibacterial, antifungal and anti-inflammatory, which can further have an immense impact in human health industry. Marine enzymes are of great importance in food, detergent, paper, leather, pharmaceutical, cosmetic and various other industries. These typical features of marine organism are obviously anchored in enzymes with concomitant unique and appealing characteristics, which have resulted in an increased interest in the prospection of marine resources for enzymes with varied applications (Arnosti et al. 2014). Hence, as we can see, metagenomics has resulted as an effective means for discovery of novel extremozymes, isolated from

EXTREMOPHILES

PSYCHROPHILES (-2-20 °C)	THERMOPHILES (55-113 °C)	ACIDOPHILES (pH < 4)	ALKALIPHILES (pH > 9)	HALOPHILES (2-5 M NaCl)
Algoriphagus, Alteromonas, Colwellia, Desulfotolea, Exiguobacterium, Flameovirga, Martelella, Planococcus, Pseudoaltermonas, Psychrobacter, Zunongwangia, etc.	*Archaeoglobus, Aquifex, Bacillus, Fervidobacteria mFlammeovirga, Geobacillus, Hydrogenobacter, Methanococcus, Microbulbifer, Pyrococcus, Pyrodictium, Pyrolobus, Staphylothermus, Sulfolobus, Thermococcus, Thermoproteus, Thermoplasma, Thermus, Thermotoga, etc.*	*Acidianus, Acidobacillus, Acidiferrobacter, Acidihalobacter, Desulfurolobus, Hydrogenobaculum, Sulfolobus, Sulfobacillus, Thiobacillus, etc.*	*Alkaliphilus, Alkalibacterium, Bacillus, Corynebacterium, Natronobacterium, Natronococcus, Pseudomonas, Streptomyces, Thioalkalivibrio, etc.*	*Bacillus, Cyanobacterium, Emericellopsis, Haloarcula, Halobacterium, Haloferax, Halorubrum, Halothermothrix, etc.*
α-Amylase, β-glucosidase, Esterase, Xylanase, Lipase, Protease, etc.	*Amylase, β-Agarase, Amylopullulanase, Cellulase, DNA Gyrase, Keratinase, Prolidase, Lipase etc.*	*Amylase, Cellulase, Esterase, Endo-gluconase, α-glucosidase, Lipase, Ligase etc.*	*Cellulase, Lipase, Protease, Pectinase, Pullulanase, Starch degrading enzymes, Xylanase, etc.*	*Amylase, Chitinase, Chitin deacetylase Xylanase, etc.*

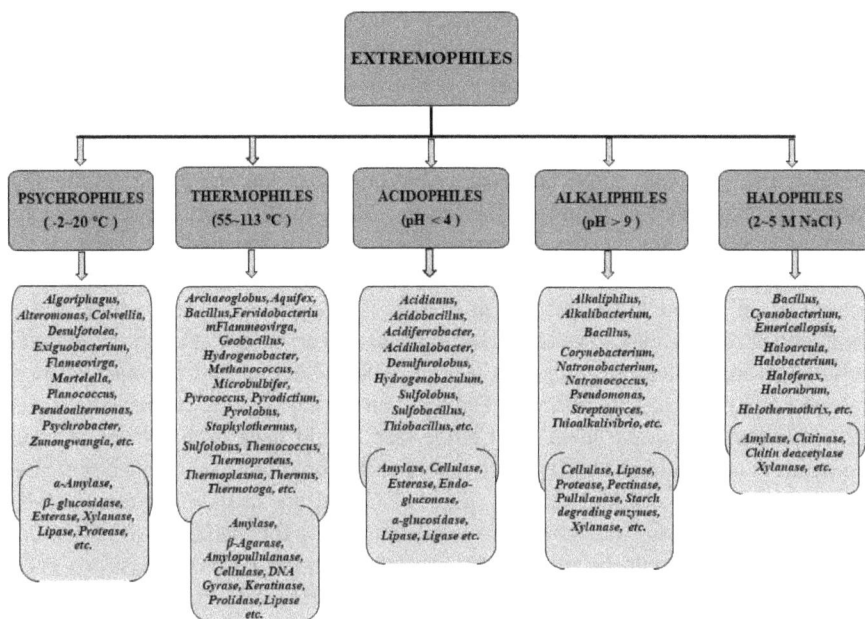

Figure 2. Examples of extremophiles and the different types of industrially important enzymes secreted by them.

extreme marine environments (Fig. 2). These can be categorized as psychrophiles, thermophiles, acidophiles, alkaliphiles and halophiles and can be much more active at those conditions compared to their conventional counterparts. These enzymes are already being used in many applications providing economic benefits and energy savings. Below is a discussion about very prominently used enzymes derived from microbes thriving in marine environment and used for varied industrial applications.

Protease

Protease makes up 60% of all industrial enzymes sold in the world with wide usage in leather, detergent, pharmaceutical and food industries and microbes believed to be the most suitable for its production. Protease hydrolyze peptide bonds in proteins and polypeptides and display wide mechanisms of action. The proteases isolated from marine microbes are heat-, cryo-, pH and metal-tolerant, stable in the presence of a broad spectrum of chemicals and hence have attracted the interest of researchers (Ahmed Abdul et al. 2019). Several proteases from marine sources have been reported, following is a review of a few of these published in past few years. High stability and compatibility with a wide range of commercial liquid and solid detergents was exhibited by two extracellular alkaline serine-proteases (BM1 and BM2) from *Bacillus mojavensis* A21 isolated from seawater (Haddar et al. 2009). Yet another protease with high applicability in detergent industry due to activity at high pH and temperature was extracted from *Engyodontium album* BTMFS10 isolated from sediment of the West coast of India (Chellappan et al. 2011). Similarly, relatively thermostable protease was pooled from a marine *Marinobacter*, isolated from the Indian Ocean (Fulzele et al. 2011). The use of proteases in food industry is intended for functions such as improved digestibility and/ or solubility; modified functionality (emulsification, fat-or water-binding, foaming and gel strengthening, whipping), improved or modified flavor, etc. Most of the proteases intended for meat tenderization are functional at either mesophilic or thermophilic temperature, have optimal temperatures of 50–80°C and retain only less than 10% of their highest activity at 20°C. The huge decrease in activity is a negative attribute when used as meat tenderizers because meat tenderization is often carried out at room temperature before cooking (Allen Foegeding and Larick 1986). Additionally, the mesophilic or thermophilic proteases are normally stable, even during cooking, which might lead to over-tenderization. An ideal meat tenderizer should have high activity at room temperature and should be easily inactivated during cooking. Cold-adapted proteases usually have relatively high activity at 0–30°C and are unstable at temperatures higher than 50°C and thus, could prove advantageous for this purpose (Khan and Sylte 2009). This could be justified by the extracellular serine protease MCP-01, produced by the deep-sea psychrophilic *Pseudoalteromonas* sp., SM9913, which proved effective in the tenderization of beef meat at 4°C. A 3–4-fold enhancement in moisture preservation resulting in fresher color and better appearance and an increase of the relative myofibrillar fragmentation index by 92% as compared to the control was an added advantage (Zhao et al. 2012a). Elucidation of elastin degradation has similarly been the focus of research, where the use of marine enzymes was involved. A pseudo alterin secreted by *Pseudoalteromonas* sp., CF6-2 from deep-sea sediment was shown to break the cross-links at structural level, which resulted in the release of filaments, droplets and spherules from elastic fibers (Zhao et al. 2012b). Tenderness of meat is also largely influenced by collagen and elastin; therefore, these structural proteins are promising targets for enzyme action (Bekhit et al. 2013). Collagenase from a deep-sea *Myroides profundi* D25, was isolated and used to further clarify the mechanism of collagen degradation (Ran et al. 2014). Overall, these efforts were expected to result in more efficient processes for meat tenderization. Psychrophilic leucyl amino peptidases, alongside with chymotrypsin and trypsin activities have been identified from heterotrophic microorganisms in the waters of Arctic fjord (Steen and Arnosti 2013). Kim et al. (2013) isolated serine protease from brown seaweed *Costaria costata* for treatment of thrombosis. Haloalkaliphilic protease from marine fish intestinal *B. subtilis* AP-MSU6 and marine crustacean shell waste *Bacillus* sp., APCMST-RS3 have also been reported (Maruthiah et al. 2013, 2015). Protease from *B. alveayuensis* CAS 5 isolated from an oil contaminated marine site near Mumbai Harbor (India) has been obtained (Annamalai et al. 2014). Ibrahim et al. (2015) investigated a halotolerant alkaline protease from *Bacillus* sp., NPST-AK15 from hyper saline soda lakes and found to be significant for laundry industry. Thao et al. (2015) reported trypsin-type enzyme and significantly high protease activity from ciliates in defined seawater microcosms when a marine heterotrophic ciliate- *Paranophrys marina* together with different concentrations of killed *Pseudomonas aeruginosa* was used as food for 10 d. A marine sediment bacterial isolate similar to *B. thuringiensis* MR-R1 was screened for bio-flocculant production. High flocculation activity indicates that the bio-flocculant may contend as an alternative to the conventionally used flocculants in water treatment (Ntozonke et al. 2017). Protease produced by *Geobacillus* sp., isolated from undersea fumaroles was anticipated to have application for skin rejuvenation, wrinkle smoothing and for producing high-quality suede (Iqbalsyah et al. 2019). A systematic investigation on the microbial community in the sediments of the Bohai Sea, Yellow Sea and South China Sea revealed the predominant protease-producing genera as *Pseudoalteromonas* and *Bacillus* (Zhang et al. 2019). The characterization of cold-active alkaline protease from an Antarctic bacterium, *Psychrobacter* sp., 94-6PB revealed that the micro diversity of this enzyme in *Psychrobacter* organisms from different cold habitats found several gene clusters that correlated with specific ecological niches (Perfumo et al. 2020). Recently, a novel extracellular protease from bacteria associated with hydrothermal vent crab from East China Sea was obtained. It was stable at a wide range of temperature, pH, solvents and detergents, making it potential for application in detergents and in peptide synthesis research (Gurunathan et al. 2021).

Lipase/Esterase

Lipases and esterases (EC 3.1.1.X) are carboxyl hydrolases able to cleave and form ester bonds. They are ubiquitous enzymes that breakdown fats and oils with eventual release of free fatty acids and glycerol and are in high demand for detergent, paper, cosmetics, food flavoring and other industries. Lipases are valuable catalysts used in various reactions such as esterification, transesterification and aminolysis and most often derived [among marine resources] from pelagic fishes which have high fat content. A novel extracellular phospholipase C was purified from *Streptomycete* among ~ 400 marine bacteria which hydrolyzed only phosphatidylcholine (Mo et al. 2009). Metagenomic libraries were constructed from deep-sea sediment collected around the summit of Edison Seamount in the New Ireland Fore-arc (Jeon et al. 2011) and Pacific Ocean (Jiang et al. 2012) and screened for lipolytic activity employing a combination of approaches like removal of signal sequence, co-expression of chaperone genes and low temperature induction. A thermophilic lipase was obtained from *Aeromonas* sp., of marine sludge, which exhibited maximum lipase activity when olive oil was used as carbon source (Charoenpanich et al. 2011). The Red Sea has long been considered as one of the most diverse and warmest regions in the world from which a thermostable lipase from *Geobacillus thermodenitrificans* was purified and characterized (Balan et al. 2012). A novel halotolerant lipase was obtained following a functional screening of a marine sponge fosmid metagenomic library. The activity and stability profile of the enzyme over a wide range of parameters suggests its utility in a variety of industrial applications (Selvin et al. 2012). Mesophilic lipases have also been obtained from marine *Vibrio* sp., isolated from the gut of a Tripod fish, with few metal salts enhancing its production (Selva Mohan et al. 2012). Lipases have been used for the hydrolysis of mono-and di-acyl glycerols, which can be used in the production of food emulsifiers. Yuan and co-workers identified cold-active lipase from *Janibacter* sp., with *sn*-1/3 specificity toward mono-and di-acyl glycerols, one of the few, if not only so far, lipase from marine bacteria with such preference (Yuan et al. 2014). Marine esterases are known to play an important role in organic carbon degradation and cycling. One such esterase was isolated from a metagenomic library containing ~ 10,500 fosmid clones from a surface sediment of the South China Sea (Li et al. 2014). According to López-lópez et al. (2014), maximum lipase activity at low temperatures is generally suitable for the cold washing process in detergents. Many enzymes sourced from the metagenome library have unique biochemical properties that make them valuable for industrial applications. A halotolerant thermostable lipase from the marine *Oceanobacillus* sp., displayed a high degree of stability over a wide range of conditions and anticipated to have utility in inhibiting biofilm formation in food processing environment (Sehgal Kiran et al. 2014). Halotolerant esterases from the sea may have good potential in industrial processes requiring high salts. A fosmid library containing 7,200 clones was constructed from a deep-sea sediment from the South China Sea and a gene *H8* encoding an esterase was identified by functional screening and expression in *Escherichia coli* (Zhang et al. 2017). Two novel esterase genes with high similarity were screened, cloned and expressed from a deep-sea sediment metagenomic library. The enzymes were characterized and showed different hydrolytic abilities toward esters with different acyl chain lengths (Huo et al. 2018). Lipase from a hyperthermophilic archae, *Pyrococcus furiosus* exhibiting lysis activity towards p-nitrophenylesters and ability to incise it from mid-chain length was reported. Additionally, higher reactivity to *p*-nitrophenylcaproate (C6), a raw material for industrial production was also revealed (Cabrera and Blamey 2018). Out of a total of 20 isolates obtained from Mediterranean Sea, the highest lipase producer was *B. cereus* HSS and hence could be exploited for economic treatment of oily wastewater (Hassan et al. 2018). Lipase production by a thermohalophilic bacterium, *Pria Laot Sabang* 80, isolated from an underwater hot spring has also recently been reported (Ulwivvay et al. 2019). In another study, the lipolytic activity of bacteria isolated from deep-sea sediments of the South Atlantic was evaluated, among which, *Halomonas sulfidaeris* LAMA 838 and *Marinobacter excellens* LAMA 842, were found to be promising (Delabary et al. 2020).

Amylase

Amylases act on starch [and its derivatives], that consists of a mixture of amylose and amylopectin, where amylopectin occurs in the 3:1 ratio to amylose. Starch hydrolysates and its derivatives are widely used in food and beverage industry requiring relatively high temperatures and is a multi-enzymatic process and hence enzymes able to cope with such environment are of interest. Amylases are classified into α-amylase, β-amylase and glucoamylase based on their three-dimensional structures, reaction mechanisms and amino acid sequences. Alkalophilic amylase with optimum activity at 40°C and pH 9.0 produced by sponge-associated marine bacterium *Halobacterium salinarum* has been reported (Santhanam et al. 2009). A novel 66 kDa α-amylase was isolated from the marine *Streptomyces* sp., (Chakraborty et al. 2009) the activity of which was retained at 85°C and was not inhibited by the presence of commercially available detergent and oxidizing agents. Promising α-amylase from *Bacillus* sp., isolated from a marine salt farm, with optimum temperature of 110°C was obtained which retained full activity at 60°C and 45% of the initial activity at 110°C (Pancha et al. 2010). Surfactant, oxidant, and detergent stable α-amylase was isolated from marine haloalkaliphilic *Saccharopolyspora* sp., A9, which was active in a wide range of NaCl concentration with maximum activity at 11% NaCl (Chakraborty et al. 2011). Reports on calcium independent α-amylases are scarce and *Streptomyces* strain isolated from marine sediment exemplifies this. The enzyme was thermostable and retained 40% of its initial activity after 48 h of incubation at 95°C, independent of the presence of Ca^{2+} (Chakraborty et al. 2012). Novel alkaline amylases from marine *B. mirini* and *Halobacillus amylus* from Andaman and Nicobar Islands, India is also available (Krishnan et al. 2011, 2013). Again, with potential application for starch processing, given the salt-tolerant and thermal stability, is the α-amylase produced by *Halomonas* sp., AAD21 (Uzyol et al. 2012). In yet another study, pH stable and salt tolerant α-amylase was isolated from marine *Zunongwangia profunda* (AmyZ2) (Wu et al. 2014). Studies on immobilization of α-amylase from marine *Nocardiopsis* sp., in gellan gum are also reported (Chakraborty et al. 2014). Geothermal vents create a stressful environment, particularly when temperature is considered. Hence, the temperature optimum of 90°C observed for an α-amylase from *Geobacillus* sp., Iso5 was therefore concomitant with the requirements to cope with such harsh conditions (Mahadevan and Neelagund 2014). A halotolerant α-amylase having the ability of digesting the insoluble raw starch was characterized from *B. subtilis*, a marine sediment isolate from Palk Bay region (Kalpana and Pandian 2014). A glycoside hydrolase family protein (AmyASS), isolated from the marine fish pathogen *Aeromonas salmonicida* sp., was expressed in *E. coli* (Peng et al. 2015). *Bacillus* sp., BCC 021-50 was found to be the best amylase-producing strain from among 50 bacterial strains, isolated from marine water and soil and was produced using agro residues (Simair et al. 2017). In another study, amylases suitable for new industrial applications from *Bacillus* sp., NRC22017 was isolated from marine samples (Elmansy et al. 2018). The *B. subtilis* SR60 isolated from coral reefs of Cabo Branco, Brazil produced thermostable and halotolerant amylase which also had the capacity to produce protease, cellulase and xylanase with similar characteristics (de Veras et al. 2018). A novel α-amylase that hydrolysed raw starch from both terrestrial and marine environments at near ambient temperature, was screened and cloned from *Pontibacillus* sp., ZY [isolated from sediment of Yongxing island] thereby suggesting its application in starch-based industrial processes (Fang et al. 2019). Enhanced α-amylase production using *Streptomyces gancidicus* ASD isolated from Andaman and Nicobar Islands (Ashwini and Shanmugam 2019) and *Nocardiopsis dassonvillei* strain KaS11 from the sea water of Kachhighadi, Dwarka coast has also been reported (Rathore et al. 2020). Recently, the antibiofilm potential of α-amylase from the marine *Pantoea agglomerans*, against food-borne pathogens was evaluated and found to efficiently disrupt the preformed biofilms (Goel et al. 2022).

Marine polysaccharide degrading enzymes

Agarases

Agarases are used in the hydrolysis of agar, a seaweed colloid, which is normally used as an additive in food processing, due to its emulsifying, gelling and stabilizing properties. Agar is a mixture of agarose and agaropectin, the former being the larger fraction (Fernandes 2014). Enzymes derived from microbes are seen as a substitute to acid degradation of agar. Agarases may display either endo, exo or both activities, where most of the agarase identified till date being the endo β-agarases (Chi et al. 2012). β-agarases have been isolated from several marine bacteria, most notable among them are species of *Alteromonas*, *Bacillus*, *Cytophaga*, *Pseudoalteromonas*, *Pseudomonas*, *Streptomyces* and *Vibrio*, where several of the agarase genes have been cloned and sequenced (Fernandes 2014). For example, in order to gain insight on the enzymatic complex responsible for agarose hydrolysis, cloning and expression of genes encoding agarases from marine *Zobellia galactanivorans* (Hehemann et al. 2012) and *aga Xa* gene from *Catenovulum* sp., X3, a sea water bacterium into *E. coli* BL21 was carried out (Xie et al. 2013). In another study, the β-agarase gene *agy1* from *Saccharophagus* sp., AG21, a marine bacterium, was cloned and over expressed as a His-tagged recombinant β-agarase in *E. coli* (Lee et al. 2013). A similar product profile for hydrolysis of agarose was observed when the *YM01-3* gene from *Catenovulum agarivorans* sp., nov., YM01T, a sea water bacterium was cloned and over expressed in *E. coli* JM109 (Cui et al. 2014) and exhibited higher thermal stability and activity, characteristics much different from other agarases. An exo-type β-agarase, *aga1*, from *Paenibacillus* sp., SSG-1, isolated from inland soil was cloned and characterized and exhibited low similarity to known glycoside hydrolases. The bioinformatic analysis showed that *aga1* may have been transferred together with its surrounding genes, from marine bacteria to soil bacteria via human microbiota. Use of seaweed as food and human feces or saliva were the most likely linkages for this gene transfer pathway (Song et al. 2016). β-agarase gene, *agaB-4*, was isolated from *Paenibacillus agarexedens* BCRC 17346 using next-generation sequencing and was further cloned and expressed in *E. coli* BL21(DE3). The purified protein found application in the recovery of DNA from agarose gels and in agar degradation for the production of neoagarotetraose (Chen et al. 2018). The secretomic analysis of *Vibrio* sp., A8, isolated from a red alga in the South China Sea, showed that only β-agarase (gene 3152) was abundantly expressed in the secretome of this species and had a good substrate specificity suggesting its potential application for agarose-oligosaccharide production (Li et al. 2020). In another study, β-agarase gene from *Cellovibrio* sp., KY-GH-1 was overexpressed as a recombinant His-tagged protein using the *E. coli* expression system, the recombinant enzyme was useful to convert agarose to NA2 (Kwon et al. 2020). Past few years have witnessed quite a number of researches on agarase from marine bacteria, a few of these and their utility have been outlined in Table 1.

Table 1. Major polysaccharide degrading enzymes derived from marine source.

Enzyme	Bacteria	Source	Activity	Refrence
Agarase	*Aeromonas* sp.	*Gracilaria* from the Kuta Coast, Central Lombok, West Nusa Tenggara	Possibility for biofuel production	Rahman et al. (2011)
	Simiduia areninigrae sp., nov.	Black sand off the shore of Jeju Island, Republic of Korea	Novel agarolytic species	Kim et al. (2012)
	Microbulbifer maritimus	Gulf of Mannar coast, India	Recovery of useful substances from seaweed and for protoplast isolation from *Gracilaria* sp.	Vijayaraghavan and Rajendran (2012)
	Halococcus sp., 197A	Solar salt sample	Aga-HC released degradation products in the order of neoagarohexose, neoagarotetraose and small quantity of neoagarobiose	Minegishi et al. (2013)
	Pseudoalteromonas sp., NJ21	Antarctic marine environment	Potential of the Antarctic agarase as a catalyst in medicine, food and cosmetic industries	Li et al. (2015)
	Vibrio sp.	Marine sediments from East coast from Pondicherry, India	Degradation of agar	Saravanan et al. (2015)
	Microbulbifer sp., Q7	Gut of sea cucumber	Recombinant agarase as an effective tool for preparing functional neoagaro-oligosaccharides	Su et al. (2017)
	Cellulophaga mitvescoria W5C	Marine samples	Breakdown of seaweed derived agar	Ramos et al. (2018)
	Vibrio, Pseudomonas and *Bacillus*	From macroalga *Ulva lactuca*	Discovery of biocatalysts and bioactive compounds of marine origin	Comba González et al. (2018)
	Gayadomonas joobiniege G7	Sea water from Gaya Island	Useful for the food, pharmaceutical, and cosmetic industries	Lee et al. (2018)
	Pseudoalteromonas ruthenica	Waters of Al-Uqair, the Arabian Gulf, Al-Ahsaa, Saudi Arabia	Recovery of DNA from agarose gels, production of high-value compounds that have various activities such as anti-inflammatory, antibacterial, and antioxidant	Khalifa and Aldayel (2019)

Table 1 contd. ...

...*Table 1 contd.*

Enzyme	Bacteria	Source	Activity	Refrence
	Shewanella algae, Microbulbifer elongatus	*Kappaphycus* sp., & *Sargassum* from Mandapam and Rameshwaram, India	Anticipated to have wide range of applications	Kandasamy et al. (2020)
	Salinicola zeshunii, Bacillus piscis, and *Bacillus licheniformis*	Associated with *Chaetomorpha* from Gunung Kidul, Indonesia	Potential for agarolytic activities	Wijaya et al. (2021)
	Pseudoalteromonas sp., MHS	Surface of *Ulva lactuca*		Sharabash et al. (2022)
Alginate lyase	*Pseudomonas fluorescens* HZJ216	Associated with *Laminaria* from the Arctic ocean	Extracellular alginate lyases with high alginate-degrading activity	Li et al. (2011)
	Psychrobacter, Winogradskyella, Psychromonas and *Polaribacter*	Associated with Laminaria from the Arctic Ocean	Study material for bacterial cold-adapted alginate lyases	Dong et al. (2012)
	Vibrio sp., QY102	Marine *Sargassum*, Ocean University of China	High alginate-degrading activity and showed polymannuronic acid specificity	Zhou et al. (2014)
	14 alginate lyase-excreting strains including one novel strain of *Marinomonas*	Associated with the brown alga *Ascophyllum nodosum*	Significance of macroalgal degrading bacteria associated with *A. nodosum*	Martin et al. (2015)
	Vibrio splendidus 12B01	Marine sample	Degrade alginate into longer chains of oligomers	Badur et al. (2015)
	Microbulbifer sp., ALW1	From rotten brown alga	Preparation of alginate oligosaccharides with low degree of polymerization (DP). Hydrolysates also exhibited the antioxidant activity	Zhu et al. (2016a)
	Bacillus litorali	Decayed kelps	Potential to ferment *Sargassum horneri* for use as biofertilizer	Wang et al. (2016)
	Paenibacillus, Bacillus, Leclercia, Isoptericola, Planomicrobium, Pseudomonas, Lysinibacillus, and *Sphingomonas*	Surface of brown algae *Laminaria japonica, Sargassum horneri* and *Sargassum siliquatrum*	*Paenibacillus* produced many kinds of extracellular enzymes which can be used in a wide range of industrial applications	Wang et al. (2017)

	Organism	Marine sample	Activity/Application	Reference
	Microbulbifer sp., Q7	Marine sample	Produced alginate oligosaccharides with specific M/G ratio and molecular weights	Yang et al. (2018)
	Flavobacterium sp., UMI-01	Marine sample	Various applications	Ojima et al. (2019)
	Pseudoalteromonas arctica M9	Isolated from *Sargassum*	Potential in preparing tri-saccharides from these polysaccharides	Xue et al. (2022)
Carrageenase	*Pseudoalteromonas porphyrae*	Decayed seaweed collected from Yellow Sea, China	Active conversion of κ-carrageenan into tetrasaccharides	Liu et al. (2011)
	Cellulosimicrobium cellulans	Seawater, sediment and algal tissue from Egyptian Mediterranean coast	Production of extracellular κ-carrageenase using free and immobilized cells	Youssef et al. (2012)
	Closely related to *Marinimicrobium* and *Microbulbifer*	Associated with *Kappaphycus alvarezii* from Philippines	Ability to degrade seaweed polysaccharides	Tayco et al. (2013)
	Cellulophaga lytica strain N5-2	Sediment of carrageenan production base	Hydrolysed κ-carrageenan into κ-neocarraoctaose-sulfate & κ-neocarrahexaose-sulfate and then broke former into κ-neocarra-biose - sulfate & κ-neocarrahexaose-sulfate	Yao et al. (2013)
	Pseudoalteromonas sp., QY203	Marine sample	Exhibited endo-κ-carrageenase activity, and depolymerizing into di & tetrasaccharide	Shangyong et al. (2013)
	Vibrio sp., NJ-2	Rotten red algae	Tool to produce κ-carrageenan oligosaccharides with various biological activities	Zhu and Ning (2016)
	Alteromonas macleodii KS62	Decaying red seaweeds	Establishes the potential of red seaweed to serve as a low-cost substrate for economic production of κ-carrageenase	Prajkta and Chandra (2019)
	Cellulophaga algicola	Various marine habitats	For industrial production of oligosaccharides	Howlader et al. (2021)
	Salinicola zeshunii, *Bacillus piscis* and *Bacillus licheniformis*	*Chaetomorpha* sp., from Sepanjang Beach, Indonesia	Carbohydrate degradation	Wijaya et al. (2021)
	Flavobacterium algicola	Marine sample	Study of κ-carrageenan, τ-carrageenan and partial λ-carrageenan catabolic pathways	Jiang et al. (2022)

Alginate lyases

Alginate is a linear complex copolymer composed of mannuronate acid (M) and guluronic acid (G)- originally extracted from the mesenchyme of kelp, gulfweed and seaweed. As a natural polysaccharide, alginate has been extensively used in food, cosmetic and biomedical industries due to its chelation, gelation and hydrophilic properties. In biomedicine, alginate is mainly used as a matrix or carrier to realize the sustained release and enhanced efficacy of a drug in targeted or localized drug delivery systems. Alginate lyases, characterized as either mannuronate lyases (EC 4.2.2.3) or guluronate lyases (EC 4.2.2.11), catalyse the degradation of alginate, a complex copolymer of α-L-guluronate and its C5 epimer β-D-mannuronate (Wong et al. 2000). Alginate lyase degradation products include oligosaccharides, which exhibit various bioactivities, such as antibiosis, anti-cancer and plant growth promotion (Wang et al. 2016). Conceivably, the marine-derived alginate lyases demonstrate salt tolerance and many are activated in the presence of salts and therefore, find applications in food industry. In the past decades, alginate lyases have been isolated and purified from various marine organisms, including bacteria (*Pseudomonas, Vibrio, Bacillus, Zobellia, Corynebacterium, Klebsiella*, etc.), fungi, algae and mollusks (Barzkar et al. 2022). Several of the alginate lyases have been successfully identified, cloned and sequenced in the last few years. The alginate lyase encoding gene (*alyPI*) form marine *Pseudoalteromonas* sp., CY24 was cloned and expressed in *E. coli* with a His-tag sequence fused at the C-terminal end and purified to electrophoretic homogeneity using Ni-sepharose affinity chromatography (Duan et al. 2009). Lee et al. (2012) cloned, purified and characterized a novel poly MG-specific alginate lyase in *Stenotrophomas maltophilia* KJ-2. The recombinant lyase preferentially degraded the glycosidic bond of poly MG block than poly M-block and poly G-block. In another study, alginate lyase gene was characterized using a metagenomic library constructed from the gut microflora of abalone. The library gave an alginate lyase positive clone which had the highest activity at pH 7.0 and 45°C in the presence of 1 mM $AgNO_3$ and preferred poly(α-d-mannuronate) as a substrate over poly(α-l-guluronate) (Jung et al. 2012). Another new alginate lyase isolated from *Flavobacterium* sp., S20, and the *Alg2A* gene was overexpressed in *E. coli*. Results indicated *Alg2A* preferred poly-α-l-guluronate as a substrate over poly-α-d-mannuronate and during saccharification of sodium alginate, the *Alg2A* yielded oligosaccharides with different degree of oligomerization (Lishuxin et al. 2013). Yet another gene *alyPM* encoding an alginate lyase of lyase family 7 (PL7) was cloned from marine *Pseudoalteromonas* sp., SM0524 and expressed in *E. coli*. AlyPM preferably degraded polymannuronate and primarily released dimers and trimers (Chen et al. 2017). In another study, *Vibrio* sp., QD-5 utilizing alginate, was isolated from rotten kelp. The genome sequencing showed that QD-5 contained four alginate lyase genes, where *Aly-IV* was anticipated to have application in large scale preparation of alginate oligosaccharides with low degree of polymerization (Chao et al. 2017). The characterization of a novel PL7 alginate lyase *AlyC3* from *Psychromonas* sp., C-3 [isolated from the Arctic *Laminaria*], in terms of its phylogenetic classification, catalytic properties and structure was carried out. The study provided a better understanding on the mechanisms of alginate degradation by alginate lyases (Xu et al. 2020). Recently, a novel alginate lyase, *Algpt*, was cloned and characterized from a marine *Paenibacillus* sp., LJ-23. The high stability and wide adaptability of the enzyme endowed it with application potential for preparation of alginate oligosaccharides (AOS) with different sizes and AOS-based products (Wang et al. 2022). A list of alginate lysases from marine bacteria with its potential usage reported during the past few years has been given in Table 1.

Carrageenases

Carrageenases are endohydrolases that hydrolyze the β-1,4 linkages in carrageenan and produce a series of homologous oligosaccharides with various biological and physiological activities including anti-tumor, anti-inflammation, anti-viral, anti-coagulation, etc. It also finds applications in the biomedical field, bioethanol production, prevention of red algal bloom, obtaining algal protoplasts, etc. (Zhu et al. 2018). Carrageenases belong to the family 16 of the glycoside hydrolases (Ghanbarzadeh et al. 2018). They are found in seaweed-associated microbes, most of which are Gram-negative, including *Pseudomonas*, *Pseudoalteromonas*, *Cytophaga*, *Vibrio*, *Tamlana*, etc (Zhu and Ning 2016), although production of carrageenases from Gram-positive bacteria such as *Cellulosimicrobium* and *Bacillus* have also been reported (Kang and Kim 2015). Studies on the cloning and sequencing of carrageenase gene from few bacteria have also been reported during the past few years. For example, the purification, cloning and characterization of a new ι-carrageenase CgiA Ce from the marine bacterium *Cellulophaga* sp., QY3 was carried out. It cleaved ι-carrageenan yielding neo-ι-carrabiose and neo-ι-carratetraose as the main end products (Ma et al. 2013). The molecular cloning, characterization and heterologous expression of a new κ-carrageenase gene from marine *Zobellia* sp., ZM-2 was reported. The κ-carrageenase production of *E. coli* was 9.0 times higher than that of ZM-2 indicating the potential use of the enzyme in the biotechnological industry (Liu et al. 2013). In another study, the gene (*Cgi82A*) of a novel GH82 family ι-carrageenase was cloned from the genome of marine bacterium *Wenyingzhuangia fucanilytica* and expressed in *E. coli*. The enzyme could reach its highest activity at 25°C, which is lower than all hitherto reported for GH82 ι-carrageenases (Shen et al. 2017). Yet another study reported the crystal structure of the catalytic module of ZgCgkA from *Z. galactanivorans* and the first substrate complex with the inactivated mutant form of *PcCgkA* at 1.7 Å resolution was also described (Matard-Mann et al. 2017). The complete catabolic pathway for carrageenans, in the marine heterotrophic *Z. galactanivorans* was elucidated. It was concluded that carrageenan catabolism relies on a multifaceted carrageenan-induced regulon, including a non-canonical polysaccharide utilization locus (PUL) and genes distal to the PUL, including a *susCD*-like pair (Ficko-Blean et al. 2017). A novel κ-carrageenase gene (*CgkB*) was cloned from *Pedobacter hainanensis* NJ-02 and expressed in *E. coli* BL21 (DE3). The high activity of the recombinant protein could be useful for producing carrageenan oligosaccharides with various activities (Zhu et al. 2018a). A putative carrageenase gene *Car3206* was obtained from the complete genome of *Polaribacter* sp., NJDZ03, isolated from the surface of Antarctic macroalgae, was cloned and expressed in *E. coli* BL21(DE3). The recombinant *Car3206* protein was characterized and the antioxidant activity of the degraded product was investigated (Gui et al. 2021). In another study, a carrageenase gene, *Car1383*, obtained from the metagenome of Antarctic macroalgae-associated bacteria was evaluated for its amino acid sequence and showed up to 33% similarity with other carrageenases and contained a GH16-family motif. The recombinant *Car1383* was heterologously expressed in *E. coli* and emerged a new candidate for the industrial preparation of bioactive algal oligosaccharides (Li et al. 2022). The details of carrageenase produced from marine bacteria during the past few years have been depicted in Table 1.

Other enzymes from marine bacteria with potential industrial applications

AHL lactonase

Quorum sensing (QS) is a population-dependent phenomenon for microorganisms that occurs via the up/down-regulation of downstream gene expression (Bassler 2004). The virulence of different pathogenic bacteria has been associated with QS, a sophisticated mechanism which coordinates gene expression by means of small signal molecules known as autoinducers (Ng and Bassler 2009). Among others, autoinducers include *N*-acylhomoserine lactones (AHLs) produced by the proteobacteria; oligopeptides produced by the firmicutes and furanosyl borate diester (AI-2) which are produced by both *proteobacteria* and *firmicutes* and are used for interspecies communication (González and Marketon 2003). Besides the ability to produce and use AHL-based communication systems, the ability to interfere with QS systems or to degrade AHLs have been reported for many marine bacteria. Table 2 cites examples of few AHL producing bacteria described in the past few years from marine environment.

L-asparaginase

L-asparaginase is an enzyme that catalyses the hydrolysis of L-asparagine to L-aspartic acid. This enzyme has an important role in medicine and food. It is an anti-cancer therapeutic enzyme and has attracted a great deal of attention from the scientific community (Chakravarty et al. 2022). These enzymes selectively target the metabolism of cancer cells by exploiting deficiencies in metabolic pathways and catalysing the degradation of L-asparagine into L-aspartic acid and ammonia, causing nutrient starvation to cancer cells eventually leading to their death (Alrumann et al. 2019). Furthermore, it is also applied for degradation of acrylamide, a carcinogenic compound in baked and fried foods (Pirhanto and Wakayama 2016). Table 2 cites examples of few L-asparaginase producing bacteria described in the past few years from marine environment.

Table 2. Other enzymes of industrial application derived from marine source.

Enzyme	Bacteria	Source	Activity	Refrence
AHL lactonase	*Pseudoalteromonas byunsanensis* strain 1A01261	Tidal flat sediment, Korea	Potential to control of Gram-negative pathogenic bacteria	Huang et al. (2012a)
	Tenacibaculum sp., 20J	Shrimp larvae	Anti-virulence strategy against important bacterial pathogens and other biotechnological applications	Mayer et al. (2015)
	Muricauda olearia	Skin of fish	MomL significantly attenuated the virulence of *Pseudomonas aeruginosa* in a *Caenorhabditis elegans* infection model, which suggests that MomL has the potential to be used as a therapeutic agent	Tang et al. (2015)
	Staphylococcus hominis	Coral	Coral associated microbes as a resource of OS inhibitors and could facilitate discovery of new biotechnologically relevant compounds	Ma et al. (2018)
	Oceanobacillus sp.	Seawater	Three active compounds exhibited quorum sensing inhibitory activities against *C. violaceum* 026 and *P. aeruginosa*	Chen et al. (2018a)
	Ruegeria mobilis YJ3	Healthy shrimp	RmmL might be used as a therapeutic agent in aquaculture	Cai et al. (2018)
	Vibrio mediterranei, *V. owensii* and *V. corallilyticus*	*Oculina patagonica* and *Cladocora caespitosa* healthy & diseased corals, seawater	Development of future therapies based on AHL disruption, the most promising alternatives for fighting infectious diseases in aquaculture	Torres et al. (2018)
	Planococcus versutus L10	Lagoon Island (Ryder Bay, Adelaide Island, maritime Antarctic)	Effective in attenuating the pathogenicity of *P. carotovorum*, a plant pathogen that causes soft-rot disease	See-Too et al. (2018)
	Streptomyces griseoincarnatus	*Callyspongia* sp.	Prospective safer alternative for treatment of antibiotic resistant respiratory pathogens, in particular, *P. aeruginosa*	Kamarudheen et al. (2019)

Table 2 contd. ...

...*Table 2 contd.*

Enzyme	Bacteria	Source	Activity	Refrence
	Phaeobacter inhibens	Oyster shell	Probiotic bacteria can exert their host protection by using a multipronged array of behaviours that limit the ability of pathogens to become established and cause infection	Zhao et al. (2019)
	Psychrobacter sp.	Palk bay sediment, India	Controlling QS and biofilm mediated multidrug resistant infection	Packiavathy et al. (2021)
L-asparaginase	*Strepyomyces noursei*	*Callyspongia diffusa* (sponge)	Potential as anti-neo plastic agent used in chemotherapy of lymphoblastic leukaemia	Dharmaraj (2011)
	Bacillus cereus	Mangrove soil	Useful to produce maximum enzyme for treatment of various infections and diseases	Thenmozhi et al. (2011)
	Bacillus pumilus	Sediment	–	Sindhwad and Desai (2015)
	Lactobacillus salivarius	Seawater	The bacteria may produce high yield of enzyme that can be used as anti-leukemic agent	Bhargavi and Jayamadhuri (2016)
	Bacillus tequilensis PV9W	Gulf of Mannar, Rameswaram, India	Potential source for glutaminase free L-asparaginase with industrial as well as pharmaceutical applications	Shakambari et al. (2016)
	Enterobacter hormaechei	Fishes	Halotolerant and sustained enzyme production at elevated temperature for commercial scale	Sudha et al. (2017)
	Bacillus firmus AVP 18	Nizampatnam mangrove sample	Clinical acceptable antitumor agent for the effective treatment of lymphosarcoma and lymphoblastic leukemia	Rudrapati and Audipudi (2017)
	Bacillus lichenformis	Isolated from Red Sea, Saudi Arabia	Potential candidate for further pharmaceutical use as an anticancer drug	Alrumman et al. (2019)

Organism	Source	Application	Reference
Bacillus australimaris	Marine sediment	Potentially significant and novel therapeutic drug	Chakravarty et al. (2021)
Bacillus altitudinis	Seawater	This asparaginase can be utilised for biotechnology applications	Hadi et al. (2021)
Pseudomonas aeruginosa HR03	Fish intestine	Potential usage in pharmaceutical and food industries	Qeshmi et al. (2022)
Chitiniphilus shinanonensis strain SAY3	Moat water of Ueda Castle in Nagano Prefecture, Japan	Capable to completely split the *N*-acetyl-D-glucosamine dimer into GlcNAc monomers	Huang et al. (2012b)
Alcaligenes faecalis AU02	Seafood effluent	For generation of high-value products	Annamalai et al. (2011)
Paenibacillus pasadenensis NCIM 5434	Alkaline littoral soil of Lonar Lake	Anti-fungal activity suggests its use as powerful biocontrol agent	Loni et al. (2014)
Pseudoalteromonas sp., DL-6	Marine sediments	GenechiA, suggested that it is a non-processive endo-type chitinase	Wang et al. (2014)
Pseudoalteromonas tunicata CCUG 44952T	Marine samples	Activity against phyto and human pathogenic fungi, proposed its used as bio- fungicide	Gracia Fraga et al. (2015)
C. prasina (strain LY03)	Marine samples	Algicidal mechanism on diatom *Thalassiosira pseudonana*	Li et al. (2016)
Acinetobacter ASK18 strain	Marine waste along the coastal regions in Chennai	Effective as pharmacological drug in anticancer and antibacterial properties	Krithika and Chellaram (2016)

Table 2 contd. ...

...*Table 2 contd.*

Enzyme	Bacteria	Source	Activity	Refrence
	Pseudoalteromonadaceae, Vibrionaceae	Galathea 3 expedition	Exemplified the industrial perspective	Paulsen et al. (2016)
	Bacillus sp., R2	Red sea	Ability to produce multiple enzymes other than chitinase	Cheba et al. (2017)
	Pseudoalteromonas sp.,	Marine samples	Untapped resource of secondary metabolites with chitinolytic machinery	Paulsen et al. (2019)
	Bacillus sp.,	Marine samples	Potential use as biocontrol agents	Kurniawan et al. (2019)
	Achromobacter xylosoxidans	Shrimp waste disposal site	Chitin degradation with concomitant production of N-acetyl-D-glucosamine	Subramanian et al. (2020)
	Microbulbifer sp., BN3	Marine samples	Adaptation to a wide pH range favourable for applications in industrial processes	Li et al. (2021)
	Bacillus aryabhattai	Arabian Sea	Activity against plant pathogenic fungi suggested use as biocontrol agent	Subramani et al. (2022)
	Thermophilic chitinase, Chi304	Marine metagenome	Efficient chitin biodegradation	Zhang et al. (2022)
	Bacillus subtilis JCM 1465 and *Bacillus tequilensis* 10b	Hot Springs of Penen Village, North Sumatera, Indonesia		
Glucosidase	*Martelella mediterranea*	Deep sea water, China	Hydrolysis of p-nitrophenyl-β-D-galactopyranoside, potential for industrial applications	Mao et al. (2010)
	Aeromonas sp., HC11e-3	Marine samples	References for investigating other glucosidases in the glycosyl family 3 and developing glucosidases in industrial area	Huang et al. (2012c)
	Proteus mirabilis VIT117	Shellfish wastes (prawn shells from local fish market), Vellore, India	Data may help to ease process of bioethanol production from lignocellulosic feedstock	Mahapatra et al. (2016)

Croceicoccus marinus E4A9	Deep-sea sediment sample from the East pacific polymetallic nodule region	Understanding the different catalytic mechanism during evolution for β-glucosidases	Shen et al. (2019)
Microbulbifer thermotolerans	Suruga Bay sediment samples in Japan	Could be used in the cosmetic, food, medical, and various biotechnological industries	Pyeon et al. (2019)
87 coral symbiotic β-glucosidase-producing bacteria	Corals from Luhuitou Coral Reef, China	Bacteria and genes associated with scleractinian corals with great potential for applications in the food and agriculture	Su et al. (2021)
Inulinase *Marinimicrobium* sp., LS-A18	Marine solar saltern of the Yellow Sea, China	A large number of monosaccharides was released after inulin hydrolysis by the inulinase from strain LS-A18	Li et al. (2012)
Bacillus cereus MU-31	Sediments of Vellar estuary, Perangipettai	Candidate for industrial production of fructose syrup and other biotechnological applications	Meenakshi et al. (2013)
Nocardiopsis sp., DN-K15	Marine sediment of Jiaozhou Bay, China	Potential candidate in biotechnological and industrial applications	Lu et al. (2014)
Bacillus subtilis	Water and sediment samples from island of São Miguel, the Azores	Useful candidates in various biotechnological process	Rodrigues et al. (2017)
Bacillus sp., SG7	Thermal water samples from Velingrad (Kostandovo, Bulgaria)	Suitability for use in the food industry	Ivanova et al. (2018)
Pullulanase *Thermoanaerobium* TOK6-B1	Thermal vent	Potential applications in the food, pharmaceutical, pulp, textile and other industries	Tomasik and Horton (2012)
Bacillus acidopullulyticus	Seawater	Potential applications in the food, pharmaceutical, pulp, textile and other industries	Tomasik and Horton (2012)
Thermococcus aggregans	Deep sea water	Potential applications in the food, pharmaceutical, pulp, textile and other industries	Tomasik and Horton (2012)

Table 2 contd. ...

...Table 2 contd.

Enzyme	Bacteria	Source	Activity	Refrence
	Auerobasidium pullulans CJ001	Isolated from sea mud	Preparation of maltotriose by hydrolysis of pullulan with pullulanase	Wu and Chen (2014)
	Shewanella arctica	Arctic sea ice	Complete conversion of pullulan to maltotriose as the sole product	Qoura et al. (2014)
	Altermonas macleodii	Seawater	Pullalan hydrolysis	Neumann et al. (2015)
	Shewanella arctica	Artic seawater	Potential candidate for industrial applications such as starch degradation for ethanol-based biofuel production	Elleuche et al. (2015)
	Pyrococcus yayanosii CH1	Deep-sea hydrothermal site	Mutant enzyme could realize the combined use of pullulanase with α-amylase during the starch liquefaction process to improve hydrolysis efficiency	Pang et al. (2019)

Chitinase/Chitosanase

Chitin is an N-acetyl-glucosamine homopolymer linked by a β-1,4-glucosidic bond that is structurally identical to cellulose, except that the hydroxyl group in cellulose at C2 is replaced by an acetamide group (Younes and Rinaudo 2015). It is among the most common polysaccharides found in nature and is the major structural component of most fungal cell walls and abundant in the crust of insects and crustaceans (Samuel et al. 2012). Till now, researchers have found a wide range of microbes that can produce chitinase or chitosanase [enzymes which hydrolyze chitosan-partially deacylated form of chitin] with potential applications in the food, medical and agricultural sectors (Eijsink et al. 2010). These are potential, natural fungicides, that could replace the chemical fungicides in plant biocontrol, and since bacterial chitinases can inhibit fungal growth, they are of particular interest for this purpose (Paulsen et al. 2016). Chitin-degrading enzymes have been isolated from several marine bacteria from different genus such as *Bacillus*, *Vibrio*, *Paenibacillus*, *Pseudoalteromonas*, *Aeromonas*, *Micrococcus*, *Streptomyces*, *Alteromonas*, *Actinomyces*, *Chitiniphilus*, *Achromobacter*, etc (Beygmoradi et al. 2018). A list of chitinase from marine source with its potential usage reported during the past few years have been detailed in Table 2.

Glucosidases

β-glucosidase is capable of cleaving β-glucosidic linkages of conjugated glucosides and disaccharides. It acts upon β, 1→4 bonds linking two glucose or glucose-substituted molecules. β-glucosidases can be divided into glycoside hydrolase family 1, GH1, and glycoside hydrolase family 3, GH3, depending on their amino acid sequences (Fernandes 2014). β- glucosidases are widely used for the processing of fruit juices, wine and beer, aiming to improve organoleptic properties and release aroma and flavor from compounds (Trincone 2013). They can also be used as additives in cellulose-based feeds for single-stomach animals, viz., chicken or pigs, since they enhance the digestibility of the feed (Zhang et al. 1996); and to increase the content of essential oil in tea beverages (Su et al. 2010). Pioneering screening of β-glucosidases from marine sources have been reported in the last few years; some of these are shown in Table 2.

Inulinases

Inulinases are enzymes with hydrolytic activity towards inulin, a poly-fructan present in several plants, viz., Jerusalem artichoke and chicory, where fructosyl units are bound together by β-(2,1) linkages (Das et al. 2019). Total or partial hydrolysis leads to the production of fructose or fructo oligosaccharides, respectively. These are widely used in the food industry, either as sweetener or as prebiotic (Lima et al. 2011). While fructose is a GRAS (generally recognized as safe) sweetner, which may replace sucrose in foods and beverages or be used for biofuel production. Oligofructose and inulin are emerging prebiotics since they promote the growth of specific bacteria such as *Bifidobacteria* and *Lactobacillus* (Chi et al. 2011). Few marine bacteria reported for the production of inulinase during the past few years are listed in Table 2.

Pullulanase

Pullulanase is an important debranching enzyme and has been widely utilized for hydrolysis of the α-1,6 glucosidic linkages in the saccharification process for glucose, maltose, maltotriose and fructose production, mainly in starch/sugar industry. Pullulanase are often combined with glucoamylase for efficient saccharification of starch, thus allowing for high glucose yield (El Shishtawy et al. 2014). These are also used in the fermentation industry to produce low carbohydrate "light beer" by adding them with α-amylase or glucoamylase to the wheat during fermentation and bioethanol production (Doman-Pytka and Bardowski 2004). Marine sources have been naturally tapped as providers of such pullulanases. Accordingly, efforts are being made to produce pullulanase variants from hyperthermophiles that display the same or higher thermal stability, while presenting a more favourable temperature dependent activity profile (Table 2).

Conclusion and outlook

This chapter is intended to focus the reader's attention on the potential of studying metagenomics from unique marine environments to successfully exploit the largely "untapped" resources within this unexplored ecological niche area. The existence of metagenomics has helped researchers unravel novel enzymes from nature that are beneficial to various industries and a thorough understanding of its application is expected to have a huge impact on humanity. There is a continuing need for a wide variety of novel biocatalysts for the improvisation of current industrial processes and for the development of newer and cleaner production processes, to reduce energy and raw material consumption. A new library of ocean-sourced enzymes tough enough to perform under the harshest of industrial conditions could provide an invaluable cost-saving shortcut for a range of enzyme-dependent sectors and the future of the "Marine Biocatalysts" looks very promising.

Acknowledgement

The author thanks the organization for providing the facility to compile this work. CSIR-CLRI communication number 1747.

References

Ahmed Abdul, B.A., S. Alijani, S. Thanabal and S.K. Kim. 2019. Marine enzymes production tools to the pharmaceutical industry. Indian J. Geo. Mar. Sci. 48: 1656–1666.

Allen Foegeding, E. and D.K. Larick. 1986. Tenderization of beef with bacterial collagenase. Meat Sci. 18(3): 201–214.

Alrumman, S.A., Y.S. Mostafa, K.A. Al-izran, M.Y. Alfaifi, T.H. Taha and S.E. Elbehairi. 2019. Production and anticancer activity of an L-asparaginase from *Bacillus licheniformis* isolated from the Red Sea, Saudi Arabia. Sci. Rep. 9: 3756.

Annamalai, N., M.V. Rajeswari, S. Vijayalakshmi and T. Balasubramanian. 2011. Purification and characterization of chitinase from *Alcaligenes faecalis* AU02 by utilizing marine wastes and its antioxidant activity. Ann. Microbiol. 61: 801–807.

Annamalai, N., M.V. Rajeswari and T. Balasubramanian. 2014. Extraction, purification and application of thermostable and halostable alkaline protease from *Bacillus alveayuensis* CAS 5 using marine wastes. Food Bioprod. Process 92: 335–342.

Arnosti, C., C. Bell, D.L. Moorhead, R.L. Sensibaugh, A.D. Steen, M. Stromberg et al. 2014. Extracellular enzymes in terrestrial, freshwater, and marine environments: Perspectives on system variability and common research needs. Biogeochemistry 117: 5–21.

Ashwini, K. and S. Shanmugam. 2019. Enhanced alpha-amylase production using *Streptomyces gancidicus* ASD by process optimization. Indian J. Geo. Mar. Sci. 48: 845–852.

Badur, A.H., S.S. Jagtap, G. Yalamanchili, J.K. Lee, H. Zhao and C.V. Rao. 2015. Alginate lyases from alginate-degrading *Vibrio splendidus* 12B01 are endolytic. Appl. Environ. Microbiol. 81: 1865–1873.

Balan, A., D. Ibrahim, R. Abdul Rahim and F.A. Ahmad Rashid. 2012. Purification and characterization of a thermostable lipase from *Geobacillus thermodenitrificans* IBRL-nra. Enzyme Res. 987523.

Barone, R., C. De Santi, F.P. Esposito, P. Tedesco, F. Galati, M. Visone et al. 2014. Marine metagenomics, a valuable tool for enzymes and bioactive compounds discovery. Front. Mar. Sci. 1: Article 38.

Barzkar, N., R. Sheng, M. Sohail, S.T. Jahromi, O. Babich, S. Sukhikh et al. 2022. Alginate lyases from marine bacteria: An enzyme ocean for sustainable future. Molecules 27: 3375.

Bassler, B.L. 2004. Cell-to-cell communication in bacteria: A chemical discourse. Harvey Lect. 100: 123–142.

Bekhit, A.A., D. Hopkins, G. Geesink, A.A. Bekhit and P. Franks. 2013. Exogenous proteases for meat tenderization. Crit. Rev. Food Sci. Nutr. 54: 1012–1031.

Beygmoradi, A., A. Homaei, R. Hemmati, P. Santos-Moriano, D. Hormigo and J. Fernández-Lucas. 2018. Marine chitinolytic enzymes, a biotechnological treasure hidden in the ocean? Appl. Microbiol. Biotechnol. 102: 9937–9948.

Bhargavi, M. and R. Jayamadhuri. 2016. Isolation and screening of marine bacteria producing anti-cancer enzyme L-asparaginase. Am. J. Mar. Sci. 4(1): 1–3.

Borchert, M., M. Gjermansen, S. Clark, B. Henrissat, M.B. Silow and P.F. Hallin. 2011. Pullulanase variants and uses thereof. *WIPO Patent Application* WO/2011/087836 A2.

Boubakri, H., M. Beuf, P. Simonet and T.M. Vogel. 2006. Development of metagenomic DNA shuffling for the construction of a xenobiotic gene. Gene 375: 87–94.

Cabrera, M.Á. and J.M. Blamey. 2018. Biotechnological applications of archaeal enzymes from extreme environments. Biol. Res. 51(1): 37.

Cai, X., M. Yu, H. Shan, X. Tian, Y. Zheng, C. Xue et al. 2018. Characterization of a novel N-acylhomoserine lactonase RmmL from *Ruegeria mobilis* YJ3. Mar. Drugs 16: 370.

Chakraborty, S., A. Khopade, C. Kokare, K. Mahadik and B. Chopade. 2009. Isolation and characterization of novel α-amylase from marine *Streptomyces* sp. D1. J. Mol. Cat. B: Enzy. 58: 17–23.

Chakraborty, S., A. Khopade, R. Biao, W. Jian, X.Y. Liu, K. Mahadik et al. 2011. Characterization and stability studies on surfactant, detergent and oxidant stable α-amylase from marine haloalkaliphilic *Saccharopolyspora* sp. A9. J. Mol. Cat. B: Enzy. 68: 52–58.

Chakraborty, S., G. Raut, A. Khopade, K. Mahadi and C. Kokare. 2012. Study on calcium ion independent alpha-amylase from haloalkaliphilic marine *Streptomyces* strain A3. Indian J. Biotechnol. 11: 427–437.

Chakraborty, S., S. Jana, A. Gandhi, K.K. Sen, W. Zhiang and C. Kokare. 2014. Gellan gum microspheres containing a novel alpha-amylase from marine *Nocardiopsis* sp. strain B2 for immobilization. Int. J. Biol. Macromol. 70: 292–299.

Chakravarty, N., A. Mathur and R.P. Singh. 2022. L-asparaginase: Insights into the marine sources and nanotechnological advancements in improving its therapeutics. *In*: H. Sarma, S. Gupta, M. Narayan, R. Prasad and A. Krishnan (eds.). Engineered Nanomaterials for Innovative Therapies and Biomedicine. Nanotechnology in the Life Sciences. Springer, Cham.

Chakravarty, N., J. Singh and R.P. Singh. 2021. A potential type-2 L-asparaginase from marine isolate *Bacillus* NBJ19: Statistical optimization, in silico analysis and structural modelling. Int. J. Biol. Macromol. 174: 527–539.

Chao, Y., S. Wang, S. Wu, J. Wei and H. Chen. 2017. Cloning and characterization of an alginate lyase from marine *Vibrio* sp. QD-5. Preprints 2017050055.

Charoenpanich, J., S. Suktanarag and N. Toobbucha. 2011. Production of a thermostable lipase by *Aeromonas* sp. EBB-1 isolated from marine sludge in Angsila, Thailand. Thail. Sci. Asia 37: 105–114.

Cheba, B.A., T.I. Zaghloul, M.H. EL-Massry and A.R. EL-Mahdy. 2017. Kinetics properties of marine chitinase from novel red sea strain of *Bacillus*. Procedia Eng. 181: 146–152.

Chellappan, S., C. Jasmin, S.M. Basheer, A. Kishore, K. Elyas, S.G. Bhat et al. 2011. Characterization of an extracellular alkaline serine protease from marine *Engyodontium album* BTMFS10. J. Ind. Microbiol. Biotechnol. 38(6): 743–752.

Chen, X., J. Chen, Y. Yan, S. Chen, X. Xu, H. Zhang et al. 2018a. Quorum sensing inhibitors from marine bacteria *Oceanobacillus* sp. XC22919. Nat. Prod. Res. 12: 1–5.

Chen, X.L., S. Dong, F. Xu, F. Dong, P.Y. Li, X.Y. Zhang et al. 2017. Characterization of a new cold-adapted and aalt-activated polysaccharide lyase family 7 alginate lyase from *Pseudoalteromonas* SM0524. Front. Microbiol. 7.

Chen, Z.W., H.J. Lin, W.C. Huang, S.L. Hsuan, J.H. Lin and J.P. Wang. 2018. Molecular cloning, expression, and functional characterization of the β-agarase AgaB-4 from *Paenibacillus agarexedens*. AMB Exp. 8: 49.

Chi, W.J., Y.K. Chang and S.K. Hong. 2012. Agar degradation by microorganisms and agar-degrading enzymes. Appl. Microbiol. Biotechnol. 94: 917–930.

Chi, Z.M., T. Zhang, T.S. Cao, X.Y. Liu, W. Cui and C.H. Zhao. 2011. Biotechnological potential of inulin for bioprocesses. Bioresour. Technol. 102: 4295–4303.

Comba González, N., M.L. Ramírez Hoyos, L. López Kleine and D. Montoya Castaño. 2018. Production of enzymes and siderophores by epiphytic bacteria isolated from the marine macroalga *Ulva lactuca*. Aquat. Biol. 27: 107–118.

Cowan, D., Q. Meyer, W. Stafford, S. Muyanga, R. Cameron and P. Wittwer. 2005. Metagenomic gene discovery: Past, present and future. Trends Biotechnol. 23(6): 321–329.

Cui, F., S. Dong, X. Shi, X. Zhao and X.H. Zhang. 2014. Overexpression and characterization of a novel thermostable β-agaraseYM01-3 from marine bacterium *Catenovulumagarivorans* YM01T. Mar. Drugs 12: 2731–2747.

Das, D., M.R. Bhat and R. Selvaraj. 2019. Review of inulinase production using solid-state fermentation. Ann. Microbiol. 69: 201–209.

de Veras, B.O., Y.Q. dos Santos, K.M. Diniz, G.S.C. Carelli and E.A. dos Santos. 2018. Screening of protease, cellulase, amylase and xylanase from the salt-tolerant and thermostable marine *Bacillus subtilis* strain SR60. F1000Research 7: 1704.

Delabary, G., M. Silva, C. Silva, L. Zanatta Baratieri, T. Melo, C. Stramosk et al. 2020. Influence of temperature and culture media on growth and lipolytic activity of deep-sea *Halomonas sulfidaeris* LAMA 838 and *Marinobacter excellens* LAMA 842. Ocean Coast. Res. 68.

Dharmaraj, S. 2011. Study of L-asparaginase production by *Streptomyces noursei* MTCC10464 610469, isolated from marine sponge *Callyspongia diffusa*. Iran. J. Biotechnol. 9(2): 102–108.

Doman-Pytka, M. and J. Bardowski. 2004. Pullulan degrading enzymes of bacterial origin. Crit. Rev. Microbiol. 30: 107–21.

Dong, S., J. Yang, X.Y. Zhang, M. Shi, X.Y. Song, X.L. Chen et al. 2012. Cultivable alginate lyase-excreting bacteria associated with the Arctic brown alga *Laminaria*. Mar. Drugs 10: 2481–2491.

Duan, G., F. Han and W. Yu. 2009. Cloning, sequence analysis, and expression of gene alyPI encoding an alginate lyase from marine bacterium *Pseudoalteromonas* sp., CY24. Can. J. Microbiol. 55(9): 1113–8.

Eijsink, V., I.A. Hoell and G. Vaaje-Kolstada. 2010. Structure and function of enzymes acting on chitin and chitosan. Biotechnol. Gen. Eng. Rev. 27: 331–66.

Elleuche, S., F.M. Qoura, U. Lorenz, T. Rehn, T. Brück and G. Antranikian. 2015. Cloning, expression and characterization of the recombinant cold-active type-I pullulanase from *Shewanella arctica*. J. Mol. Catal. B Enzym. 116: 70–77.

Elmansy, E.A., M.S. Asker, E.M. El-Kady, M.H. Saadia and F.M. El-Beih. 2018. Production and optimization of α-amylase from thermo-halophilic bacteria isolated from different local marine environments. Bull. Natl. Res. Cent. 42: 31.

El-Shishtawy, R.M., S.A. Mohamed, A.M. Asiri, A.B. Gomaa, I.H. Ibrahim and H.A. Al-Talhi. 2014. Solid fermentation of wheat bran for hydrolytic enzymes production and saccharification content by a local isolate *Bacillus megatherium*. BMC Biotechnol. 24: 14–29.

Eswaran, R. and L. Khandeparker. 2019. Seasonal variation in β-glucosidase-producing culturable bacterial diversity in a monsoon-influenced tropical estuary. Environ. Monit. Assess. 191: 662.

Fang, W., S. Xue, P. Deng, X. Zhang, X. Wang, Y. Xiao et al. 2019. AmyZ1: A novel α-amylase from marine bacterium *Pontibacillus* sp. ZY with high activity toward raw starches. Biotechnol. Biofuels 12: 95.

Fei, X., X.L. Chen, X.H. Sun, F. Dong, C.Y. Li, P.Y. Li et al. 2020. Structural and molecular basis for the substrate positioning mechanism of a new PL7 subfamily alginate lyase from the arctic. J. Biol. Chem. 295: 16380–16392.

Fernandes, P. 2014. Marine enzymes and food industry: Insight on existing and potential interactions. Front. Mar. Sci. 1.

Ficko-Blean, E., A. Préchoux, F. Thomas, T. Rochat, R. Larocque, Y. Zhu et al. 2017. Carrageenan catabolism is encoded by a complex regulon in marine heterotrophic bacteria. Nat. Commun. 8: 1685.

Fulzele, R., E. DeSa, A. Yadav, Y. Shouche and R. Bhadekar. 2011. Characterization of novel extracellular protease produced by marine bacterial isolated from the Indian Ocean. Braz. J. Microbiol. 42: 1364–1373.

García-Fraga, B., A.F. da Silva, J. López-Seijas and C. Sieiro. 2015. A novel family 19 chitinase from the marine-derived *Pseudoalteromonas tunicata* CCUG 44952T: Heterologous expression, characterization and antifungal activity. Biochem. Eng. J. 93: 4–93.

Ghanbarzadeh, M., A. Golmoradizadeh and A. Homaei. 2018. Carrageenans and carrageenases: Versatile polysaccharides and promising marine enzymes. Phytochem. Rev. 17(3): 535–571.

Goel, C., C. Shakir, A.Tesfaye, K.R. Sabu, A. Idhayadhulla, A. Manilal et al. 2022. Antibiofilm potential of alpha-amylase from a marine bacterium, *Pantoea agglomerans*. Can. J. Infect. Dis. Med. Microbiol. 10.

González, J.E. and M.M. Marketon. 2003 Quorum sensing in nitrogen-fixing *Rhizobia*. Microbiol. Mol. Biol. Rev. 67: 574–592.

Gui, Y., G. Xiaoqian, F. Liping, Z. Qian, Z. Peiyu and L. Jiang. 2021. Expression and characterization of a thermostable carrageenase from an Antarctic *Polaribacter* sp. NJDZ03 strain. Front. Microbiol. 12.

Gurunathan, R., B. Huang, V.K. Ponnusamy, J.S. Hwang and H.U. Dahms. 2021. Novel recombinant keratin degrading subtilisin like serine alkaline protease from *Bacillus cereus* isolated from marine hydrothermal vent crabs. Sci. Rep. 11: 12007.

Haddar, A.R., R. Agrebi, A. Bougatef, N. Hmidet, A. Sellami-Kamoun and M. Nasri. 2009. Two detergent stable alkaline serine protease from *Bacillus mojavensis* A21. Purification, characterization and potential application as a laundry detergent additive. Biresour. Technol. 100: 3366–3373.

Hadi, W.A.M., B.T. Edwin and A.J. Nair. 2021. Isolation and identification of marine *Bacillus altitudinis* KB1 from coastal kerala: Asparginase producer. J. Mar.Biol. Asso. India 63(2).

Handelsman, J. 2004. Metagenomics: Application of genomics to uncultured microorganisms. Microbiol. Mol. Biol. Rev. 68(4): 669–685.

Hassan, S.W.M., H. Abd El Latif Hala and M. Ali Safaa. 2018.. Production of cold-active lipase by free and immobilized marine *Bacillus cereus* HSS: Application in wastewater treatment. Front. Microbiol. 9.

Healy, F.G., R.M. Ray, H.C. Aldrich, A.C. Wilkie, L.O. Ingram and K.T. Shanmugam. 1995. Direct isolation of functional genes encoding cellulases from the microbial consortia in a thermophilic, anaerobic digester maintained on lignocellulose. Appl. Microbiol. Biotechnol. 43: 667–674.

Hehemann, J.H., G. Correc, F. Thomas, T. Bernard, T. Barbeyron, M. Jam et al. 2012. Biochemical and structural characterization of the complex agarolytic enzyme system from the marine bacterium *Zobellia galactanivorans*. J. Biol. Chem. 287: 30571–30584.

Howlader, M.M., J. Molz, N. Sachse and R. Tuvikene. 2021. Optimization of fermentation conditions for carrageenase production by *Cellulophaga* species: A comparative study. Biology 10(10): 971.

Huang, L., A. Shizume, M. Nogawa, G. Taguchi and M. Shimosaka. 2012b. Heterologous expression and functional characterization of a novel chitinase from the chitinolytic bacterium *Chitiniphilus shinanonensis*. Biosci. Biotechnol. Biochem. 76: 517–522.

Huang, W., Y. Lin, S. Yi, P. Liu, J. Shen, Z. Shao et al. 2012a. QsdH, a novel AHL lactonase in the RND-type inner membrane of marine *Pseudoalteromonas byunsanensis* strain 1A01261. PLoS ONE 7: e46587.

Huang, X., Y. Zhao, Y. Dai, G. Wu, Z. Shao, Q. Zeng and Z. Liu. 2012c. Cloning and biochemical characterization of a glucosidase from a marine bacterium *Aeromonas* sp. HC11e-3. World J. Microbiol. Biotechnol. 28(12): 3337–44.

Huo, Y.Y., S.L. Jian, H. Cheng, Z. Rong, H.L. Cui and X.W. Xu. 2018. Two novel deep-sea sediment metagenome-derived esterases: Residue 199 is the determinant of substrate specificity and preference. Microb. Cell Fact. 17(1): 16.

Ibrahim, A., A. Alsalamah, Y. Elbadawi, M. Eltayeb and S. Ibrahim. 2015. Production of extracellular alkaline protease by new halotolerant alkaliphilic *Bacillus* sp. NPST-AK15 isolated from hyper saline soda lakes. E-J Biotech. 46.

Imhoff, J.F., A. Labes and J. Wiese. 2011. Bio-mining the microbial treasures of the ocean: New natural products. Biotech. Adv. 29: 468–482.

Iqbalsyah, T.M., M. Malahayati, A. Atikah and F. Febriani. 2019. Purification and partial characterization of a thermo-halostable protease produced by *Geobacillus* sp. strain PLS A isolated from undersea fumaroles. J. Taibah Univ. Sci. 13: 850–857.

Ivanova, V., S. Gavrailov and V. Pashkoulova. 2018. Purification of an exo-inulinase from *Bacillus* sp. SG7. J. Microbiol. Biotechnol. Food Sci. 685–691.

Jeon, J.H., J.T. Kim, H.S. Lee, S.J. Kim, S.G. Kang, S.H. Choi et al. 2011. Novel lipolytic enzymes identified from metagenomic library of Deep-Sea sediment. Evid. Based Complement. Alternat. Med. Article ID 271419.

Jiang, C., H. Jiang, T. Zhang, Z. Lu and X. Mao. 2022. Enzymatic verification and comparative analysis of carrageenan metabolism pathways in marine bacterium *Flavobacterium algicola*. Appl. Environ. Microbiol. 88(7): e0025622.

Jiang, X., X. Xu, Y. Huo, Y. Wu, X. Zhu, X. Zhang et al. 2012. Identification and characterization of novel esterases from a deep-sea sediment metagenome. Arch. Microbiol. 194: 207–214.

Jung, S.S., B. Keun, P. Seong, C. Han, S. Chi, S. Tai-Sun et al. 2012. Characterization of alginate lyase gene using a metagenomic library constructed from the gut microflora of abalone. J. Ind. Microb. Biotech. 39(4): 585–593.

Kalpana, B.J. and S.K. Pandian. 2014. Halotolerant, acid-alkali stable, chelator resistant and raw starch digesting α-amylase from a marine bacterium *Bacillus subtilis* S8-18. J. Basic Microbiol. 54: 802–811.

Kamarudheen, N. and K.V.B. Rao. 2019. Fatty acyl compounds from marine *Streptomyces griseoincarnatus* strain HK12 against two major biofilm forming nosocomial pathogens; an *in vitro* and *in silico* approach. Microb. Pathogenesis 127: 121–130.

Kamble, P. and S. Vavilala. 2018. Discovering novel enzymes from marine ecosystems: A metagenomic approach. Bot. Mar. 2: 161–175.

Kandasamy, K.P., R.K. Subramanian, R. Srinivasan, S. Raghunath, S. Balaji et al. 2020. *Shewanella algae* and *Microbulbifer elongatus* from marine macro-algae - isolation and characterization of agar-hydrolysing bacteria. Access Microbiol. 2(11): acmi000170.

Kang, S. and J.K. Kim. 2015. Reuse of seaweed waste by a novel bacterium, *Bacillus* sp. SYR4 isolated from a sandbar. World J. Microbiol. Biotechnol. 31(1): 209–217.

Karl, D.M. 2007. Microbial oceanography: Paradigms, processes and promise. Nat. Rev. Microbiol. 5(10): 759–769.

Kennedy, J., J.R. Marchesi and A.D.W. Dobson. 2008. Marine metagenomics: Strategies for the discovery of novel enzymes with biotechnological applications from marine environments. Microb. Cell Factories 7: 27.

Khalifa, A. and M. Aldayel. 2019. Isolation and characterization of the agarolytic bacterium *Pseudoalteromonas ruthenica*. Open Life Sci. 14(1): 588–594.

Khan, M.T. and I. Sylte. 2009. Determinants for psychrophilic and thermophilic features of metallopeptidases of the M4 family. In Silico Biol. 9(3): 105–124.

Kim, B.C., H. Poo, K.H. Lee, M.N. Kim, D.S. Park, H.W. Oh et al. 2012. *Simiduia areninigrae* sp. nov. an agarolytic bacterium isolated from sea sand. Int. J. Syst. Evol. Microbiol. 62: 906–911.

Kim, D.W., K. Sapkota, J.H. Choi, Y.S. Kim, S. Kim and S.J. Kim. 2013. Direct acting anti-thrombotic serine protease from brown seaweed *Costaria costata.* Process Biochem. 48(2): 340–350.

Krishnan, A., G. Kumar, K. Loganathan and B. Rao. 2011. Optimization, production and partial purification of extracellular α-amylase from *Bacillus* sp., *marini.* Arc. Appl. Sci. Res. 3.

Krishnan, A., K. Loganathan, G. Kumar and B. Rao. 2013. Purification and activity of amylase of marine *Halobacilus* sp., *amylus* HM454199. Indian J. Geo-Mar. Sci. 42: 781–785.

Krithika, S. and C. Chellaram. 2016. Isolation, screening, and characterization of chitinase producing bacteria from marine wastes. Int. J. Pharm. Pharm. Sci. 8: 34.

Kurniawan, E., S. Panphon and M. Leelakriangsak. 2019. Potential of marine chitinolytic *Bacillus* isolates as biocontrol agents of phytopathogenic fungi. IOP Conf. Series: Earth Environ. Sci. 217: 012044.

Kwon, M., W.Y. Jang, G.M. Kim and Y.H. Kim. 2020. Characterization and application of a recombinant exolytic GH50A β-Agarase from *Cellvibrio* sp. KY-GH-1 for enzymatic production of neoagarobiose from agarose. ACS Omega 5: 29453–29464.

Lee, S.I., S.H. Choi, F.Y. Lee and H.S. Kim. 2012. Molecular cloning, purification, and characterization of a novel polyMG-specific alginate lyase responsible for alginate MG block degradation in *Stenotrophomas maltophilia* KJ2. J. Appl. Microb. Biotech. 95(6): 1643–1653.

Lee, Y., C. Oh, M. DeZoysa, H. Kim, W.D.N. Wickramaarachchi, I. Whang et al. 2013. Molecular cloning, overexpression, and enzymatic characterization of glycosyl hydrolase family16 β-agarase from marine bacterium *Saccharophagus* sp.AG21in *Escherichia coli.* J. Microbiol. Biotechnol. 23: 913–922.

Lee, Y.R., S. Jung, W.J. Chi, C.H. Bae, B.C. Jeong, S.K. Hong et al. 2018. Biochemical characterization of a novel GH86 β-agarase producing neoagarohexaose from *Gayadomonas Joobiniege* G7. J. Microbiol. Biotechnol. 28(2): 284–292.

Li, A.X., L.Z. Guo and W.D. Lu. 2012. Alkaline inulinase production by a newly isolated bacterium *Marinimicrobium* sp. LS–A18 and inulin hydrolysis by the enzyme. World J. Microbiol. Biotechnol. 28(1): 81–89.

Li, C., C. Li, L. Li, X. Yang, S. Chen, B. Qi and Y. Zhao. 2020. Comparative genomic and secretomic analysis provide insights into unique agar degradation function of marine bacterium *Vibrio fluvialis* a8 through horizontal gene transfer. Front. Microbiol. 11: Article 1934.

Li, J., Q. Hu, Y. Li and Y. Xu. 2015. Purification and characterization of cold-adapted beta-agarase from an Antarctic psychrophilic strain. Braz. J. Microbiol. 46(3): 683–690.

Li, J., X. Gu, Q. Zhang, L. Fu, J. Tan and L. Zhao. 2022. Biochemical characterization of a carrageenase, Car1383, derived from associated bacteria of Antarctic macroalgae Front. Microbiol. 13.

Li, L., X. Jiang, H. Guan, P. Wang and H. Guo. 2011. Three alginate lyases from marine bacterium *Pseudomonas fluorescens* HZJ216: Purification and characterization. Appl. Biochem. Biotechnol. 164: 305–317.

Li, P.Y., P. Ji, C.Y. Li, Y. Zhang, G.L. Wang and X.Y. Zhang. 2014.Structural basis for dimerization and catalysis of a novel esterase from the GTSAG motif subfamily of the bacterial hormone-sensitive lipase family. J. Biol. Chem. 289: 19031–19041.

Li, R.K., Y.J. Hu, Y.J. He, T.B. Ng, Z.M. Zhou and X.Y. Ye. 2021. A thermophilic chitinase 1602 from the marine bacterium *Microbulbifer* sp. BN3 and its high-level expression in *Pichia pastoris.* Biotech. Appl. Biochem. 68: 1076–1085.

Li, Y., X. Lei, H. Zhu, H. Zhang, C. Guan, Z. Chen et al. 2016. Chitinase producing bacteria with direct algicidal activity on marine diatoms. Sci. Rep. 6: 21984.

Lima, D.M., P. Fernandes, D.S. Nascimento, R.C.L.F. Ribeiro and S.A.A. de Assis. 2011. Fructose syrup: A biotechnology asset. Food Technol. Biotechnol. 49: 424–434.

Lishuxin, H., Z. Jungang, P. Qiang, L. Hong and D. Yuguang. 2013. Characterization of a new alginate lyase from newly isolated *Flavobacterium* sp. S20. J. Ind. Microb. Biotech. 40(1): 113–122.

Liu, G., Y. Li, Z. Chi and Z.M. Chi. 2011. Purification and characterization of κ-carrageenase from the marine bacterium *Pseudoalteromonas porphyrae* for hydrolysis of κ-carrageenan. Proc. Biochem. 46: 265–271.

Liu, Z., G. Li, Z. Mo and H. Mou. 2013. Molecular cloning, characterization, and heterologous expression of a new κ-carrageenase gene from marine bacterium *Zobellia* sp. ZM-2. Appl. Microbiol. Biotechnol. 97: 10057–10067.

Loni, P.P., J.U. Patil, S.S. Phugare and S.S. Bajekal. 2014. Purification and characterization of alkaline chitinase from *Paenibacillus pasadenensis* NCIM 5434. J. Basic Microbiol. 54: 1080–1089.

López-lópez, O., M.E. Cerdán and M.I.G. Siso. 2014. New extremophilic lipases and esterases from metagenomics. Curr. Protein Pept. Sci. 15: 445–455.

Lorenz, P. and J. Eck. 2005. Metagenomics and industrial applications. Nat. Rev. Microbiol. 3(6): 510–516.

Lu, W.D., A.X. Li and Q.L. Guo. 2014. Production of novel alkalitolerant and thermostable inulinase from marime actinomycete *Nocardiopsis* sp. DN-K15 and inulin hydrolysis by the enzyme. Annals Microbiol. 64(2): 441–449.

Ma, S., G. Duan, W. Chai, C. Geng, Y. Tan, L. Wang et al. 2013. Purification, cloning, characterization and essential amino acid residues analysis of a new ι-carrageenase from *Cellulophaga* sp. QY3. PLoS ONE 8(5): e64666.

Ma, Z.P., Y. Song, Z.H. Cai, Z.J. Lin, G.H. Lin, Y. Wang et al. 2018. Anti-quorum sensing activities of selected coral symbiotic bacterial extracts from the South China Sea. Front. Cell. Infect. Microbiol. 8: 144.

Mahadevan, G.D. and S.E. Neelagund. 2014. Thermostable lipase from *Geobacillus* sp. Iso5: Bioseparation, characterization and native structural studies. J. Basic Microbiol. 54(5): 386–96.

Mahapatra, S., A.S. Vickram, T.B. Sridharan, R. Parmasewari and M. Ramesh Pathy. 2016. Screening, production, optimization and characterization of β-glucosidase using microbes from shellfish waste. 3 Biotech. 6: 213.

Mao, X., Y. Hong, Z. Shao, Y. Zhao and Z.J. Liu. 2010. A novel cold-active and alkali-stable β-glucosidase gene isolated from the marine bacterium *Martelella mediterranea*. Appl. Biochem. Biotechnol. 162: 2136–2148.

Martin, M., T. Barbeyron, R. Martin, D. Portetelle, G. Michel and M. Vandenbol. 2015. The cultivable surface microbiota of the brown alga *Ascophyllum nodosum* is enriched in macroalgal-polysaccharide-degrading bacteria. Front. Microbiol. 6: 1487.

Maruthiah, T., B. Somanath, G. Immanuel and A. Palavesam. 2015. Deproteinization potential and antioxidant property of haloalkalophilic organic solvent tolerant protease from marine *Bacillus* sp. APCMSTRS3 using marine shell wastes. Biotechnol. Rep. 8: 124–132.

Maruthiah, T., P. Esakkiraj, G. Prabakaran, A. Palavesam and G. Immanuel. 2013. Purification and characterization of moderately halophilic alkaline serine protease from marine *Bacillus subtilis* AP-MSU 6. Biocatal. Agric. Biotechnol. 2: 116–119.

Matard-Mann, M., T. Bernard, C. Leroux, T. Barbeyron, R. Larocque, A. Préchoux et al. 2017. Structural insights into marine carbohydrate degradation by family GH16 κ-carrageenases. J. Biol. Chem. 292(48): 19919–19934.

Mayer, C., M. Romero, A. Muras and A. Otero. 2015. Aii20J, a wide-spectrum thermostable N-acylhomoserine lactonase from the marine bacterium *Tenacibaculum* sp. 20J, can quench AHL-mediated acid resistance in *Escherichia coli*. Appl. Microbiol. Biotechnol. 99: 9523–9539.

Meenakshi, S., S. Umayaparvathi, P. Manivasagan, M. Arumugam and T. Balasubramanian. 2013. Purification and characterization of inulinase from marine bacterium, *Bacillus cereus* MU-31. Indian J. Geo-Mar. Sci. 42(4): 510–515.

Metzker, M.L. 2010. Sequencing technologies-the next generation. Nat. Rev. Genet. 11: 31–46.

Minegishi, H., Y. Shimane, A. Echigo, Y. Ohta, Y. Hatada, M. Kamekura et al. 2013. Thermophilic and halophilic β-agarase from a halophilic archaeon *Halococcus* sp. 197A. Extremophiles 17(6): 931–9.

Mo, S., J.H. Kim and K.W. Cho. 2009. Enzymatic properties of an extracellular phospholipase C purified from a marine *Streptomycete*. Biosci. Biotechnol. Biochem. 73(9): 2136–7.

Nazir, A. 2016. Review on metagenomics and its applications. IJIR 2: 277–286.

Neumann, A.M., J.P. Balmonte, M. Berger, H.A. Giebel, C. Arnosti, S. Vogct et al. 2015. Different utilization of alginate and other algal polysaccharides by marine *Alteromonas macleodii* ecotypes. Environ. Microbiol. 17(10): 3857–3868.

Newman, D.J. and R.T. Hill. 2006. New drugs from marine microbes: the tide is turning. J. Ind. Microbiol. Biotechnol. 33: 539–544.

Ng, W. and B. Bassler. 2009. Bacterial quorum-sensing network architectures. Annu. Rev. Genet. 3: 197–222.

Ntozonke, N., K. Okaiyeto, A.S. Okoli, A.O. Olaniran, U.U. Nwodo and A.I. OkohInt. 2017. A marine bacterium, *Bacillus* sp., isolated from the sediment samples of Algoa Bay in South Africa produces a polysaccharide-bioflocculant. J. Environ. Res. Public Health 14: 1149.

Ojima, T., R. Nishiyama and A. Inoue. 2019. Production of value-added materials from alginate using alginate lyases and 4-deoxy-l-erythro-5-hexoseulose uronic acid–metabolic enzymes from alginolytic bacteria and marine gastropods. *In*: Enzymatic Technologies for Marine Polysaccharides, 1st Edition, Imprint CRC Press, Pages 16.

Pace, N.R., D.A. Stahl, D.J. Lane and G.J. Oslen, 1985. Analyzing natural microbial populations by rRNA sequences. ASM News 51: 4–12.

Packiavathy, I.A.S.V., A. Kannappan, S. Thiyagarajan, R. Srinivasan, D. Jeyapragash, J. Bosco et al. 2021. AHL-lactonase producing *Psychrobacter* sp. from Palk bay sediment mitigates quorum sensing-mediated virulence production in Gram negative bacterial pathogens. Front. Microbiol. 12: 748.

Pancha, I., D. Jain, A. Shrivastav, S.K. Mishra, B. Shethia, S. Mishra et al. 2010. A thermoactive α-amylase from a *Bacillus* sp. isolated from CSMCRI salt farm. Int. J. Biol. Macromol. 47(2): 288–291.

Pang, B., L. Zhou, W. Cui, Z. Liu, S. Zhou, J. Xu et al. 2019. A hyperthermostable type II pullulanase from a deep-sea microorganism *Pyrococcus yayanosii* CH1. J. Agri. Food Chem. 67(34): 9611–9617.

Paulsen, S.S., B. Andersen, L. Gram and H. Machado. 2016. Biological potential of chitinolytic marine bacteria. Mar. Drugs 14: 230.

Paulsen, S.S., M.L. Strube, P.K. Bech, L. Gram and E.C. Sonnensshein. 2019. Marine chitinolytic *Pseudoalteromonas* represents an untapped reservoir of bioactive potential. Appl. Environ. Sci. 4: 00060–19.

Peng, H., M. Chen, L. Yi, X. Zhang, M. Wang, Y. Xiao et al. 2015. Identification and characterization of a novel raw-starch-degrading α-amylase (AmyASS) from the marine fish pathogen *Aeromonas salmonicida* sp., *salmonicida*. J. Mol. Cat. B: Enzy. 119: 71–77.

Perfumo, A., G.J.F. von Sass, E.L. Nordmann, N. Budisa and D. Wagner. 2020. Discovery and characterization of a new cold-active protease from an extremophilic bacterium via comparative genome analysis and *in vitro* expression. Front. Microbiol. 11.

Poli, A., I. Finore, I. Romano, A. Gioiello, L. Lama and B. Nicolaus. 2017. Microbial diversity in extreme marine habitats and their biomolecules. Microorganisms 5(2): 25.

Prajakta, N. and S. Chandra. 2019. *Alteromonas macleodii* KS62 (MTCC 12606): A novel κ-carrageenase producing microorganism. Int. J. Microbiol. Biotechnol. 4(4): 128–132.

Prihanto, A. and M. Wakayama. 2016. Marine microorganism: An underexplored source of L-asparaginase. pp. 1–25. *In*: S.K. Kim and F. Toldrá (eds.). Advances in Food and Nutrition Research. Academic Press, 79.

Pyeon, H.M., Y.S. Lee and Y.L. Choi. 2019. Cloning, purification, and characterization of GH3 β-glucosidase, MtBgl85, from *Microbulbifer thermotolerans* DAU221. Peer J 7: e7106.

Qeshmi, F.I., A. Homaei, K. Khajeh, E. Kamrani and P. Fernandes. 2022. Production of a novel marine *Pseudomonas aeruginosa* recombinant L-asparginase: Insight on the structure and biochemical characterization. Mar. Biotechnol. 1–15.

Qoura, F., S. Elleuche, T. Brück and A. Garabed. 2014. Purification and characterization of a cold-adapted pullulanase from a psychrophilic bacterial isolate. Extremophiles: Life Under Extreme Conditions 18: 1095–1102.

Rahman, F., A. Meryandini, M. Zairin and I. Rusmana. 2011. Isolation and identification of an agar-liquefying marine bacterium and some properties of its extracellular agarases. Biodiversitas 12: 1412–33.

Ramos, K.R.M., K.N.G. Valdehuesa, G.M. Nisola, W.K. Lee and W.J. Chung. 2018. Identification and characterization of a thermostable endolytic β-agarase Aga2 from a newly isolated marine agarolytic bacteria *Cellulophaga omniscoria* W5C. New Biotech. 40: 261–267.

Ran, L.Y., H.N. Su, M.Y. Zhou, L. Wang, X.L. Chen, B.B. Xie et al. 2014. Characterization of a novel subtilisin-like protease myroicolsin from deep sea bacterium *Myroides profundi* D25 and molecular insight into its collagenolytic mechanism. J. Biol. Chem. 289: 6041–6053.

Rathore, D.S., K. Malaviy, A. Dobariya and S.P. Singh. 2020. Optimization of the production of an amylase from a marine actinomycetes *Nocardiopsis Dassonvillei* strain Kas11. Proc. Natl. Conf. Innov. Biol. Sci. (NCIBS).

Rodrigues, C.J.C., R.F. Pereira, P. Fernandes, J.M. Cabral and C.C.C.R. de Carvalho. 2017. Cultivation-based strategies to find efficient marine biocatalysts. Biotechnol. J. 12: 1700036.

Rondon, M.R., P.R. August, A.D. Bettermann, S.F. Brady, T.H. Grossman, M.R. Liles et al. 2000. Cloning the soil metagenome: A strategy for accessing the genetic and functional diversity of uncultured microorganisms. Appl. Environ. Microbiol. 66: 2541–2547.

Rudrapati, P. and A.V. Audipudi. 2017. Production and purifcation of anticancer enzyme L-asparaginase from *Bacillus frmus* AVP 18 of mangrove sample through submerged fermentation. Int. J. Curr. Microbiol. App. Sci. 5: 1–18.

Saleena, S.K., J.I. Johnson, J.K. Joseph, K.K. Padinchati and M.H.A. Abdulla. 2022. Production and optimization of L-asparginase by *Streptomyces koyangensis* SK4 isolated from Arctic sediment. J. Basic Microbiol. 62: 1–10.

Samuel, P., A. Raja and P. Prabakaran. 2012. Investigation and application of marine derived microbial enzymes: Status and prospects. Int. J. Oceanograpy Mar. Ecol. Sys. 1(1): 1–10.

Santhanam, S., G. Kiran Seghal, J. Selvin, R. Gandhimathi, T. Baskar, A. Manilal et al. 2009. Optimization, production, and partial characterization of an alkalophilic amylase produced by sponge associated marine bacterium *Halobacterium salinarum* MMD047. Biotechnol. Bioprocess. Eng. 14: 67–75.

Saravanan, D., V. Suresh Kumar and M. Radhakrishnan. 2015. Isolation and optimization of agarase producing bacteria from marine sediments. Int. J. Chem. Tech. Res. 8(4): 1701–1705.

See-Too, W.S., P. Convey, D.A. Pearce and K.G. Chan. 2018.Characterization of a novel *N*-acylhomoserine lactonase, AidP, from Antarctic *Planococcus* sp. Microb. Cell Fact. 17: 179.

Seghal Kiran, G., A. Nishanth Lipton, J. Kennedy, A.D. Dobson and J. Selvin. 2014. A halotolerant thermostable lipase from the marine bacterium *Oceanobacillus* sp. PUMB02 with ability to disrupt bacterial biofilms. Bioeng. 5(5): 305–18.

Selva Mohan, T., A. Palavesam and R.L. Ajitha. 2012. Optimization of lipase production by *Vibrio* sp.—A fish gut isolate. Eur. J. Zool. Res. 1: 23–25.

Selvin, J., J. Kennedy, D.P. Lejon, G. Sehgal Kiran and A.D.W. Dobson. 2012. Isolation, identification and biochemical characterization of a novel halo-tolerant lipase from the metagenome of the marine sponge *Haliclona simulans*. Microb. Cell Fact. 11: 72.

Shakambari, G., A.K. Birendranarayan, M.J.A. Lincy, S.K. Rai, Q.T. Ahmad, B. Ashokkumar et al. 2016. Hemocompatible glutaminase free L-asparaginase from marine *Bacillus tequilensis* PV9W with anticancer potential modulating p53 expression. RSC Adv. 31.

Shangyong, L.I., J.L.A. Panpan, W. Linna, Y.U. Wengong and H. Feng. 2013. Purification and characterization of a new thermostable κ-carrageenase from the marine bacterium *Pseudoalteromonas* sp. QY203. J. Ocean Univ. China 12(1): 155–159.

Sharabash, M.M., S.S. Abouelkheir, M.E.M. Mabrouk, H.A. Ghozian and S.A. Sabry. 2022. Agarase production by marine *Pseudoalteromonas* sp. MHS: Optimization, and purification. J. Mar. Sci. 4(1).

Sharma, R., R. Ravi, K. Raj and G. Amit. 2005. Unculturable' bacterial diversity: An untapped resource. Curr. Sci. 89: 72.

Sharpton, T.J. 2014. An introduction to the analysis of shotgun metagenomic data. Front. Plant Sci. 5.

Shen, J., Y. Chang, S. Dong and F. Chen. 2017. Cloning, expression and characterization of a ι-carrageenase from marine bacterium *Wenyingzhuangia fucanilytica*: A biocatalyst for producing ι-carrageenan oligosaccharides. J. Biotech. 259: 103–109.

Shen, Y., Z. Li, Y.Y. Huo, L. Bao, L. Gao, P. Xiao et al. 2019. Structural and functional insights into CmGH1, a novel GH39 family β-glucosidase from deep-sea bacterium. Front. Microbiol. 10.

Shendure, J. and H. Ji. 2008. Next-generation DNA sequencing. Nat. Biotechnol. 26: 1135–1145.

Simair, A.A., I. Khushk, A.S. Qureshi, M.A. Bhutto, H.A. Chaudhry, K.A. Ansari et al. 2017. Amylase production from thermophilic *Bacillus* sp. BCC 021-50 isolated from a marine environment. Fermentation 3(2): 25.

Sindhwad, P. and K. Desai. 2015. Media optimization, isolation and purification of L-asparaginase from marine isolate. Asian Pac. J. Health Sci. 293: 72–82.

Song, T., H. Xu, C. Wei, T. Jiang, S. Qin, W. Zhang et al. 2016. Horizontal transfer of a novel soil agarase gene from marine bacteria to soil bacteria via human microbiota. Sci. Rep. 6: 34103.

Steen, A.D. and C. Arnosti. 2013. Extracellular peptidase and carbohydrate hydrolase activities in an Arctic fjord (Smeerenburg fjord, Svalbard). Aquat. Microb. Ecol. 69: 93–99.

Su, E., T. Xia, L. Gao, Q. Dai and Z. Zhang. 2010. Immobilization of β-glucosidase and its aroma-increasing effect on tea beverage. Food Bioprod. Process. 88: 83–89.

Su, H., Z. Xiao, K. Yu, Q. Zhang, C. Lu, G. Wang et al. 2021. High diversity of β-glucosidase-producing bacteria and their genes associated with Scleractinian corals. Int. J. Mol. Sci. 22: 3523.

Su, Q., T. Jin, Y. Yu, M. Yang, H. Mou and L. Li. 2017. Extracellular expression of a novel β-agarase from *Microbulbifer* sp. Q7, isolated from the gut of sea cucumber. AMB Expr. 7: 220.

Subramani, A.K., R. Raval, S. Sundareshan, R. Sivasengh and K. Raval. 2022. A marine chitinase from *Bacillus aryabhattai* with antifungal activity and broad specificity toward crystalline chitin degradation. Prep. Biochem. Biotechnol. 52: 1160–1172.

Subramanian, K., B. Sadaiappan, W. Aruni, A. Kumarappan, R. Thirunavukarasu, G.P. Srinivasan et al. 2020. Bioconversion of chitin and concomitant production of chitinase and N-acetylglucosamine by novel *Achromobacter xylosoxidans* isolated from shrimp waste disposal area. Sci. Rep. 10: 11898.

Sudha, K., J. Rani and T.R. Sivansankari. 2017. Extracellular l-asparginase production by halotolerant strain of *Enterobacter hormaechei* isolated from marine fishes. Int. J. Adv. Res. 5(1): 2619–2625.

Tang, K., Y. Su, G. Brackman, F. Cui, Y. Zhang, X. Shi et al. 2015. MomL, a novel marine-derived N-acyl homoserine lactonase from *Muricauda olearia*. Appl. Environ. Microbiol. 81: 774–782.

Tayco, C.C., F. Tablizo, R.S. Regalia and A.O. Lluisma. 2013. Characterization of a k-carrageenase-producing marine bacterium, isolate ALAB-001. Philippine J. Sci. 142: 45–54.

Thao, N.V., A. Nozawa, Y. Yumiko Obayashi, S.-I. Kitamura, T. Yokokawa and S. Suzuki. 2015. Extracellular proteases are released by ciliates in defined seawater microcosms. Mar. Environ. Res. 109: 95–102.

Thenmozhi, C., R. Sankar, V. Karuppiah and P. Sampathkumar. 2011. L-asparaginase production by mangrove derived *Bacillus cereus* MAB5: Optimization by response surface methodology. Asian Pac. J. Trop. Med. 4(6): 486–491.

Tomasik, P. and D. Horton. 2012. Enzymatic conversions of starch. *In*: Derek Horton (ed.). Advances in Carbohydrate Chemistry and Biochemistry, Academic Press, 68: 59–436.

Torres, M., J.C. Reina, J.C. Fuentes-Monteverde, G. Fernández, J. Rodríguez, C. Jiménez et al. 2018. AHL-lactonase expression in three marine emerging pathogenic *Vibrio* spp. reduces virulence and mortality in brine shrimp (*Artemia salina*) and Manila clam (Venerupis philippinarum).

Trincone, A. 2013. Biocatalytic processes using marine biocatalysts: Ten cases in point. Curr. Org. Chem. 17: 1058–1066.

Uchiyama, T., T. Abe, T. Ikemura and K. Watanabe. 2005. Substrate-induced gene-expression screening of environmental metagenome libraries for isolation of catabolic genes. Nat. Biotechnol. 23(1): 88–93.

Ulwiyyah, F.N.H., N. Saidi and T.M. Iqbalsyah. 2019. Screening and production of lipase from a thermo halophilic bacterial isolate of Pria Laot Sabang 80 isolated from under water hot spring. KnE Eng. 4(2): 105.115.

Uzyol, K.S., B.S. Akbulut and D.D. Kazan. 2012. Thermostable α-amylase from moderately halophilic *Halomonas* sp., AAD21. Turk. J. Biol. 36: 327–338.

Vijayaraghavan, R. and S. Rajendran. 2012. Identification of a novel agarolytic γ-proteobacterium *Microbulbifer maritimus* and characterization of its agarase. J. Basic Microbiol. 52: 705–712.

Wang, M., L. Chen, Z. Liu, Z. Zhang, S. Qin and P. Yan. 2016. Isolation of a novel alginate lyase-producing *Bacillus litoralis* strain and its potential to ferment *Sargassum horneri* for biofertilizer. Microbiol. Open 5(6): 1038–1049.

Wang, M., L. Chen, Z. Lou, X. Yuan, G. Pan, X. Ren et al. 2022. Cloning and characterization of a novel alginate lyase from *Paenibacillus* sp. LJ-23. Mar. Drugs 20(1): 66.

Wang, M., L. Chen, Z. Zhang, X. Wang, S. Qin and P. Yan. 2017. Screening of alginate lyase-excreting microorganisms from the surface of brown algae. AMB Exp. 7(1): 74.

Wang, S.L., T.W. Liang, B.S. Lin, C.L. Wang, P.C. Wu and J.R. Liu. 2010. Purification and characterization of chitinase from a new species strain *Pseudomonas* sp. TKU008. J. Microbiol. Biotechnol. 20: 1001–1005.

Wang, X., Y. Zhao, H. Tan, N. Chi, Q. Zhang, Y. Du et al. 2014. Characterisation of a chitinase from *Pseudoalteromonas* sp. DL-6, a marine psychrophilic bacterium. Int. J. Biol. Macromol. 70: 455–462.

Whitman, W.B., D.C. Coleman and W.J. Wiebe. 1998. Prokaryotes: The unseen majority. Proc. Natl. Acad. Sci. USA 95(12): 6578–6583.

Wijaya, A.P., M.T. Sibero, D.S. Zilda, A.N. Windiyana, A. Wijayanto, E.H. Frederick et al. 2021. Preliminary screening of carbohydrase-producing bacteria from *Chaetomorpha* sp. in Sepanjang Beach, Yogyakarta, Indonesia. IOP Conf. Series: Earth and Environmental Science 750: 012027.

Wong, T.Y., L.A. Preston and N.L. Schiller. 2000. Alginate Lyase: Review of major sources and enzyme characteristics, structure-function analysis, biological roles, and applications. Annu. Rev. Microbiol. 54: 289–340.

Wu, G., Y. Qin, Q. Cheng and Z. Liu. 2014. Characterization of a novel alkali-stable and salt-tolerant α-amylase from marine bacterium *Zunongwangia profunda*. J. Mol. Catal. B Enzym. 110: 8–15.

Wu, S.J. and J. Chen. 2014. Preparation of maltotriose from fermentation broth by hydrolysis of pullulan using pullulanase. Carbohydrate Pol. 107: 94–97.

Xie, W., B. Lin, Z. Zhou, G. Lu, J. Lun, C. Xia et al. 2013. Characterization of a novel β-agarase from an agar-degrading bacterium *Catenovulum* sp. X3. Appl. Microbol. Biotechnol. 97: 4907–4915.

Xu, F., X.L. Chen, X.H. Sun, F. Dong, C.Y. Li, P.Y. Li et al. 2020. Structural and molecular basis for the substrate positioning mechanism of a new PL7 subfamily alginate lyase from the arctic. J. Biol. Chem. 27: 16380–16392.

Xu, M., D. Fujita and N. Hanagata. 2009. Perspectives and challenges of emerging single-molecule DNA sequencing technologies. Small 5: 2638–2649.

Xue, Z., X.M. Sun, C. Chen, X.Y. Zhang, X.L. Chen, Y.Z. Zhang et al. 2022. A novel alginate lyase: Identification, characterization, and potential application in alginate trisaccharide preparation. Mar. Drugs 20(3): 159.

Yamada, K., T. Terahara, S. Kurata, T. Yokomaku, S. Tsuneda and S. Harayama. 2008. Retrieval of entire genes from environmental DNA by inverse PCR with pre-amplification of target genes using primers containing locked nucleic acids. Environ. Microbiol. 10(4): 978–987.

Yang, M., Y. Yuan, Y. Suxiao, S. Xiaohui, M. Haijin and L. Li. 2018. Expression and characterization of a new polyg-specific alginate lyase from marine bacterium *Microbulbifer* sp. q7. Front. Microbiol. 9.

Yao, Z., F. Wang, Z. Gao, L. Jin and H. Wu. 2013. Characterization of a κ-carrageenase from marine *Cellulophaga lytica* strain N5-2 and analysis of its degradation products. Int. J. Mol. Sci. 14: 24592–24602.

Younes, I. and M. Rinaudo. 2015. Chitin and chitosan preparation from marine sources. Structure, properties and applications. Mar. Drugs 13: 1133–1174.

Youssef, A., E. Beltagy, M. El-Shenawy and A. El-Aassar Samy. 2012. Production of k-carrageenase by *Cellulosimicrobium cellulans* isolated from Egyptian Mediterranean coast. Afr. J. Microbiol. Res. 6: 6618–6628.

Yuan, D., D. Lan, R. Xin, B. Yang and Y. Wang. 2014. Biochemical properties of a new cold-active mono- and di-acyl glycerol lipase from marine member *Janibacter* sp., strain HTCC2649. Int. J. Mol. Sci. 15: 10554–10566.

Zhang, J., M. Chen, J. Huang, X. Guo, Y. Zhang, D. Liu et al. 2019. Diversity of the microbial community and cultivable protease-producing bacteria in the sediments of the Bohai Sea, Yellow Sea and South China Sea. PLoS ONE 14(4): e0215328.

Zhang, Y., F. Guan, G. Xu, X. Liu, Y. Zhang, J. Sun et al. 2022. A novel thermophilic chitinase directly mined from the marine metagenome using the deep learning tool Preoptem. Bioresour. Bioprocess 9: 54.

Zhang, Y., J. Hao, Y.Q. Zhang, X.L. Chen, B.B. Xie, M. Shi et al. 2017. Identification and characterization of a novel salt-tolerant esterase from the Deep-Sea sediment of the South China Sea. Front. Microbiol. 8.

Zhang, Z., R.R. Marquardt, G. Wang, W. Guenter, G.H. Crow, Z. Han et al. 1996. A simple model for predicting the response of chicks to dietary enzyme supplementation. J. Anim. Sci. 74: 394–402.

Zhao, G.Y., M.Y. Zhou, H.L. Zhao, X.L. Chen, B.B. Xie, X.Y. Zhang et al. 2012a. Tenderization effect of cold-adapted collagenolytic cprotease MCP-01 on beef meat at low temperature and its mechanism. Food Chem. 134: 1738–1744.

Zhao, H.L., X.L. Chen, B.B. Xie, M.Y. Zhou, X. Gao and X.Y. Zhang. 2012b. Elastolytic mechanism of an ovel M23 metalloprotease pseudo alterin from deep-sea *Pseudoalteromonas* sp. CF6-2: Cleaving not only glycyl bonds in the hydrophobic regions but also peptide bonds in the hydrophilic regions involved in crosslinking. J. Biol. Chem. 287: 39710–39720.

Zhou, J., M. Cai, T. Jiang, W. Zhou, W. Shen, X. Zhou et al. 2014. Mixed carbon source control strategy for enhancing alginate lyase production by marine *Vibrio* sp. QY102. Bioprocess Biosyst. Eng. 37: 575–584.

Zhao, X., C. Chen, L. Chen, Y. Wang and X. Geng. 2015. Genome-based studies of marine microorganisms. pp. 231–236. *In*: K.E. Nelson (ed.). Encyclopedia of Metagenomics: Genes, Genomes and Metagenomes: Basics, methods, databases and tools, Springer, New York.

Zhao, W., T. Yuan, C. Piva, E.J. Spinard, C. Schuttert, D.C. Rowley et al. 2019. The probiotic bacterium *Phaeobacter inhibens* downregulates virulence factor transcription in the shellfish pathogen, *Vibrio coralliilyticus*, by N-acyl homoserine lactone production. Appl. Environ. Microb. 85: e01545–18.

Zhu, B. and L. Ning. 2016. Purification and characterization of a new-carrageenase from the marine bacterium *Vibrio* sp. NJ-2. J. Microbiol. Biotech. 26: 255–262.

Zhu, B., F. Ni, Y. Sun, X. Zhu, H. Yin, Z. Yao et al. 2018. Insight into carrageenases: Major review of sources, category, property, purification method, structure, and applications. Crit. Rev. Biotechnol. 38: 1261–1276.

Zhu, B.W., Q. Xiong, F. Ni, Y. Sun and Z. Yao. 2018a. High-level expression and characterization of a new κ-carrageenase from marine bacterium *Pedobacter hainanensis* NJ-02. Lett. Appl. Microbiol. 66: 409–415.

Zhu, Y., L. Wu, Y. Chen, H. Ni, A. Xiao and H. Cai. 2016a. Characterization of an extracellular biofunctional alginate lyase from marine *Microbulbifer* sp. ALW1 and antioxidant activity of enzymatic hydrolysates. Microbiol. Res. 182: 49–58.

Chapter 7

Quantum Dots in Biological Sciences

Kamla Rawat[1] and *Himadri B. Bohidar*[2,3],*

Introduction

Nanoparticles are typically defined as structures having a dimension less than a 100 nm (Owens and Pools Jr. 2007). As their size is further reduced a limit is approached when their energy bandgap starts to have a strong dependence on size which then translates into size governing fluorescent emission. Typically, this is manifested in semiconductor and carbonaceous nanomaterials of size in the range of 1–10 nm. As the size becomes comparable to the de Broglie wavelength, quantum effects start dominating. Therefore, such particles are called quantum dots (QDs) due to their zero-dimensionality. From the historical perspectives the credit for inventing quantum dots, and providing a theoretical understanding of the same goes to Alexei Ekimov and Louis Brus, and Alexander Efros, respectively. The advent of QDs has revolutionized the field of imaging biomaterials and living cells in real time. This is further extended to high resolution optical sensing of pathological biomarkers at the point of care and pesticides, organic toxins, and heavy metals in food and environment (Ikanovic et al. 2007, Choi et al. 2006, Niu et al. 2018, Ma et al. 2018).

In comparison to conventional dyes and fluorophores, these quantum dots are brighter and can provide multicolor emission with excitation from a single light source (Kir'yanov et al. 2015, Hoa et al. 2012, Chen and Wu 2012, Peng and Peng 2001, Deng et al. 2005, Rogach et al. 1999). Luminescent semiconductor quantum dots (QD) has been the pivot of most research endeavours in recent times (Soheyli et al. 2019, Zhou et al. 2017). These have proven to be excellent functional materials due to their remarkable opto-electronic properties, fluorescent nature and stability

---
[1] Department of Chemistry, School of Chemical and Life Sciences, Jamia Hamdard, New Delhi, India.
[2] School of Physical Sciences, Jawaharlal Nehru University, New Delhi, India.
[3] TERI-Deakin Nanobiotechnology Center, TERI Gram, Gual Pahari, Gurgaon, India.
* Corresponding author: h.bohidar@teri.res.in; bohidarjnu@gmail.com

against photobleaching properties compared to conventional fluorophores (Maluleke et al. 2020, Jawhar et al. 2020, Yang et al. 2015, Guo et al. 2014). The change in band-gap that remarkably correlates with the size increase allows the emission of photons from the visible to deep infrared region (Gil et al. 2021, Toufanian et al. 2018, Geng et al. 2018).

More recently ternary [containing three elements] QDs have been preferred to the conventional binary [containing just two elements] QDs due to the absence of toxic heavy metals such as Cd, Se and Pb in composition. These are less toxic and environmental friendly. Besides the low toxicity, ternary QDs retain excellent optical and electronic properties such as tunable absorption band to the near infra-red region, wide Stoke shift, large photo-fluorescence lifetime, etc. (Mohan et al. 2017, Tsolekile et al. 2019). These sets of QDs have also been used as drug conjugates for effective and safe delivery of the drug to the target point. A lot of studies have been carried out on these QD/drug conjugates, especially in relation to photodynamic therapy, a treatment of diverse forms of cancers, infections and inflammations which is an alternative to the conventional surgery and chemotherapy (Hsu et al. 2011).

As indicated earlier, QDs may be classified into primary, secondary, tertiary, and quaternary nanomaterials depending on the number of elements present in their composition. Carbon and graphene QDs will belong to the primary class, while CdSe, CdS, MnSe, MoS, ZnS, ZnSe, etc., will be called secondary QDs because they contain two elements each. CuInS and AgInS will belong to the tertiary while CuInS@ZnS and AgInS@ZnS to the quaternary class. This is purely an empirical classification which is used in this review to distinguish between different QD materials. Under this definition ZnSe@ZnS, will be a tertiary QD material which is a core-shell structure as per nanoscience literature. Here, ZnS is the core and ZnSe is the shell material. Similarly, in the QD literature, the CdSe, CdS, MnSe, MoS, ZnS, ZnSe, etc., are called II–VI semiconductor nanomaterials because the elements concerned belong to group 12 and 16 of the modern periodic table [and to IIB and VI of the classical version of the periodic table]. Note that to engineer the bandgap of QDs multiple elements are selectively coordinated during the synthesis protocol, which facilitates tailoring of their fluorescence emissions, enhancement in quantum yield, elimination of blinking, and reduction in photo-bleaching, thereby making these suitable for high resolution imaging applications. In this review, we discuss recent developments in the synthesis and bioconjugation of some selected QDs with focus on how these ultrafine nanoparticles influence protein secondary structure, and enzymatic activity. Cellular uptake and cytotoxicity with their applications in cellular imaging further expands the scope of this review.

Synthesis methods

The generalized synthesis of quantum dot entails choosing a suitable precursor which can be any organic matter or an inorganic compound. In the case of carbon dots the precursor is then carbonized under controlled pyrolysis. This may be achieved in one of the following ways: (i) microwave, (ii) hydrothermal or (iii) direct heating in a furnace. The synthesis of inorganic QDs follows a structured pathway where the nucleation and growth of the nanomaterial is controlled through regulated chemical reactions. These methods have their respective pros and cons. For achieving control over the quantum dot morphology, and its size distribution the hydrothermal or microwave pyrolysis is normally preferred. The nucleation and growth of a stable QD structure critically depends on the processing temperature, duration of pyrolysis, reaction pressure, and the reaction medium. Based on extensive optimization protocols, several prescribed synthesis methods have been established in the literature which are described in this section.

(a) Primary QDs: CdSe and ZnSe

Initial QDs synthesized belong to the CdSe family of nanocrystals where often Cd was replaced by Zn. This synthesis was largely inspired from the kinetic growth method of preparation reported by Peng et al. and others (Peng et al. 2001, Yu et al. 2002, Peng et al. 2002, Qu et al. 2001).

In comparison to aforesaid method, another approach was adopted which makes it easy to handle and to reproduce QDs with better size and shape tunability where the chemicals used were less toxic, inexpensive, and relatively green (Kippeny et al. 2002, Bunge et al. 2003, Peng 2002, Tyagi et al. 2016). The synthesis protocol in finer details is described in ref. (Mir et al. 2016). In brief, a stock solution of Se precursor is prepared by dissolving 30 mg of Se in 5 mL of 1-octadecene (ODE, 90%) in a 10 mL flask clamped on a hot plate. A micro-syringe is used to draw 0.4 mL of trioctylphosphine (TOP) and added to the flask, which is stirred for about 15 min at 50°C. This Se stock solution is mixed with the different surfactant ligands (SDS, CTAB, TX-100, etc.) (0.05 mM in 1 mL of Se stock sol) in separate vials and again stirred for 1 h. This stock solution is stored at room temperature (25°C) in an airtight container.

In the next step Cd precursor is prepared by the addition of 13 mg of CdO to a 25 mL flask containing 0.6 mL of oleic acid and 10 mL of octadecene. This flask is heated to the set temperature of 200°C for size group QD1, and 230°C for size group QD2 to which 1 mL of previously prepared selenium-surfactant solution is added. The physical morphology of the products depends on reaction time and temperature in the kinetic growth model. For getting rid of unreacted precursors and by-products, samples are mixed with hexane in the extraction solvent, which is an equal volume mixture of $CHCl_3$ and $CH3OH$ in the ratio 1:1:1. This process is repeated in multiple cycles. The reacted solution now has several segregated liquid layers. Unreacted precursors and excess amines are extracted into the methanol layer, and QDs remain in the 1-octadecene (ODE)/hexanes layer. Clearly, this generated hydrophobic nanocrystals.

For the transformation from the hydrophobic to hydrophilic phase, 10 mL of oleic acid capped nanoparticle dispersion is dispersed in 25 mL of chloroform to which 2 mL of a basic methanolic solution (pH ~ 10) of MPA (3-mercaptopropionic (MPA) acid) with 2.5 mL of deionized water is added. After stirring, MPA-capped QDs are successfully transferred into the water phase. By repeated centrifugation and decantation steps, excess MPA is removed from the aqueous dispersion of QDs (Priyadarshini et al. 2022, Ding et al. 2013).

(b) MnSe

The preparation of MnSe QDs of various size with oleic acid capping performed by the hot injection method is described in ref. (Das et al. 2016). This approach offers many advantages, like narrow particle-size distribution of the product, good control of shape and higher yield. The following describes the protocol. Briefly, Se precursor is first prepared by mixing 60 mg of Se and 1 mL of trioctylphosphine (TOP) in 5 mL of 1-octadecene (ODE) (90%) and maintained at a temperature of 60°C under stirring for 30 min, and then cooled to room temperature 25°C. Solution of Mn precursor is prepared by dissolving 13 mg of $(CH_3COO)_2Mn \bullet 4H_2O$ in 0.6 mL of oleic acid (OA) and 10 mL of octadecene. 1 mL of Se precursor was injected into this hot solution. Reaction temperature and time were at 200, 240 and 280°C, and for 30 min each to complete the synthesis which yielded 3 different sizes of MnSe QDs. After that it is left to cool naturally to room temperature. For the removal of unreacted precursors and by-products, the cooled sample was mixed with equal volume of hexane in extraction solvent (an equal volume mixture of $CHCl_3$/ CH_3OH) in the ratio of 1:1:1, and then the aliquot was extracted from it. The unreacted precursors and excess OA are extracted into the methanol layer and only nanoparticles remained in the ODE/ hexane layer. All samples are isolated using methanol and excess acetone, followed by centrifugation for 10 min at 1000 xg and the aliquots are collected at the bottom (Das et al. 2016).

(c) MoS₂

Room temperature stable MoS_2 quantum dots are prepared by using ammonium tetrathiomolybdate $((NH_4)_2MoS_4)$ as a precursor which is dissolved in 50 mL of deionized water and ultrasonicated for 15 min (Pandey et al. 2020). To this solution 0.25 g of L-cysteine is added followed by ultrasonication for 10 min. This preparation is then loaded into a Teflon-lined stainless steel hydrothermal cell and placed inside an autoclave maintained at 200°C for 24 h. The obtained aliquot is purified by repeated centrifugation, followed by dialysis against water using a membrane of 10 kDa cut off.

(d) Carbon dots and conjugates

Synthesis of carbon dots (CDs) is described in finer details in ref. (Pandey et al. 2020). This method uses PEG and citric acid as the precursors, and synthesis was performed via microwave pyrolysis. This is a one-step high yield protocol.

Synthesis of Au@CD/Ag@CD Nanoconjugates. Synthesized CDs are appropriately diluted and used for the preparation of gold@-carbon dots (Au@CDs) and silver@ carbon dots (Ag@CDs). Au@CDs conjugates were prepared from $HAuCl_4$ precursor (Priyadarshini et al. 2017, Priyadarshini et al. 2020). For the synthesis of Ag@CD nanoconjugates, an aqueous solution of precursor silver nitrate is heated to boiling at 100°C to which CD solution (5.3 mg mL^{-1}) is added. A change in color of the reaction solution to reddish-brown is noticed. For optimization of the synthesis, the concentrations of CD and that of precursor $AgNO_3$ were varied so as to obtain well-dispersed and stable particles (Priyadarshini and Rawat 2017, Priyadarshini et al. 2020, Priyadarshini et al. 2022).

Secondary quantum dots: Core@Shell structures

ZnSe quantum dots were synthesized via an aqueous route by Ding et al. (2013) in a one-pot method, and at room temperature with good regulation over size and morphology (Ding et al. 2013). Glutathione (GSH) was replaced with thioglycolic acid (TGA) and thiourea (as a sulphur source) in a modified technique, might produce the same results. This method produced ZnSe QDs of mean size of 3.60 ± 0.12 nm (TEM data) (Mir et al. 2017).

For the core-shell structures, typically 20 mL of ZnSe dispersion (0.08% w/v) is loaded into a 100 mL three-neck flask to which 10 mL of mixture solution containing 0.18 mM of TGA, 0.12 mM of $Zn(OAc)_2$, and 0.12 mM of thiourea are added. The pH is adjusted to 10 by adding NaOH solution. The reaction mixture is then heated to 90°C, and monitored at different time intervals with recording of UV–vis, and fluorescence emission spectra. The reaction is terminated by cooling to room temperature, then the ZnSe@ZnS nanomaterial is purified by repeated centrifugation, and decantation in ethanol. This method produces core-shell structure of average size of 4.8 ± 0.20 nm (TEM data) (Mir et al. 2017).

Ternary quantum dots

Ternary quantum dots are optically superior nanomaterials with excellent quantum yield very well suited for biological applications (Mir et al. 2018).

(a) AgInS and AgInS @ZnS quantum dots

In a typical method, 0.02 mmol of $Ag(NO_3)$, 0.08 mmol of $In(NO_3)_3 \cdot xH_2O$, and 0.4 mmol of GSH are added to 42 mL of deionized water in a 200 mL reaction vessel. The solution turbidity changes to transparent under vigorous stirring on pH adjustment to 8.5 by adding 1.0 M NaOH. Next, a freshly prepared Na_2S solution (0.05 M) is added to this solution and then microwaved (at 160 W) for 5.30 min. The production of AgInS QDs is inferred from the appearance of a golden color of the dispersion. After cooling to room temperature, 0.1 M of $Zn(Ac)_2$ and 0.05 M of Na_2S solutions are added and further microwaved at 160 W for 5.30 min which produced AgInS @ZnS core-shell structures (Mir et al. 2018). It may be noted that AgInS and AIS refer to the same material.

(b) CuInS (CIS) and CIS@ZnS quantum dots

In a typical synthesis method, 0.02 mM of $Cu(NO_3)$, 0.08 mM of $In(NO_3)_3 xH_2O$, and 0.4 mmol of GSH is added to of 40 mL deionized water in a of 200 mL flask. The solution turbidity reduced under vigorous stirring almost to transparent when adjusted to pH 8.5 with 1.0 M NaOH followed by addition of freshly prepared Na_2S solution (0.05 M) to the reaction. CIS QDs are generated after microwaving the reaction solution for 5 min followed by cooling to room temperature. The colour of the obtained dispersion is orange. For further details refer to (Mir et al. 2018).

For the deposition of ZnS shell on the CIS QDs, the reaction is performed at a room temperature of 25°C. 0.1 M Zn(Ac)$_2$ and 0.05 M Na$_2$S solutions are added dropwise into the freshly made CIS QD dispersion. This mixture is microwaved at 100°C for 5 min to produce a fine core-shell QDs of light orange color. Following the synthesis, the aliquots are treated with an excess of ethanol followed by centrifugation at ~ 500 xg and redispersion in deionized water (Mir et al. 2018). This cycle is repeated multiple times and the final product is dispersed in deionized water/buffer for storage and future use.

Morphological and optical properties

(a) CdSe and ZnSe

Yu et al. (2003) reported the synthesis of hydrophilic CdTe QDs (Yu et al. 2003) and the same prepared with five different types of surfactant coatings: OA (oleic acid), CTAB (Cyltrimethylammonium bromide), DTAB (Dodecyltrimethylammonium bromide), SDS (Sodium dodecyl sulfate) and TX-100(Triton X-100) was also explored (Das et al. 2016).

TEM probed the dried particles while the DLS measured the size of the hydrated structures that were prone to clustering (Fig. 1). TEM data allowed observation of large changes in the structure of quantum dots with temperature for a given reaction time.

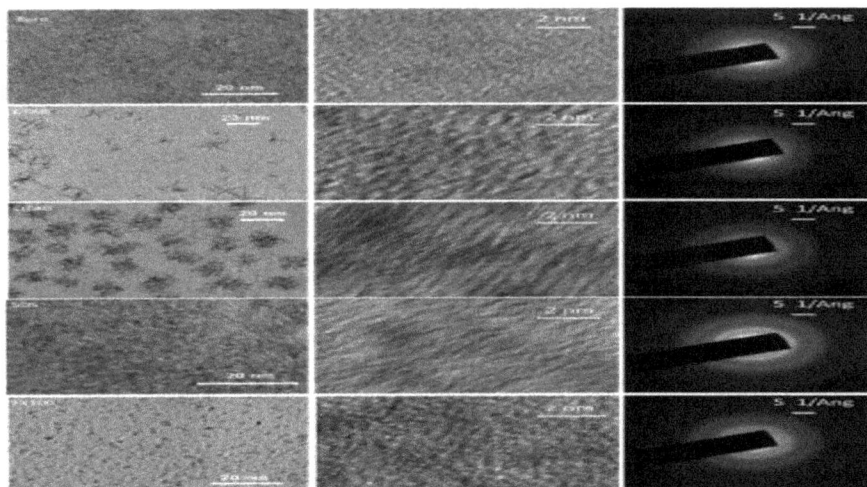

Figure 1. TEM, HRTEM images and SAED pattern of CdSe QDs of first size group (core = 2.5 nm) without and with different surfactant capping. (Reproduced from Das et al. 2016 with permission from Elsevier).

The hydrodynamic radius and zeta potential of these QDs are shown in Fig. 2, which reveals that the higher surface charge was observed for bare particles given by −64 and −52 mV for 3.5 and 2.5 nm core size QDs respectively. Surface capping did not change the charge polarity of these QDs.

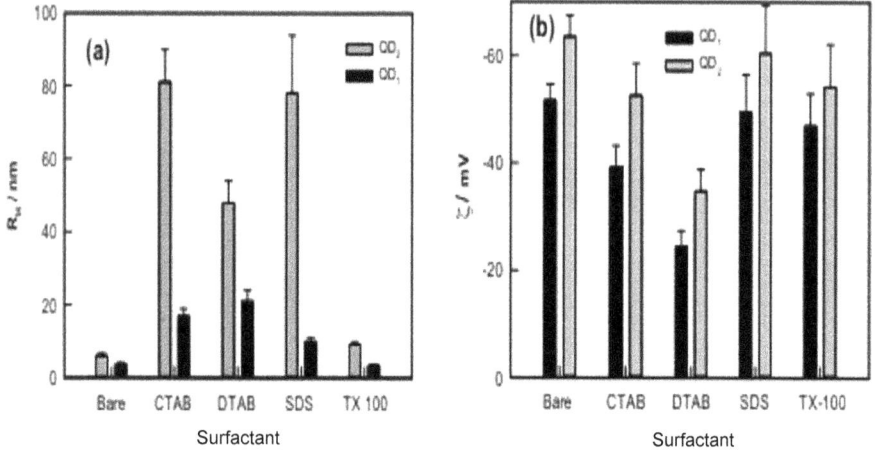

Figure 2. (a) Average hydrodynamic radius R_H and (b) zeta potential ζ of QDs capped with different ligands. Two different samples of CdSe (QD_1 and QD_2) are compared in these diagrams. (Reproduced from Das et al. 2016 with permission from Elsevier).

The empirical evaluation of QD diameter D (in nm) is possible from the first absorption maxima data given by (Jasieniak et al. 2009)

$$D = 59.60816 - 0.54736\lambda + 1.8873 \times 10^{-3}\lambda^2 - 2.85743 \times 10^{-6}\lambda^3 + 1.62974 \times 10^{-9}\lambda^4 \ (1)$$

where λ is the wavelength of the first excitonic absorption peak of the sample under probe.

Figures 3 and 4 depict the absorbance, and the fluorescence spectra of CdSe quantum dots synthesized with t surfactant capping. Any change in the spectra is

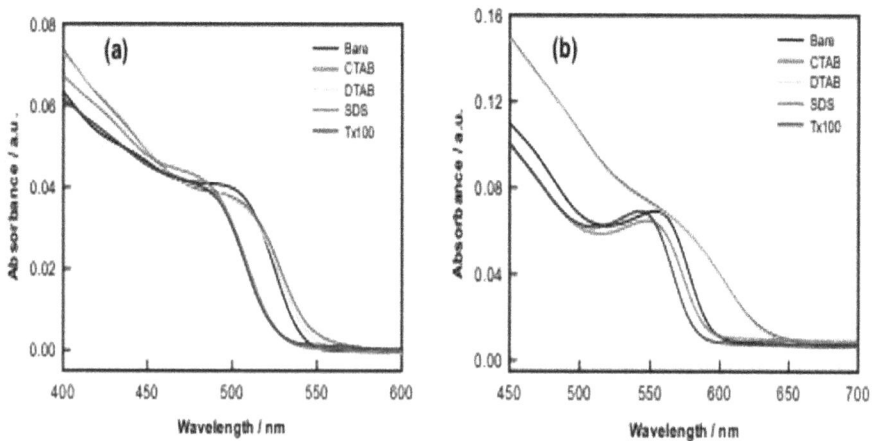

Figure 3. Absorbance spectra of quantum dots of core diameter (a) 2.5 nm and (b) 3.5 nm capped with four different surfactants. Reproduced from Das et al. 2016 with permission from Elsevier.

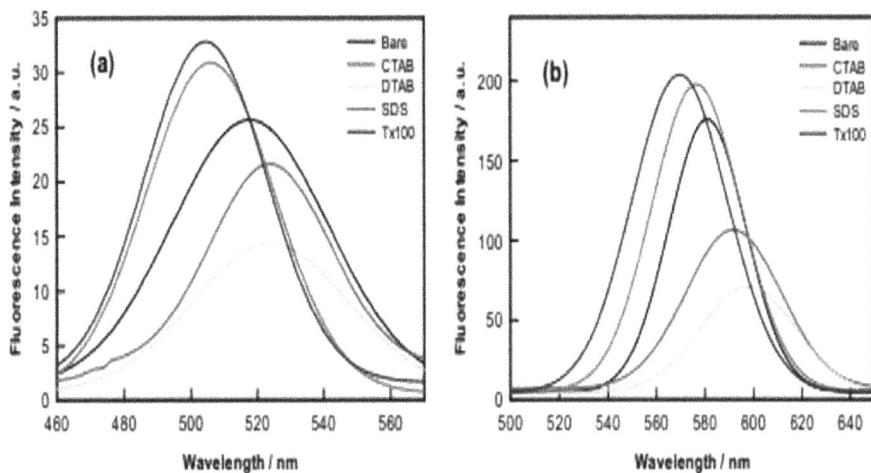

Figure 4. Comparison of Photoluminescence (PL) em1ss1on measurements for CdSe quantum dots with core diameter (a) 2.5 nm and (b) 3.5 nm with different surface ligands. (Reproduced from Das et al. 2016 with permission from Elsevier).

attributed to the choice of capping ligand, and the consequent alteration of surface defects.

The functionalization with surfactants resulted in a substantial change in the PL emission intensity relative to the OA sample. The highest enhancement in fluorescence was seen in TX-100 functionalized then in SDS functionalized QDs, whereas CTAB and DTAB surfactants caused loss in the PL intensity compared to OA capped samples.

(b) MnSe quantum dots

Monodisperse and spherical MnSe nanocrystals of different size with fluorescent and magnetic properties were successfully synthesized by hot injection method (Das et al. 2018). TEM images shown in Fig. 5 reveal spherical particles of mean size of 7, 12 and 16 nm for samples designated as a, b and c, respectively made under different temperature conditions. High resolution TEM images showed crystal lattice fringes with d-spacing of about 0.276 nm that refers to the dominant [200] crystallographic planes of the cubic lattice of MnSe (Fig. 5d). Particle size histograms are shown in the insets.

Figure 6 shows the absorbance, and the fluorescence spectra of samples prepared at different reaction temperatures and capped with oleic acid (OA). Note that the prominent first absorption edge and fluorescent peak are red shifted with particle size from 311 to 344 nm for absorption, and 411 to 432 nm (blue light) for emission. It was concluded that with proper surface passivation it is possible to get well defined emission peaks in the blue light region which create applications in short wavelength optoelectronics and magneto-optical devices (Sarma et al. 2014).

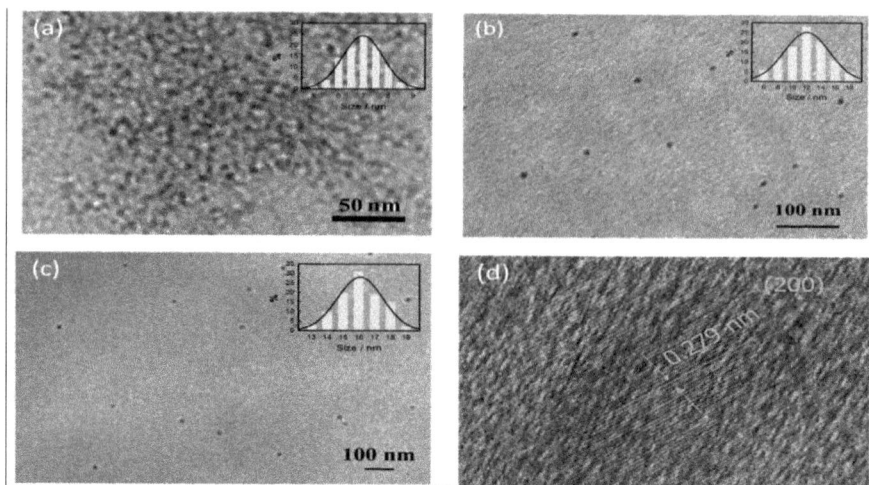

Figure 5. TEM images and size-distribution histogram (inset) of MnSe QDs: (a) 7, (b) 12 and (c) 16 nm made at different temperatures. (d) HRTEM image of size 16 nm sample. Reproduced from Das et al. 2018 with permission from IOP Publishing Ltd, UK.

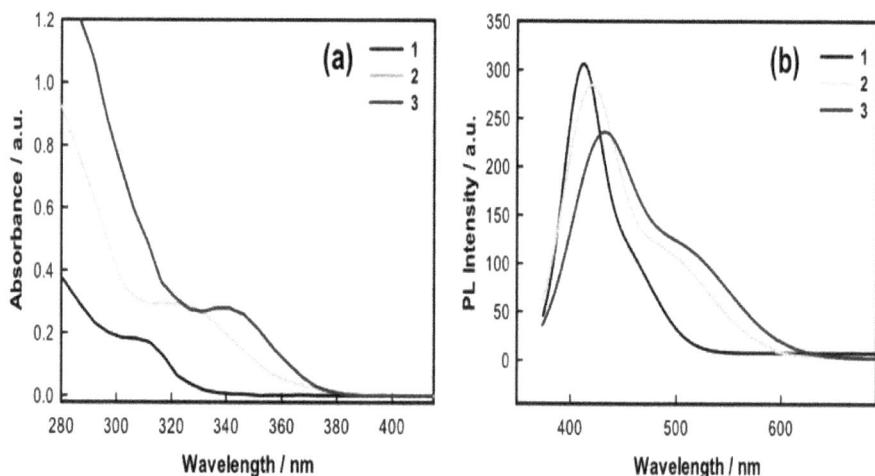

Figure 6. (a) Absorbance and (b) emission spectra (λ_{max} = 300 nm) of MnSe QDs: (1) 7, (2) 12 and (3) 16 nm, synthesized at different temperature. Reproduced from Das et al. 2018 with permission from IOP Publishing Ltd, UK.

(c) MoS₂ quantum dots

The mean size of quantum dots estimated from TEM analysis was 3.1 nm as clearly seen from Fig. 7, and the inset shows a size histogram (Pandey et al. 2020). The crystalline profile obtained from the XRD pattern is depicted in Fig. 7c consistent HRTEM data (Fig. 7d) image shows diffraction fringes (Fig. 7e) that infer the crystal plane spacing information, which was ~ 0.19 nm. The X-ray diffraction pattern

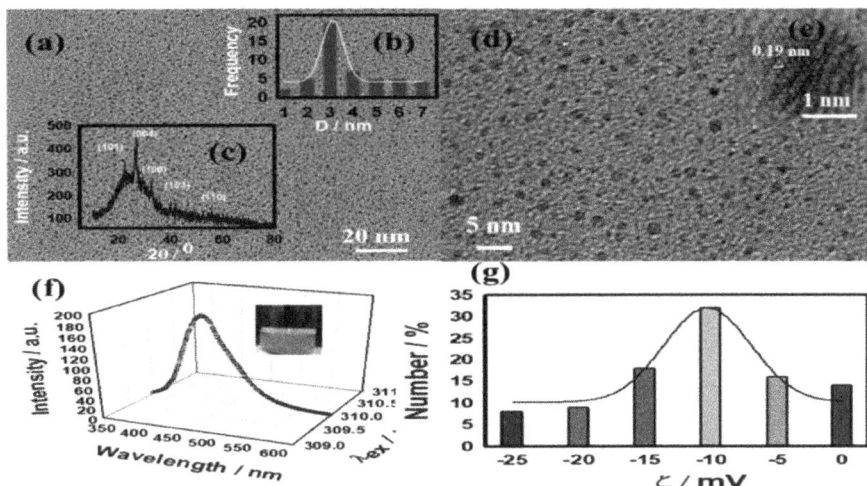

Figure 7. (a) TEM image on the 20 nm scale, (b) size distribution histogram, (c) XRD data of synthesized MoS_2 QDs, (d) TEM image on the 5 nm scale, (e) HRTEM image, (f) fluorescence at 400 nm under 310 nm UV excitation, and (g) zeta potential distribution. Adapted with permission from Pandey et al. 2020, American Chemical Society.

shows clear diffraction peaks assigned to the (002), (101), (004), (100), (103), and (110) planes of the crystalline structure (Fig. 7c). Debye-Scherrer formula (Patterson 1939) applied to the XRD pattern provides the average crystallite size of ~ 7.4 nm. The surface charge of these determined from the zeta potential analysis was –10 mV (Fig. 7g). For the excitation at the absorption edge wavelength (310 nm), the emission peak was found at 400 nm. The optical band gap energy calculated with the Tauc equation which was 4.25 eV (Tauc et al. 1966).

(d) Carbon dots and organo-metallic nanoconjugates

Carbon dots (CDs) are a family of carbonaceous materials easily synthesized by pyrolytic carbonization of a variety of precursors (Priyadarshini and Rawat 2017, Priyadarshini et al. 2020). A strong surface plasmon resonance peak is observed close to 530 and 420 nm for Au@CDs and Ag@CDs, respectively. Quenching of the fluorescence intensity confirms the formation of nanoconjugates (Priyadarshini et al. 2022). The crystalline nature of the particles was confirmed by XRD analysis. Table 1 provides a summary of the physical parameters of the CDs and their nanoconjugates.

Table 1. Physical parameters of the synthesized nanoconjugates. Reproduced from Priyadarshini et al. (2022).

Sample	Maximum absorbance (nm)	Maximum emission (nm)	DLS Rh (nm)
CDs	305	454	13 ± 1
Au@CDs	520	454	47 ± 2
Ag@CDs	415	454	65 ± 2

Dynamic light scattering determined hydrodynamic radius Rh normally refers to the hydrated particle size.

It is worth noting that the PL emission remained fixed at 454 nm regardless of the fact that the CD and its organo-metallic conjugates had absorbance maxima at characteristically different wavelengths. Further, the DLS estimated size refer to the solvated, and possibly aggregated nanostructures.

Core-Shell (ZnSe@ZnS) structures

The physical properties of the ZnSe@ZnS core-shell structures are depicted in Fig. 8. The ZnSe QDs had a mean TEM size of ~ 3.6 nm, and zeta potential of –20 mV which changed to ~ 4.8 nm and –38 mV on shell capping indicating clear change in size and shape characteristics. The DLS determined size was much larger as expected. Size histogram revealed similar polydispersity for ZnSe and ZnSe@ZnS samples (Mir et al. 2017).

Complete particle size information is given in Fig. 9 based on the TEM data. One clearly sees ZnSe QDs having a size of 3.6 nm and for the ZnSe@ZnS, it was 4.8 nm with some amount of polydispersity. The HRTEM determined crystallite size was 0. 245 nm for the core and 0.225 nm for the core-shell structure (Mir et al. 2017).

The absorption edge for the samples were observed at ~ 375 nm, and the PL emission was found at 490 and 510 nm respectively (Fig. 10). Remarkably, the absorbance decreased while the PL intensity increased when a shell was put on the ZnSe core (Mir et al. 2018).

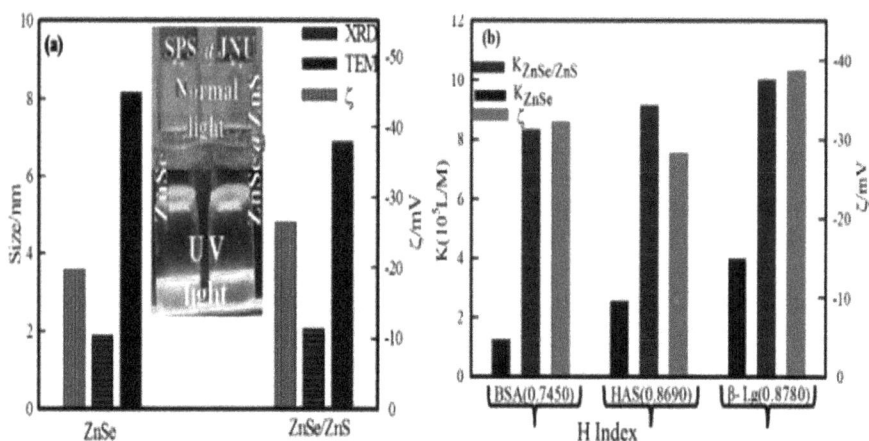

Figure 8. (a) Plot of TEM, and XRD, and surface charge of ZnSe and ZnSe@ZnS quantum dots, inset photo is visual under normal light and UV illumination. (b) Plot of binding constant and surface charge of proteins with Hydrophobicity-index (H-index) of proteins (BSA, HSA and β-Lg). This will be discussed later. Reproduced from Mir et al. 2017 with permission from Elsevier.

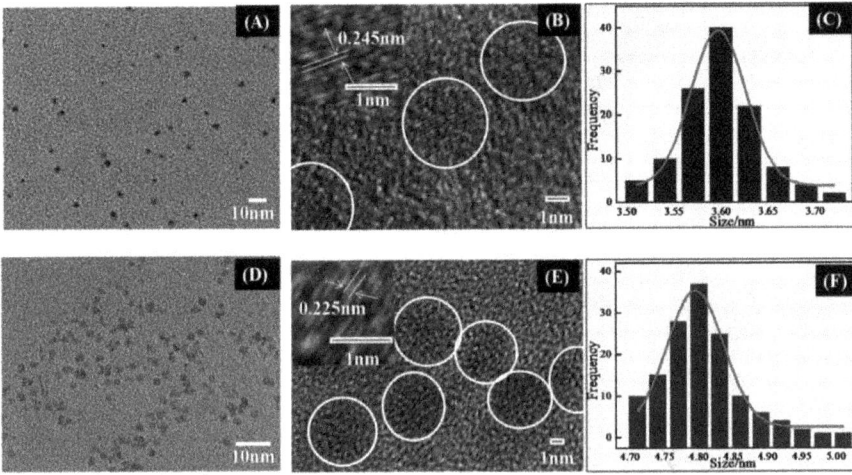

Figure 9. TEM and HRTEM images (A and B) of the ZnSe core, (D and E) the as-prepared ZnSe@ZnS core-shell QDs. The average size for ZnSe was 3.6 ± 0.12 nm, and for ZnSe@ZnS it was 4.8 ± 0.20 nm, shown in the histograms (C and F), respectively. Note that the crystallite size was 0.245 and 0.225 nm, (B) and (E) for ZnSe and ZnSe@ZnS. Reproduced from Mir et al. 2017 with permission from Elsevier.

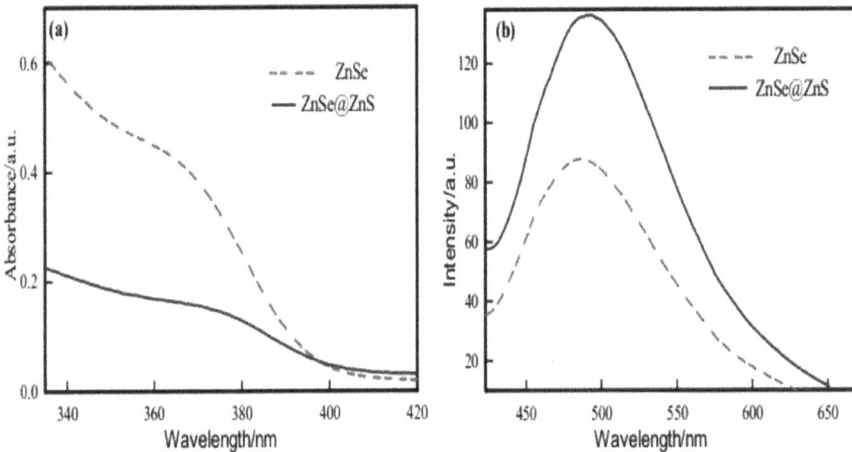

Figure 10. Comparison of (a) absorption, and (b) fluorescence emission spectra of the ZnSe and ZnSe@ ZnS samples. Notice the decrease in absorbance and enhancement in fluorescence intensity due to the presence of shell. Reproduced from Mir et al. 2017 with permission from Elsevier.

Quaternary quantum dots

(a) CuInS@ZnS

The hydrophilic CIS core-only, and CIS@ZnS core–shell dots in normal and under a 365 nm UV illumination show good fluorescence emission (Mir et al. 2018). These

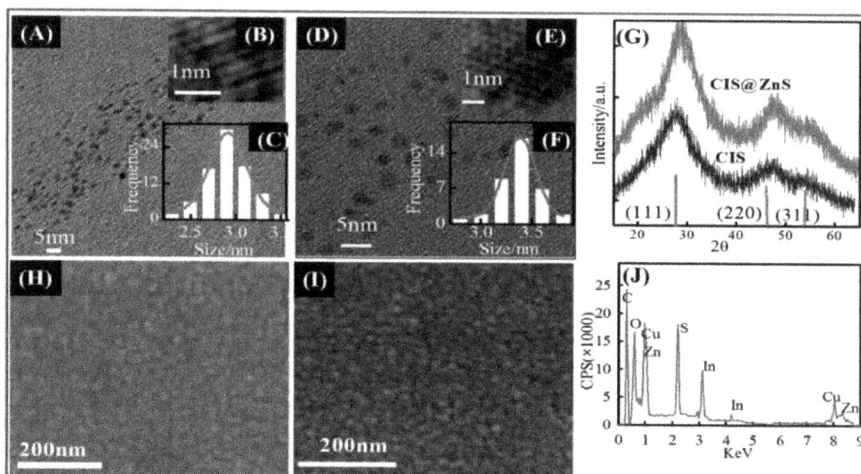

Figure 11. A and D are TEM, inset, B and E are HRTEM images of CIS and CIS@ZnS core-shell QDs, and C and F are size histograms. G is XRD profile of these QDs. H and I are FESEM images of CIS CIS@ZnS core shell QDs respectively. (J) EDX spectra of CIS and CIS@ZnS core-shell QDs. Reproduced from Mir et al. 2018. RSC Adv. 8: 30589.

monodisperse CIS and CIS@ZnS quantum dots had a typical size of 2.9 and 3.5 nm respectively (Fig. 11). American Chemical Society monograph on Green Chemistry mandates that methods must require minimum energy for a given reaction that must occur in a green solvent. Thus, for the synthesis to be designated green use of safer solvents, and minimum time and energy consumption are a definite requirement (Ryan and Tinnes 2002, Dallinger and Kappe 2007). Synthesis of these nanomaterials using water as the reaction medium and microwave heating used for less than 5 min for thermal activation has been reported (Tang et al. 2012). It has been possible to make highly stable quaternary quantum dots under facile synthesis conditions. Figure 11 elucidates the complete morphology of CIS and CIS@ZnS quantum dots determined using TEM and XRD techniques (Mir et al. 2018).

Figure 11 depicts the HRTEM images of CIS and CIS@ZnS quantum dots. These reveal a narrow particle size distribution and with a mean size of about 2.9 nm for core as against the core-shell size of 3.5 nm. The XRD diffraction in peaks (Fig. 11G) indicate these have a small crystallite size. Three major peaks of CIS QDs could be indexed to (112), (024), and (116) lattice planes of the tetragonal crystal structure for the CIS nanocrystals (JCPDS no. 47-1372). It is pertinent to note that the core-shell QDs exhibited almost similar diffraction peaks, but peaks were shifted to higher angular values. This implied that a fraction of Zn^{2+} ions diffused into the crystal structure of CIS, widening and distorting it. FESEM images shown in Fig. 11(H and I) illustrate composition. Zeta potential (ζ) values were found to be –32 and –40 mV for core and core-shell structures, respectively. The higher negative value (> –30 mv) of zeta potential attributes higher stability.

(b) AIS and AIS@ZnS quantum dots

AIS based core-shell QDs exhibit very high luminescence that could be visualized directly under ultraviolet lamp. Figure 12A shows the AIS QDs with the mean size of ~ 2.5 nm, and clear lattice fringes, which after the ZnS shell formation increased to 3.25 nm (Fig. 12E). HRTEM (Figs. 12B and F) images of AIS and AIS@ZnS QDs assigned values of 0.22 and 0.23 nm to their corresponding crystallite size. Uniform spherical structure of these quantum dots and their size histogram is inferred from Fig. 15B. The core-only QDs show a negative zeta potential of about –40 mV (Fig. 12J) while the core-shell QDs have a marginally higher value of the same polarity.

The crystalline morphology of the QDs determined from the X-ray diffraction (XRD) studies reveal three broad peaks located at $2\theta \approx 27.2°$, 45.4° and 54.3° representing (112), (204) and (312) diffraction planes, respectively (JCPDS NO32–0483) (Fig. 12D). Close to similar results were observed from the AIS@ZnS samples, but with the peaks marginally shifting towards higher 2θ values (Dallinger et al. 2007). A certain amount of Zn^{2+} ion diffusion into the AIS lattice was not ruled out here. XRD profiles and TEM images of the samples made with different Ag:In ratios which indicate the crystal structure and particle size remained invariant of Ag:In ratio (Mir et al. 2018).

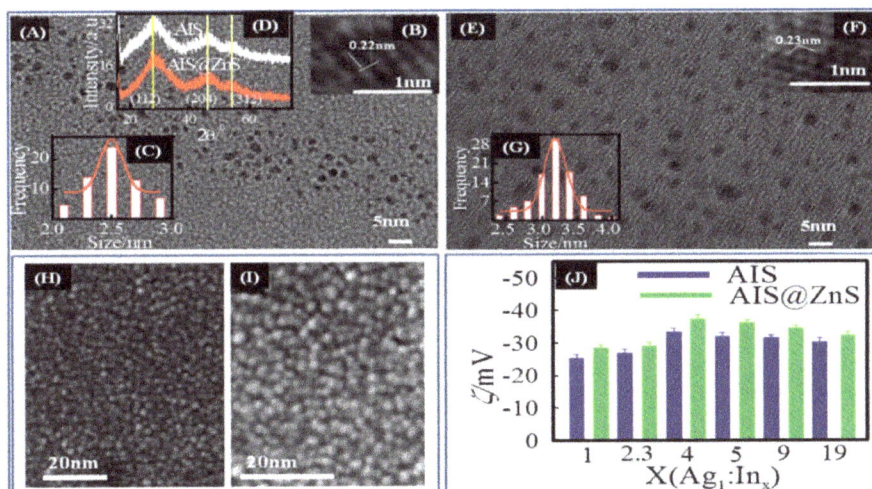

Figure 12. (A) TEM, (B) HRTEM images and (C) size histogram of AIS core only. (D) XRD of AIS core and AIS@ZnS core-shell. (E) TEM, (F) HRTEM images and (G) size histogram of core-shell QDs. These images and XRD are for 1:4 {Ag:In} ratio samples. (H and I) are FESEM images of AIS and AIS@ZnS QDs, respectively. (J) Zeta potential distribution plot of core and core-shell quantum dots with different ratio of Ag:In. Reproduced from Mir, I.A., V.S. Radhakrishanan, K. Rawat, T. Prasad and H.B. Bohidar. 2018. Sci. Rep. 8: 9322.

Applications in Biology

(a) Interaction of CdSe with BSA

Protein-QD interactions often cause conformational changes in the native state of the protein at the secondary structure level. Since BSA is a good model protein its binding to the CdSe QDs was explored (Das et al. 2016). The circular dichroism (CD) spectra provide direct information on the secondary structure changes. This data for BSA-QD complexes reveals signature peaks (two negative doublet peaks) arising from α-helix content (of proteins) that become shallower. Hence, it was concluded that the helix content of BSA continuously reduced inferring disruption of the protein secondary structure (Fig. 13). Several other studies have reported similar observations (Tang et al. 2012, Shang et al. 2001, Cui et al. 2003, Cheng et al. 2009).

It is necessary to note the size and coating-dependent differential binding for quantum dots of size 2.5 and 3.5 nm to BSA, which has an apparent hydrodynamic size of 3.5 nm. The zeta potential of both type of QDs was close to within 10% for any given surfactant (Fig. 2). The differential binding must be decided by the protein surface charge anisotropy, different ligand surface area, size and ionic nature of coating present on the QD surface. BSA when dissolved in water has a pH ~ 5.5, hence surface charge is approximately −10 mV (Rawat and Bohidar 2012). Hence, the quantum dots having less negative surface charge density would permit the proteins to adsorb more onto their surfaces which would again be biased towards bigger size and DTAB capped QDs. Note that the similar polarity binding is governed by the surface patch binding mechanism which in turn is facilitated by anisotropic protein surface charge distribution. This is a very unique binding protocol normally seen in proteins which is elaborately discussed in ref. (Pathak et al. 2017). Intuitively,

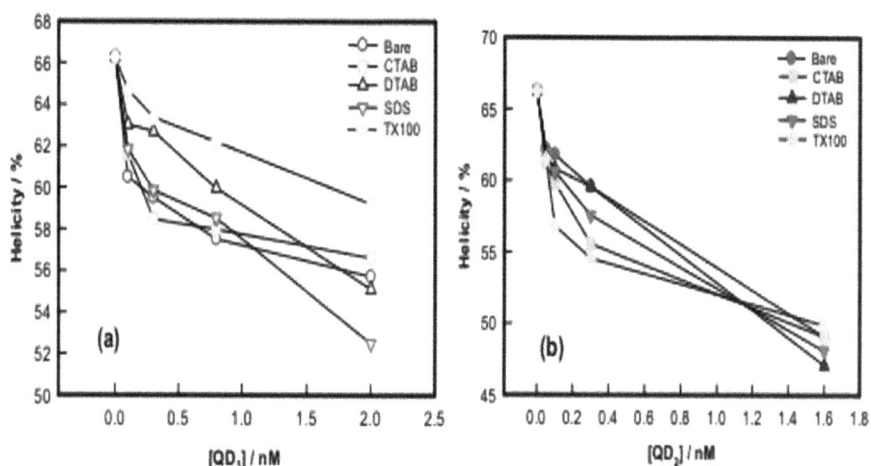

Figure 13. Variation of secondary structure (helicity) of proteins BSA on the concentration of (a) QD1 (2.5 nm) and (b) QD2 (3.5 nm) coated with different surfactants. Note the drastic loss in the secondary structure for bigger size quantum dots (3.5 nm). Reproduced from Das et al. 2016 with permission from Elsevier.

it may be argued that an increase in favorable interaction causes the water molecules bound to BSA to redistribute, and give protein a hydrophobic environment which would destroy the secondary structure of the protein.

Based upon CD spectra of BSA and those of its quantum dot bound complexes it can be concluded that the helix content of BSA continuously decreased inferring strong associative interaction with quantum dots of both size (Shang et al. 2007, Cui et al. 2003, Cheng et al. 2009).

(b) QD-DNA Hydrogel

Synthesis and formation of biocompatible fluorescent hydrogels is relevant to biology due to its application potential. Remarkably, this was achieved in a DNA-quantum dot system (Pandey et al. 2020). The mean zeta potential of MoS_2 quantum dots and DNA were measured to be -10 and -60 mV, respectively. In an aqueous sol of DNA strands, colloidal MoS_2 QDs were dispersed and interestingly gelation ensued. Interestingly, though both entities had similar charge polarity, a self-assembled network of interconnected and hierarchical porous structures could be observed for these hydrogels implying that these QDs assisted in the evolution of the gel structure from a solution that would otherwise fail to gel (Fig. 14). Notice their differential microstructures.

Because of non-existent crosslinks, DNA strands cannot assemble to form a network gel, but in the presence of MoS_2 quantum dots gelation did occur. It is pertinent to discuss in what capacity these QDs could play the role of pseudo-crosslinks between predominantly negatively charged DNA strands. The science of Wigner glasses and gels, where the self-organization in a repulsive electrostatic environment was reported long ago (Bonn et al. 1999, Angelini et al. 2014).

The DNA-MoS_2 system is similar to the colloidal dispersion case discussed in the Wigner glass except that there is a small content of negatively charged colloidal particles co-dispersed in a heterogeneously distributed large DNA strands of the same polarity. Logically, these QDs would locate themselves in pockets where there

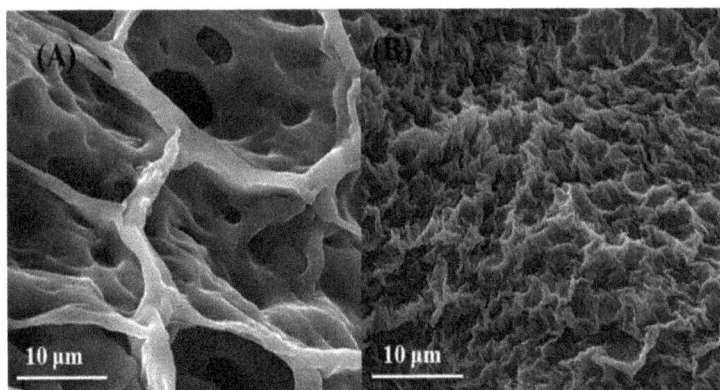

Figure 14. Morphology of the (A) DNA hydrogel, and (B) DNA-MoS_2 hydrogel obtained from FESEM. Adapted from Pandey et al. 2020 with permission from American Chemical Society.

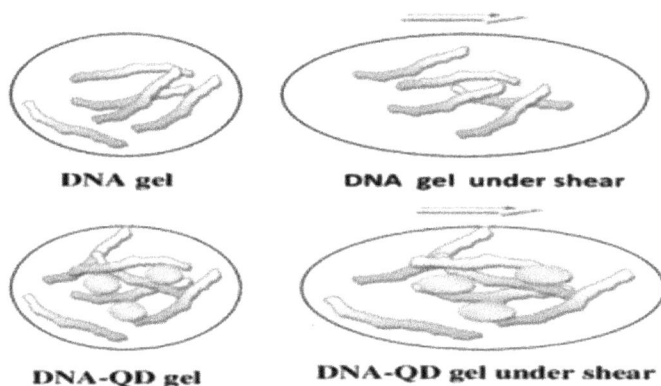

Figure 15. DNA hydrogel Microstructure (Top) and the same in the presence of MoS$_2$ QD (Bottom). Adapted from Pandey et al. 2020 with permission from American Chemical Society.

is minimum repulsion field. When a given number of colloidal particles are added to this dispersion, then these will occupy the free volumes available between the randomly distributed rods in the medium and, reduce the overall free-volume in the sample. However, the overall environment is still repulsive, hence, it is reasonable to describe this hydrogel as a Wigner-like gel. This is depicted schematically in Fig. 15. Such fluorescent hydrogels will find applications in a variety of biological applications like real time imaging of cells, tissues, targeting and internalization of drugs to specific sites, etc.

(c) Cellular uptake and toxicity of carbon dots and conjugates

It is worth realizing that carbon happens to be one of the organic materials that can be converted into a quantum dot like structure. The anti-proliferative effect of carbon dot nanoconjugates was studied against HeLa cells with concentration of conjugates varying in the range of 25–200 µg/ mL (Priyadarshini et al. 2022). Ag@CDs showed the highest inhibition of HeLa cells with an IC50 value of 50 µg/ mL, while bare CDs gave the least toxicity with IC50 value of 180 µg/mL, for the Au@CDs samples it was 150 µg/mL. In the literature, the toxicity of silver NPs with an almost 60% decrease in cell viability at a concentration of 65 µg/mL in L929 cells is reported. In the present case, the highest inactivation of cell proliferation was observed in Ag@CDs, which can be due to the Ag+ ions forming the particle core. Similarly, gold NPs are reported to inhibit the proliferation of dalton lymphoma cells with 40–50% viability in the concentration range of 80–100 µl. The low toxicity of Au@CDs observed might be due to the CD shell over the gold particles that renders these nontoxic.

Clear changes in the morphology as well as in the cell density was noticed compared to the control (Fig. 17). While the control set (untreated HeLa cells) showed intact morphology, the treated cells showed disrupted cell organization and shrinkage. The cells, post shrinking were round in shape. After cell death and those under stress normally appear round. Marked reductions in the number of surviving cells suggested the high toxicity and induction of apoptosis and necrosis at

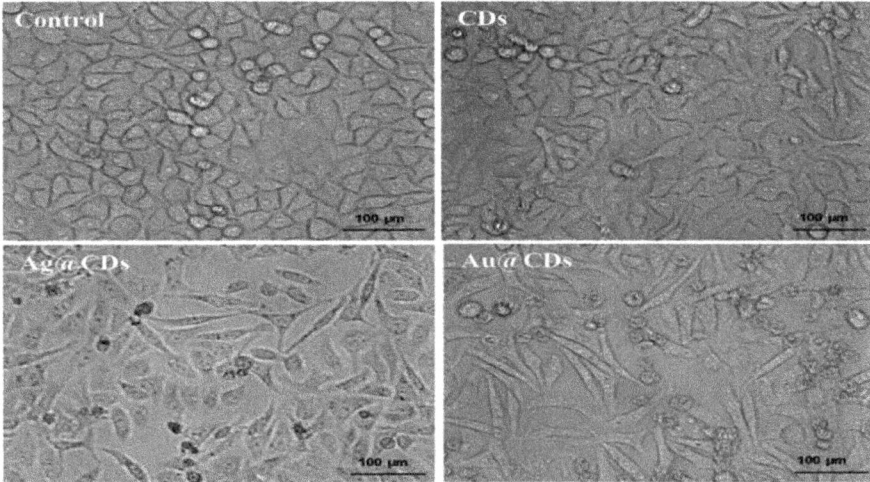

Figure 16. Morphological changes in Hela cells after treatment with the nanoconjugates (100 µg/ml). No morphological alterations were found in cells treated with CDs and control set. Images were captured at 20x magnification. Adapted from Priyadarshini et al 2022.

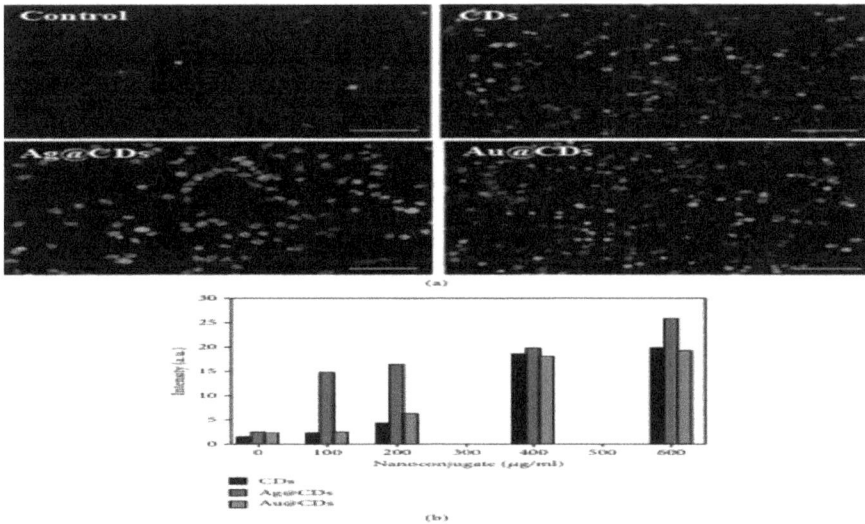

Figure 17. (a) Fluorescence microscope images of DCFDA-stained He La cells including control cells, CD-treated cells, Ag@CD-treated cells, and Au@CD-treated cells. (b) ROS level in treated cells after incubation with the nanoconjugates. Adapted from Priyadarshini et al 2022.

100 µg/mL of Ag@CD conjugates. However, cells treated with CDs had morphology same as of the control cells suggested the nontoxicity of CDs. Likewise, the cells treated with Au@CDs showed few round cells with intact morphology signifying their lesser toxicity in comparison to silver conjugates. The respective TEM images are depicted in Fig. 18.

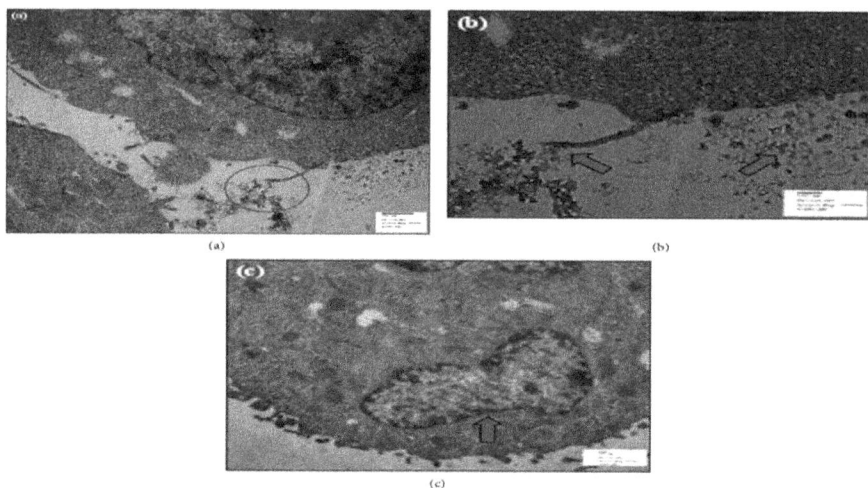

Figure 18. TEM images of Hela cells treated with carbon dots. (a) and (b) correspond to different magnifications depicting the interaction with the cell microvilli. (c) represents the internalization in the cytoplasmic vesicles. Adapted from Priyadarshini et al (2022).

Studies suggest that the induction of toxicity is generally mediated by apoptosis, mitochondrial damage, metabolic inactivity, and oxidative stress (Li et al. 2010, Clark et al. 2012, Kim et al. 2010, Pan et al. 2009, Lakowicz 2006). These processes are known to be assisted by the production of ROS. A fluorescent dye DCFDA was used for analysis of ROS generation, where a direct correlation was found between the ROS amount, and the green fluorescence intensity (Priyadarshini et al. 2022). There was no observation of any fluorescence in the control experiments, while for the CD treated HeLa cells, a weak and diffused green fluorescence was detected. The HeLa cells treated with Ag@CDs showed a high intensity of green fluorescence, and a good reduction in the number of cells, which suggested cell death due to high toxicity. ROS analysis inferred that the Au@CDs were less toxic compared to Ag@CDs, signified by lower fluorescence intensity (Fig. 19(a)). Similar analysis of ROS showed relatively high intensity of DCFDA in Ag@CD treated cells compared to CDs and Au@CD treated samples (Fig. 19(b)). The potential of the nanoconjugates in live cell imaging studies and other biomedical applications was ascertained from the uptake of CDs and its nanoconjugates by treating the cells with 50 μg/mL concentrations, which was almost half the concentration of toxicity limit. Figure 19 shows the confocal imaging data indicating the cellular uptake of the nanoconjugates (internalization). The bare cells were taken as control. The data suggested that the scattering intensity was the highest for Au@CD-treated samples. Interestingly, all the three particles could be internalized in the HeLa cells. The fluorescence intensity followed the trend Au@CD > Ag@CD > CD.

These quantum dots are easily internalized and get uniformly distributed inside the cell as is revealed in Fig. 19. Therefore, these CDs and its conjugates possess the potential to be used as superior fluorescent probes for biomedical and theragnostic applications.

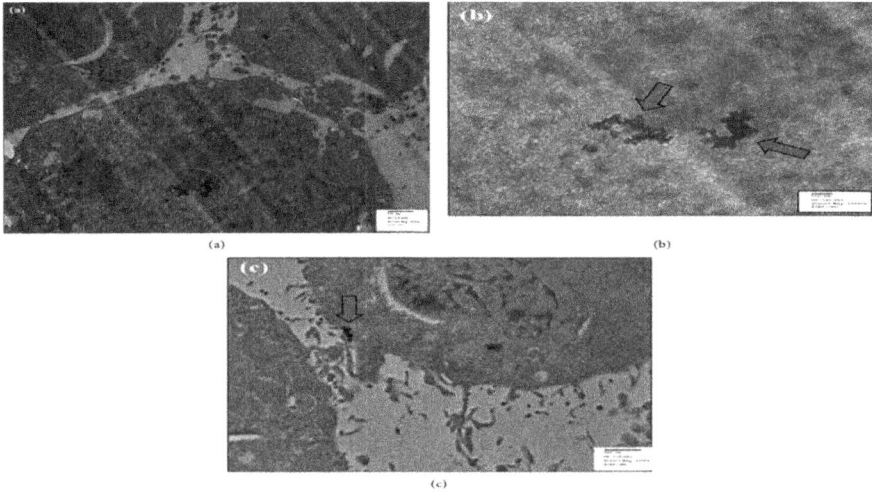

Figure 19. TEM images of Hela cells treated with Au@CDs. (a) shows the localization within the cells (arrows). (b) is the magnified image of the same cell. (c) depicts the interaction of the Au@CDs at the cell surface. Adapted from Priyadarshini et al (2022).

(d) Interaction of proteins with ZnSe@ZnS quantum dots

The structural morphology of core-shell quantum dots is very different from that of their core-only counterparts. In order to map this a systematic study of the interaction between the core-shell QDs and a few serum proteins like BSA, HSA, and β-Lg was explored (Mir et al. 2017). The observed results may be discussed in terms of specific interaction between these QDs and plasma proteins through protein-QD complex formation. Both the absorbance and emission properties of the protein-QD complexes are distinctively different from their individual components. This is clearly illustrated in Fig. 20 for the particular case of BSA. Two observations are made here: (i) absorbance increases, and (ii) fluorescence intensity gets quenched with QD content for a fixed protein concentration.

The observed interaction can be visualized as flows. The equilibrium binding between protein (P) and quantum dot (Q_d) may be expressed as

$$P + Q_d \leftrightarrow P . . Q_d \tag{2}$$

$$K_{app} = \frac{[P...Q_d]}{[P][Q_d]} \tag{3}$$

The apparent association constant (K_{app}) can be given by (Lakowicz 2006),

$$\frac{1}{(A - A_0)} = \frac{1}{(A_C - A_0)} + \left(\frac{1}{K_{app}}\right)\left(\frac{1}{Q_d}\right)\frac{1}{(A_C - A_0)} \tag{4}$$

The absorbance in the absence and presence of QDs is represented as A_0 and A. A plot of $1/(A - A_0)$ versus $1/[Q_d]$ is linear with a slope of

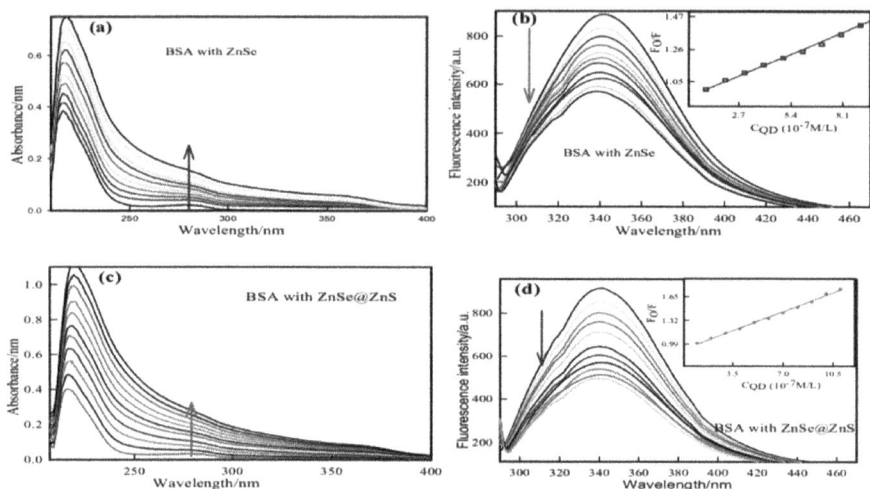

Figure 20. (a) Absorbance and (b) Fluorescence spectra of BSA-ZnSe complex at pH = 7.0. BSA concentration was 0.4 M and ZnSe QD concentration was varied from 0–10 M in steps of one unit. (c) and (d) show similar data for BSA-ZnSe@ZnS complexes. Reproduced from Mir et al. (2017) with permission from Elsevier.

$(1/K_{ap})(1/(A_c - A_0))$ and intercept of $1/(A_c - A_0)$. The slope to intercept ratio will yield the binding parameter K_{app} and the results obtained indicated a binding trend given by BSA > HSA > β-Lg. The fluorescence emission spectra can be similarly analyzed to get a linear dependence plot between the reciprocal of fluorescence intensity versus QD concentration.

The protein-QD binding, which was primarily electrostatic, had no strong impact on the protein secondary structure, and hence, on their bioactivity (Moriyama and Takeda 2005, Parker and Song 1992). Overall, it was seen that the QDs quenched the fluorescence emission of the samples in a size and structure dependent manner. Thus, it is concluded that these core-shell structures may be used in biological applications including cell and tissue imaging.

Conclusion and future prospects

The breakthroughs achieved in the synthesis methods to generate designed nanomaterials with tailored properties has provided sufficient impetus to the field of nanotechnology for growth. The pluralistic application of these in medicine, agriculture, paint and pulp industry, opto-electronics, personal care products, and in manufacturing has proved the wide acceptance of these materials. Within the family of nanomaterials, quantum dots occupy a special place due to their remarkably dominant spectroscopic signatures, which remains to be exploited to its full potential. Clearly, this new class of fluorophores which bear high optical quantum yield need to be exploited more in solving problems related to bio-imaging, sensor designing, smart farming and diagnostics at the point of care.

For instance, the targeted delivery of drugs to the effected tissues would require real time monitoring of the drug trajectory inside the patient. In principle, a drug molecule tagged with a suitable quantum dot will adequately meet this requirement, which has been proved in model cell experiments. Organic quantum dots have proven successful as bio-stimulants in promoting growth of plants through enhancement of photosynthesis activity. Chloroplast photosynthesis normally uses the blue and red region of the optical spectrum to convert light into chemical energy. This can be augmented by internalization of quantum dots in the plant leaves. Many quantum dots absorb light in the UV region and emit in the blue, the UV component of solar radiation, which is typically ~ 8–10 % of total energy, may be thus harvested and used in photosynthesis enhancement.

Quantum dots can be functionalized to selectively adsorb heavy metal ions present in the industrial effluents and contaminated ground water. The QD-metal ion complex normally leads to the change in the photoluminescence intensity, which may be used as an indicator for quantitative detection and mitigation of this class of pollutants. Such functionalization may be altered to capture, detect, and monitor the organic pollutants like phenols, dioxane, dyes, pesticides, etc., present in the environment. The hybrid quantum dots have shown promise as efficient light emitting diodes, multiplexers, optical gates, and relays, highspeed switches, etc. Thus, the possibilities are limitless.

References

Angelini, R., M. Sztucki and B. Ruzicka. 2014. Glass-glass transition during aging of a colloidal clay. Nat. Commun. 5(4049).

Bonn, D., H. Tanaka, G. Wegdam and H. Kellay. 1999. Aging of a colloidal and quot; Wigner and quot; glass. Europhys. Lett. 45: 52–57.

Bunge, S.D., K.M. Krueger, T.J. Boyle, M.A. Rodriguez, T.J. Headly and V.L. Colvin. 2003. Growth and morphology of cadmium chalcogenides: The synthesis of nanorods, tetrapods, and spheres from CdO and Cd (O2CCH3)2. J. Mater. Chem. 13: 1705–1709.

Chen, Z. and D. Wu. 2012. Colloidal ZnSe quantum dot as pH probes for study of enzyme reaction kinetics by fluorescence spectroscopic technique. Colloids Surf. A: Physicochem. Eng. 414: 174–179.

Cheng, X.X., Y. Lui, B. Zhou, X.H. Xiao and Y. Liu. 2009. Probing the binding sites and the effect of berbamine on the structure of bovine serum albumin. Spectrochim. Acta. A 72: 922–928.

Choi, J.H., K.H. Chen and M.S. Strano. 2006. Aptamer-capped nanocrystal quantum dots: A new method for label-free protein detection. J. Am. Chem. Soc. 128: 15584–15585.

Clark, K.A., C. O'Driscoll, C.A. Cooke, B.A. Smith, K. Wepasnick, D.H. Fairbrother, P.S. Lees and J.P. Bressler. 2012. Evaluation of the interactions between multiwalled carbon nanotubes and Caco-2 cells. J. Toxicol. Environ. Health - A: Curr. Issues 75(1): 25–35.

Cui, F.L., J. Fan, D.L. Ma, M.C. Liu, X.G. Chen and Z.D. Hu. 2003. A study of the interaction between a new reagent and serum albumin by fluorescence spectroscopy. Anal. Lett. 36: 2151–2166.

Dallinger, D. and C.O. Kappe. 2007. Microwave-assisted synthesis in water as solvent. Chem. Rev. 107: 2563–2591.

Das, K., I.A. Mir, R. Ranjan and H.B. Bohidar. 2018. Size-dependent magnetic properties of cubic-phase MnSe nanospheres emitting blue-violet fluorescence. Mater. Res. Express 5: 056106.

Das, K., S. Sanwlani, K. Rawat, C.R. Haughn, M.F. Doty and H.B. Bohidar. 2016. Spectroscopic profile of surfactant functionalized CdSe quantum dots and their interaction with globular plasma protein BSA. Colloids Surf. Physicochem. Eng. Asp. 506: 495–506.

Deng, Z., L. Cao, F. Tang and B. Zou. 2005. A new route to zinc-blende CdSe nanocrystals: Mechanism and synthesis. J. Phys. Chem. B 109(35): 16671–16675.

Ding, Y., H. Sun, D. Liu, F. Liu, D. Wang and Q. Jiang. 2013. Water-soluble, high-quality ZnSe@ ZnS core/shell structure nanocrystals. J. Chin. Adv. Mater. Soc. 1(1): 56–64.

Gautam, P., S. Kumar, M. Tomar, R. Singh and A. Acharya. 2017. Biologically synthesized gold nanoparticles using Ocimum sanctum (Tulsi leaf extract) induced anti-tumor response in a T cell daltons lymphoma. J. Cell Sci. 8(2).

Geng, B., D. Yang, D. Pan, L. Wang, F. Zheng, W. Shen, C. Zhang and X. Li. 2018. NIR responsive carbon dots for efficient photothermal cancer therapy at low power densities. Carbon 134: 153–162.

Gil, H.M., T.W. Price, K. Chelani, J.G. Bouillard, S.D.J. Calaminus and G.J. Stasiuk. 2021. NIR-quantum dots in biomedical imaging and their future. iScience 24: 102189.

Guo, W.S., W.T. Yang, Y. Wang, X.L. Sun, Z.Y. Liu Zhang, J. Chang and X.Y. Chen. 2014. Color-tunable Gd-ZnCu-in-S/ZnS quantum dots for dual modality magnetic resonance and fluorescence imaging. Nano Res. 7: 1581–1591.

Hoa, T.T.Q., N.N. Long, V.T.H. Hanh, V.D. Chinh and P.T. Nga. 2012. Luminescent ZnS: Mn/thioglycerol and ZnS: Mn/ZnS core/shell nanocrystals: Synthesis and characterization. Opt. Mater. 35(2): 136–140.

Hsu, J.C., C.C. Huang, K.L. Ou, N. Lu, F.D. Mai, J.K. Chen and J.Y. Chang. 2011. Silica nanohybrids integrated with CuInS2/ZnS quantum dots and magnetite nanocrystals: Multifunctional agents for dual-modality imaging and drug delivery. J. Mater. Chem. 21: 19257–19266.

Ikanovic, M., W.E. Rudzinski, J.G. Bruno, A. Allman and M.P. Carrillo. 2007. Fluorescence assay based on aptamer–quantum dot binding to Bacillus thuringiensis spores. J. Fluoresc. 17: 193–199.

Jasieniak, J., L. Smith, J.V. Embden and P. Mulvaney. 2009. Re-examination of the size-dependent absorption properties of CdSe quantum dots. J. Phys. Chem. C 113: 19468–19474.

Jawhar, N.N., E. Soheyli, A.F. Yazici, E. Mutlugun and R. Sahraei. 2020. Preparation of highly emissive and reproducible Cu–In–S/ZnS core/shell quantum dots with a mid-gap emission character. J. Alloys Compd. 824: 67–72.

Kim, J.S., K.S. Song, H.J. Joo, J.H. Lee and I.J. Yu. 2010. Determination of cytotoxicity attributed to multiwall carbon nanotubes (MWCNT) in normal human embryonic lung cell (WI-38) line. J. Toxicol. Environ. Health - A: Curr. Issues 73(21-22): 1521–1529.

Kippeny, T., L.A. Swafford and S.J. Rosenthal. 2002. Semiconductor nanoparticles: A powerful visual aid for introducing the particle in a box. J. Chem. Educ. 79: 1094–1100.

Kir'yanov, A.V., N.N. Il'ichev, E.S. Gulyamova, A.S. Nasibov and P.V. Shapkin. 2015. Nonlinear change in refractive index and transmission coefficient of ZnSe: Fe2+ at long-pulse 2.94-m excitation. Opt. Photon. J. 5(01): 15–27.

Lakowicz, J.R. 2006. Principles of Fluorescence Spectroscopy, 3rd ed., Springer, New York.

Li, J.J., D. Hartono, C.-N. Ong, B.-H. Bay and L.-Y.L. Yung. 2010. Autophagy and oxidative stress associated with gold nanoparticles. Biomaterials 31(23): 5996–6003.

Ma, F., C.C. Li and C.Y. Zhang. 2018. Development of quantum dot-based biosensors: Principles and applications. J. Mater. Chem. B 6: 6173–6190.

Maluleke, R., E.H.M. Sakho and O.S. Oluwafemi. 2020. Aqueous synthesis of glutathione-capped CuInS2/ZnS quantum dots-graphene oxide nanocomposite as fluorescence "switch OFF" for explosive detection. Mater. Lett. 269: 127669.

Mir, I.A., K. Das, K. Rawat and H.B. Bohidar. 2016. Hot injection versus room temperature synthesis of CdSe quantum dots: A differential spectroscopic and bioanalyte sensing efficacy evaluation. Colloid Surf. A 494: 162–169.

Mir, I.A., K. Das, T. Akhter, R. Ranjan, R. Patel and H.B. Bohidar. 2018. Eco-friendly synthesis of CuInS2 and CuInS2@ZnS quantum dots and their effect on enzyme activity of lysozyme. RSC Adv. 8: 30589.

Mir, I.A., K. Rawat and H.B. Bohidar. 2017. Interaction of plasma proteins with ZnSe and ZnSe@ZnS core-shell quantum dots. Colloids and Surfaces A: Physicochem. Eng. Aspects 520: 131–137.

Mir, I.A., V.S. Radhakrishanan, K. Rawat, T. Prasad and H.B. Bohidar. 2018. Bandgap tunable AgInS based quantum dots for high contrast cell imaging with enhanced photodynamic and antifungal applications. Sci. Rep. 8: 9322.

Mohan, C.N., V. Renuga and A. Manikandan. 2017. Influence of silver precursor concentration on structural, optical and morphological properties of Cu1-xAgxInS2 semiconductor nanocrystals. J. Alloys Compd. 729: 407–417.

Moriyama, Y. and K. Takeda. 2005. Protective effects of small amounts of bis (2-ethylhexyl) sulfosuccinate on the helical structures of human and bovine serum albumins in their thermal denaturations. Langmuir 21(12): 5524–5528.

Niu, X.F., Y. Zhong, R. Chen, F. Wang, Y. Liu and D. Luo. 2018. A "turn-on" fluorescence sensor for Pb2+ detection based on graphene quantum dots and gold nanoparticles. Sens. Actuators B Chem. 255: 1577–1581.

Owens, F. and C. Poole Jr. 2007. Introduction to Nanotechnology, Wiley, New York; Kindle Edn., 2022.

Pan, Y., A. Leifert, D. Ruau, S. Neuss, J. Bornemann, G. Schmid, W. Brandau, U. Simon and W. Jahnen-Dechent. 2009. Gold nanoparticles of diameter 1.4 nm trigger necrosis by oxidative stress and mitochondrial damage. Small 5(18): 2067–2076.

Pandey, P., K. Hidayath Ulla, M.N. Satyanarayan, K. Rawat, A. Gaur, S. Gawali, P.A. Hassan and H.B. Bohidar. 2020. Fluorescent MoS2 Quantum Dot-DNA nanocomposite hydrogels for organic light-emitting diodes. ACS Appl. Nano Mater. 3: 1289–1297.

Parker, W. and P.-S. Song. 1992. Protein structures in SDS micelle-protein complexes. Biophys. J. 61(5): 1435–1439.

Pathak, J., E. Priyadarshini, K. Rawat and H.B. Bohidar. 2017. Complex coacervation in charge complementary biopolymers: Electrostatic versus surface patch binding. Adv. Colloid Interface Sci. 250: 40–53.

Patterson, A.L. 1939. The Scherrer formula for X-Ray particle size determination. Phys. Rev. 56: 978–982.

Peng, X. 2002. Green chemical approaches toward high-quality semiconductor nanoparticles. Chem. Eur. J. 8: 335–339.

Peng, Z.A. and X. Peng. 2001. Formation of high-quality CdTe, CdSe, and CdS nanocrystals using CdO as precursor. J. Am. Chem. Soc. 123(1): 183–184.

Peng, Z.A. and X. Peng. 2002. Nearly monodisperse and shape-controlled CdSe nanoparticles via alternative routes: Nucleation and growth. J. Am. Chem. Soc. 124: 3343–3353.

Priyadarshini, E. and K. Rawat. 2017. Au@ carbon dot nanoconjugates as a dual mode enzyme-free sensing platform for cholesterol. J. Mater. Chem. B 5(27): 5425–5432.

Priyadarshini, E., K. Rawat and H.B. Bohidar, 2020. Multimode sensing of riboflavin via Ag@ carbon dot conjugates. Appl. Nanosci. 10(1): 281–291.

Priyadarshini, E., R. Meena, H.B. Bohidar, S.K. Sharma, M.H. Abdellattif, M. Saravanan and P. Rajamani. 2022. Oxid. Med. Cell. Longev. 2022: Article ID 3483073.

Qu, L., A. Peng and X. Peng. 2001. Alternative routes toward high quality CdSe nanoparticles. Nano Lett. 1: 333–337.

Rawat, K. and H.B. Bohidar. 2012. Universal charge quenching and stability of proteins in 1-methyl-3-alkyl (hexyl/octyl) imidazolium chloride ionic liquid solutions. J. Phys. Chem. B 116: 11065–11074.

Rogach, A.L., A. Kornowski, M. Gao, A. Eychmüller and H. Weller. 1999. Synthesis and characterization of a size series of extremely small thiol-stabilized CdSe nanocrystals. J. Phys. Chem. B 103(16): 3065–3069.

Ryan, M.A. and M. Tinnes. 2002. Introduction to green chemistry, instructional activities for introductory chemistry. American Chemical Society.

Sarma, R., Q. Das, A. Hussain, A. Ramteke, A. Choudhury and D. Mohanta. 2014. Physical and biophysical assessment of highly fluorescent, magnetic quantum dots of a wurtzite-phase manganese selenide system. Nanotechnology 25: 275101.

Shang, L., Y. Wang, J. Jiang and S. Dong. 2007. pH-dependent protein conformational changes in albumin: Gold nanoparticle bioconjugates: A spectroscopic study. Langmuir 23: 2714–2721.

Shang, Z.C., P.G. Yi, Q.L. Yu and R.L. Lin. 2001. Reaction mechanism between ciprofloxacin hydrochloride and bovine serum albumin. Acta Phys. Chim. Sin. 17: 48–52.

Soheyli, E., D. Azad, R. Sahraei, A.A. Hatmnia and M. Alinazari. 2019. Synthesis and optimization of emission characteristics of water-dispersible ag-in-s quantum dots and their bactericidal activity. Colloids Surf. B. Biointerfaces 182: 110389.

Tang, X., W.B.A. Ho and J.M. Xue. 2012. Synthesis of Zn-doped AgInS2 nanocrystals and their fluorescence properties. J. Phys. Chem. C 116(17): 9769–9773.

Tauc, J., R. Grigorovici and A. Vancu. 1966. Optical properties and electronic Structure of amorphous germaniun. Phys. Status Solidi B 15: 627–637.

Toufanian, R., A. Piryatinski, A.H. Mahler, R. Iyer, J.A. Hollingsworth and A.M. Dennis. 2018. Bandgap engineering of indium phosphide-based core/shell heterostructures through shell composition and thickness. Front. Chem. 6: 567.

Tsolekile, N., S. Nelena and O.S. Oluwafemi. 2019. Porphyrin as diagnostic and therapeutic agent. Molecules 24: 2669.

Tyagi, A., K. Rawat, A.K. Verma and H.B. Bohidar. 2016. Mechanistic evaluation of the size dependent antimicrobial activity of water soluble QDs. Anal. Methods 8: 1060–1068.

Yang, W.T., W.S. Guo, X.Q. Gong, B.B. Zhang, S. Wang, N. Chen, W.T. Yang, Y. Tu, X.M. Fang and J. Chang. 2015. Facile synthesis of Gd–Cu–In–S/ZnS bimodal quantum dots with optimized properties for tumor targeted fluorescence/MR *in vivo* imaging. ACS Appl. Mater. Interfaces 7: 18759–18768.

Yu, W.W. and X. Peng. 2002. Formation of high-quality CdS and other II–VI semiconductor nanoparticles in noncoordinating solvents: Tunable reactivity of monomers. Angew. Chem. Int. Ed. Engl. 41: 2368–2371.

Yu, W.W., Y.A. Wang and X. Peng. 2003. Formation and stability of size-, shape-, and structure-controlled CdTe nanocrystals: Ligand effects on monomers and nanocrystals. Chem. Mater. 15: 4300–4308.

Zhou, J., Y. Liu, J. Tang and W. Tang. 2017. Surface ligands engineering of semiconductor quantum dots for chemosensory and biological applications. Mater. Today 20: 360–376.

Chapter 8

Chirality in Nanomaterials

Occurrence, Methods of Determination and Biochemical Significance

Wells Utembe[1,2,3]

Introduction

Nanomaterials (NMs) are materials with structural components smaller than 100 nm in at least one dimension (Buzea et al. 2007). These materials have been shown to possess unique biochemical, physico-chemical, electrical, and mechanical properties when compared to their related conventional or bulk substances (Murty el. 2013, Sudha et al 2018). Therefore, NMs are used in many applications in medicines, electronics, food, pesticides, textiles, construction materials and many others (Sharma et al., 2018). However, there are human and environmental safety concerns arising from toxicity of some NMs, including ZnO nanoparticles (NPs) (Subashkumar and Selvanayagam 2014), Ag NPs (Kovrižnych et al. 2013), C_{60} NPs (fullerenes) (Lovern and Klaper 2006), and many others. The toxicity of NMs is affected by many factors, such as material composition (Li et al, 2010), size (Al-Khedhairy and Wahab 2022), functional groups (Hansjosten et al. 2022), surface charge (Li et al. 2022), and in some cases chirality (Xu et al. 2022).

Chirality occurs in objects that contain an asymmetric point, resulting in two non-superimposable mirror images in a three-dimensional space (Smith 2009). Often recognized in organic molecules, an asymmetric point is often a carbon atom with four different functional groups. The resulting molecules are isomers as they bear identical molecular formulae and connectivity of atoms but have different orientation of atoms in space (Fig. 1).

[1] Toxicology and Biochemistry Department, National Institute for Occupational Health, National Health Laboratory Service, 2000 South Africa.
[2] Department of Environmental Health, University of Johannesburg, Johannesburg, 2000, South Africa.
[3] Environmental Health Division, School of Public Health and Family Medicine University of Cape Town, 7925, South Africa.
Email: wellsu@nioh.ac.za; wutembe@cartafrica.org

Mirror

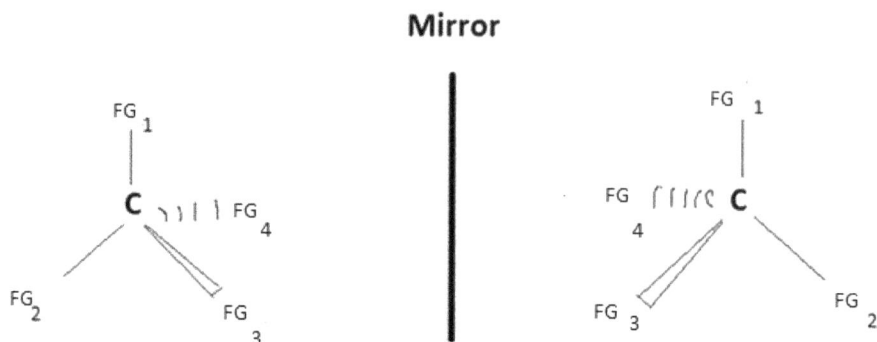

Figure 1. Two non-superimposable images of a molecule made of one central asymmetric carbon atom that is connected to four different functional groups, FG_1, FG_2, FG_3 and FG_4.

Chirality also exists in inorganic compounds, especially in coordination metal complexes that possess a central metal ion and 4 different ligands (von Zelewsky 1996, King 2001). Moreover, chirality can also exist in crystals and NMs where two or more non-superimposable images arise from an asymmetric point (Oila and Koskinen 2006, Gautier and Bürgi 2008, 2009).

Chiral substances exist as pairs of enantiomers, one of which interacts with and rotates the plane of polarized light to the left, while the other rotates it to the right. Consequently, enantiomers are said to be optically active. These optical isomers are part of stereoisomers, together with cis-trans isomers and diastereoisomers (diastereomers). The latter possess more than one chiral centre that result in non-mirror images (Smith 2009). While enantiomeric pairs are related to each other as mirror images and have all centres of chirality in reversed configurations, diastereomers are not related to each other as objects and their corresponding mirror image (Jozwiak 2012).

Chirality is conceptualised in terms of handedness, where chiral objects are either right-handed or left-handed, even though substances can exhibit non-handed chirality (King 2003). Handedness and chirality can exist at all length scales (Gellman 2010), as shown by the right hand which cannot be superimposed on the left hand. Most chiral molecules in nature are based on asymmetric tetrahedral carbon atoms that are handed. However, non-handed chirality can exist in molecules based on other polyhedral structures, especially octahedral structures, for which left-hand and right-hand designations are not possible (King 2003). King (2001) presents a comprehensive analysis of non-handed chiral octahedral complexes.

Although enantiomers show differences in their optical activity, they have identical physico-chemical properties such as melting points, boiling points and chemical reactivity (Baddeley and Richardson 2010). The enantiomer that rotates the plane of polarised light to the left is said to be levo-rotary (−) or lowercase l- whereas the other enantiomer is said to be dextro-rotary (+) or lowercase d-. The enantiomers can also be portrayed using Fischer projections (capital D/L) as well as the Cahn-Ingold-Prelog System (R for rectus and S for sinister). In Fischer Projections, a molecule is assigned a D- or L-configuration based on its relationship to an arbitrary standard D-glyceraldehyde, whereas in the Cahn-Ingold-Prelog convention rules are used to assign the priority of substituents about the chiral center, from the heaviest to the lightest groups, as viewed along the carbon to the lowest priority bond. The molecule is designated an "R" configuration if the atomic priority decreases in clockwise, right-to-left order, and it is conferred an "S" configuration if the atomic priority decreases in counterclockwise, left-to-right order (Jozwiak 2012). Since the dextro- or levo-rotation of each enantiomer is not related to the actual assignment of its atoms in space or the D/L or R/S naming conventions, dextro- or levo-rotation of a substance (±) is often provided alongside its D/L or R/S name.

A mixture of equal amounts of two enantiomers is not optically active and is referred to as a racemic mixture or racemate (Smith 2009). Racemic mixtures are designated the prefix rac- (D,L)-, or (R,S)- in order to distinguish them from single enantiomer. Some physicochemical properties of a racemate may differ significantly from those of the respective single enantiomers (Jozwiak 2012).

Many molecules in nature possess specific enantiomeric configurations. For example, amino acids in humans are of an L-configuration whereas carbohydrates in humans have D-configuration (Smith 2009). Consequently, most physiological processes involve chiral receptor molecules that can distinguish two enantiomeric substrate molecules, resulting in selective different pharmacological and toxicological profiles for enantiomers. Chirality also affects many important environmental processes, such as bioaccumulation, persistence degradation rates, and toxicity (Wong 2006, Ye et al. 2010, 2015). This chapter discusses the occurrence of chirality in NMs, methods of its detection and its implications for biomedical sciences.

Chirality in nanomaterials

Chirality induced by adsorbing ligands

Chirality can be induced on achiral particles by the adsorption of chiral molecules as shown in Fig. 2 below (Shemer et al. 2006, Govorov et al. 2011, Kumar et al. 2016) or by an asymmetric arrangement of achiral ligands around the NP (Kumar et al. 2016). The transfer of chirality to the particle surface largely depends on the structure of the adsorbing molecules.

For example, Shukla et al. (2010) could bestow chirality on chemically synthesized 5 nm gold (Au) NPs by adsorbing chiral thiols (D- or L-cysteine). Chirality in the D- or L-cysteine-capped Au NPs was confirmed by circular dichroism (CD) spectroscopy. Moreover, the D-cysteine-capped Au NPs could adsorb S-propylene whereas the L-cysteine-capped Au NPs could selectively adsorb (*R*)-propylene oxide. Also using CD spectroscopy, Yao et al. (2005) confirmed chirality in D- and L-penicillamine-capped 1–2 nm Au nanoclusters (Au NCs). The CD signals were not observed in Au NCs capped with racemic penicillamine. Chiral Au NPs have also been synthesized by other authors including Gautier and Bürgi (2006), Gautier et al. (2008), Huang et al. (2012) and Gautier and Bürgi (2009). Most interestingly, exchange of a chiral ligand with its opposite enantiomer was shown to reverse the optical activity, while

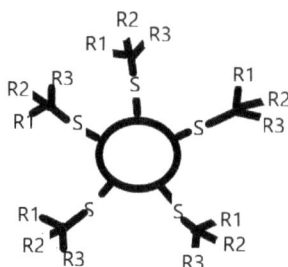

Figure 2. Chiral molecules attached to a NP core through covalent bonds with sulphur (S) atoms.

leaving the form of the CD spectra largely unaffected (Gautier and Bürgi 2008). The reversed optical activity arises from the influence of the opposite enantiomer, which should result in the mirror image CD spectra.

Ligand-induced chirality has also been observed in other NMs. For example, chirality was observed in 23.5 nm silver (Ag) NPs functionalised with glutathione (GSH), cysteine and penicillamine (Li et al. 2004), Ag NPs (5–20 nm) synthesized in mucin glycoprotein (Hendler et al. 2011), as well as Ag NPs (2 nm and 5 nm) synthesized in deoxyribonucleic acid (DNA) (Shemer et al. 2006, Molotsky et al. 2010). Furthermore, optically active Ag NCs (of size less than 2 nm) were prepared using several optically active thiols (such as GSH, cysteine, and captopril) (Cathcart et al. 2009), as well as L- and D-penicillamine (Nishida et al. 2008).

Chirality was also observed after replacing achiral trioctylphosphine oxide/ oleic acid ligands on achiral 4.4 nm cadmium selenide (CdSe) quantum dots (QD) with chiral D- and L-cysteines (Tohgha et al. 2013a, Tohgha et al. 2013b). Most importantly, opposite CD signals were observed for the CdSe QDs capped with L- and D-cysteine. Chirality has also been induced in other NPs including 30–50 nm TiO_2 NPs (Cleary et al. 2017) and silica NPs (> 300 nm) (Li et al. 2016).

The mechanistic reasons for optical activity of functionalized NPs has been discussed extensively by Hidalgo and Noguez (2016), Guerrero-Martínez et al. (2011), Noguez and Garzón (2009) and Sánchez-Castillo et al. (2010). Various mechanisms, including the *dissymmetric field effect* and *the chiral footprint model*, have been postulated to explain ligand-induced chirality in NMs (Kumar et al. 2016). In the *dissymmetric field effect model*, chirality is induced to an achiral core either through a chiral adsorption pattern or through vicinal effects by chiral ligands. In this model, an electrostatic field from the dissymmetric environment of the ligands perturbs and breaks down the symmetry of the electronic state in the NMs, whereas in the *chiral footprint model*, the chirality of metal NPs arises from chiral distortion of the NP surface atoms involved in the adsorption of the ligand (Karimova and Aikens 2019).

Some ligands that are used to improve specific properties of NPs are chiral. The ability of the ligands to induce chirality in NMs largely depends on numerous factors including the distance of the chiral center from the NP surface, ability to undergo self-assembly mediated by hydrogen bonds, and the ability to interact with the cluster surface through at least two functional groups (Gautier and Bürgi 2009).

The CD response of NPs depends largely on the coupling strength between the NPs and the chiral molecules, the size of NPs, the orientation of chiral molecules on the NP surfaces and the distance between centres of the molecules and NP (Li et al. 2017). Indeed, the CD signals or the anisotropy factor increases with decreasing size, because in smaller NPs most of the metal atoms are located at the surface where they can easily interact with the chiral ligands. This makes NCs, which are a special class of NMs that typically contain few hundred atoms, the best candidates for chiral NMs (Gautier and Bürgi 2009).

Chiral NMs can also be synthesized from achiral ligands that are arranged in an asymmetric manner. For example, intrinsically chiral Au20 NPs were prepared from achiral polydentate phosphine ligand tris (2-(diphenylphosphino)-ethyl) phosphine as well as Au102 from para-mercaptobenzoic acid (Knoppe et al. 2014). According to Hidalgo et al. (2016), chirality depends on the position and orientation of the adsorbed ligands, the asymmetric arrangement of the ligands around the NP and the distortion of the surface atoms upon ligand adsorption.

The effect on chirality of the nature of the bond/interaction between the ligands and the NP core appear not to have been investigated. Chiral or achiral ligands appear to be largely covalently bonded to the NP core, mostly through a sulphur atom in the case of sulphur-bearing ligands. However, the interplay between van der Waals and chemical forces in sulphur-bearing ligands and NMs has been discussed by Reimers et al. (2017).

Intrinsic chirality in nanomaterials

Although pure metals comprise of achiral bulk crystalline structures, they can have chiral surfaces when the surface lies outside the bulk crystal mirror planes (Gellman 2021). Indeed, according to Gellman et al. (2001), it is possible to create from a single crystalline material high Miller index surfaces "with structures that consist of kinked steps separated by flat low Miller index terraces". Miller indices are a set of numbers that represent vectors normal to the surface. In this regard, low Miller indices represent areas of high symmetry, and thus show lack of chirality, while high Miller indices represent low symmetry, and do therefore signify chirality (Kühnle et al. 2006, Sholl and Gellman 2009).

Chirality in inorganic NMs can also be intrinsically induced through designs of crystals that exposes these "kinked and stepped surface structures" (McFadden et al. 1996). Similar to the chirality in bulk crystals, the kink sites lack symmetry and are not superimposable on their mirror images when the step lengths on either side of the kink site are unequal, and thereby can be thought of as chiral, with an "R" or "S" configuration. The process for assigning R or S configurations for such surfaces is beyond this text but has been explained by Sholl et al. (2001) and Baddeley and Richardson (2010).

A number of intrinsically chiral NMs have been prepared including Au NCs (Knoppe and Bürgi 2014, Dolamic et al. 2015), fullerenes (Maroto et al. 2014), carbon nanotubes (CNTs) (Imtani and Jindal 2009, Liu et al. 2011), and silicon nanotubes (Wang et al. 2017). In the special case of CNTs, intrinsic chirality arises from the folding patterns of the graphene layer, which can be defined by pairs of integer indices that denote the chiral angle θ, and the chiral vector (Rao and Govindaraj 2011, Pérez and Martín 2012).

Intrinsic chirality can be explained in terms of chiral core model where chirality arises from inherently chiral or twisting atom-packing disordered structure (Yao et al. 2005). Mechanistic origins of intrinsic chirality in NCs have been discussed in detail by Zeng and Jin (2017), Zhu et al. (2020) and Karimova and Aikens (2019).

Analytical techniques

It is difficult to distinguish pairs of enantiomers from each other since they possess exactly the same physiochemical properties such as molecular weights, melting points, boiling points, solubilities and reactivities. However, enantiomers have some vectorial properties in which opposite enantiomers exhibit distinguishable properties, especially those based on differential absorption of left- and right-circularly polarised light (circular dichroism, CD) as well as the rotation of the plane of circularly polarised light (optical rotary dispersion (ORD)) (Smith et al. 2017, Gogoi et al. 2021). These two phenomena are the basis of many advanced analytical techniques that provide stereochemical, conformational and three-dimensional structural information of chiral substances. However, characterization of chirality in NMs is quite challenging because of the small sizes of chiroptical signals, which result in low signal-to-noise ratios.

Circular dichroism (CD) is one of the most widely used techniques for assessing the chirality of materials. Different enantiomers absorb polarised light at different wavelengths and to different extents. Consequently, CD is expressed as the difference in the absorption of left circularly polarized light (A_L) and right circularly polarized light (A_R), when light that is circularly polarized both to the left and to the right passes through a sample. An optically active substance in the sample absorbs left- and right-circularly polarized light differently, while at the same time obeying the Beer-Lambert law. The optical activity can be measured using the anisotropy factor, also known as g-factor, shown in equation 1, where ε and Δε are the molar extinction and molar circular dichroism, respectively (Guerrero-Martínez et al. 2011):

$$g = \frac{\Delta \in}{\in} \tag{1}$$

Depending on the wavelength of the radiation applied, CD can be based on electronic, vibrational, reflection or fluorescence absorptions, resulting in electronic circular dichroism (ECD), vibrational circular dichroism (VCD), Raman optical activity (ROA), diffuse reflection circular dichroism (DRCD) and circularly polarized luminescence (CPL) (Kumar et al. 2016). ECD involves electronic transitions and is a chiroptical counterpart of UV-Visible absorption spectroscopy, while VCD, ROA, and CPL are counterparts of the non-chiral infrared (IR), Raman and fluorescence spectroscopic techniques, respectively. Although two enantiomers will have identical UV-visible, IR and Raman spectra, they will have distinguishable, mirror-image ECD, VCD and ROA spectra that contain the structural (conformation and configuration) information of the enantiomers. These methods have successfully been applied in the determination of chirality of many NMs. For example, VCD has been utilised to determine the chirality of many NMs including N-acetyl-L-cysteine protected Au NPs (Gautier and Bürgi 2005), 1,1′-binaphthyl-2,2′-dithiol (BINAS)-stabilised Au NPs (Gautier and Bürgi 2010). Tohgha et al. (2013) successfully obtained size-dependent ECD and CPL spectra for L- and D-cysteine-capped CdSe QDs.

While CD assesses chirality in energy transitions, optical rotation refers to the extent to which a beam of linearly polarized light is rotated as it passes through a sample (Valev 2016, Collins et al. 2017). The most widely used optical rotation techniques are polarimetry and optical rotatory dispersion (ORD), two very similar methods that differ only in the fact that polarimetry uses a limited number of pre-selected wavelengths, while ORD is measured across the spectrum (Purdie and Swallows 1989, Valev 2016). Shukla et al. (2010) and Shukla et al. (2015) used polarimetric measurements to demonstrate the enantioselective adsorption of R- or S-propylene oxide by Au NPs coated with D- or L-cysteine from a solution of racemic propylene oxide. In this study, larger optical rotations could be obtained by increasing the temperature, decreasing wavelength and decreasing NP size.

The spectroscopic methods for assessing NMs are complemented by chiral separation techniques such as direct chiral chromatographic systems, which contain chiral selectors that can be incorporated into the stationary phase or the surface of the column packing material, or indirect chiral derivatization techniques. The later involves formation of diastereoisomers that can be separated using different separation techniques. The application of these separation techniques to NMs has been said to be tedious and of low efficiency (Yang et al. 2017). As a comprehensive assessment of the methods used to assess chirality are beyond the scope of this chapter, readers are referred to other texts, such as those by Davankov (1997), Yanez et al. (2007) and Rocco et al. (2013).

Chiral nanomaterials and biochemical systems: interactions and applications

Chiroptic effects in NMs can be stronger than those observed in chiral molecules, and similar to chirality in molecules, chiral receptors can differentiate between enantiomeric pairs of chiral Nms. For example, blood plasma proteins had a stronger interaction with L-Lysine that was attached to TiO_2 films than D-Lysine whereas more proteins were adsorbed to D-tartaric acid as compared to L-tartaric acid (Fan et al. 2017). Similarly, chirality determined the interaction and recognition of transferrin with its receptor on the cellular membrane of 14 nm gold (Au) NPs (Wang et al. 2017). Although the mechanistic effects of chirality on protein adsorption kinetics are yet to be elucidated, these enantioselective adsorption processes may have many significant consequences on biochemical processes. Firstly, chirality may affect the cell uptake of NMs which occurs in a number of steps, including adsorption of NPs to cells (Wilhelm et al. 2003, Cho et al. 2009) and the interaction of the NPs with components of the cell membrane. In that regard, chirality affected the *in vitro* uptake of 16-nm Au NPs that were modified with 2-mercaptoacetyl-valine

(MAV) and poly(acryloyl-valine) (PAV) of different chiralities (Deng et al. 2016). A greater uptake was observed for D-PAV-Au NPs than for L-PAV-Au NPs by A549 cells and HepG2 cells, although there were no differences in the internalisation of the L- and D-MAV-Au NPs. Zhang et al. (2022) observed a greater uptake in GL261 and bEnd.3 cells for D-GSH Au nanooctopods (NOPs) than L-GSH and racemic NOPs. Similarly, CNTs with lower chirality translocated faster through lipid bilayer cell membranes than CNTs with higher chirality (Skandani et al. 2012), while L-Cysteine QDs was translocated at almost twice the rate of D-Cysteine QDs (Martynenko et al. 2016).

Chirality may not only affect the uptake rates of NMs but also their intracellular distribution and consequently their toxicity. For example, 16-nm L-PAV-Au NPs were located in lysosomes of A549 cells while D-PAV-Au NPs were located in the cytoplasm and closer to the cell nuclei (Deng et al. 2016). On the other hand, both enantiomers were mainly distributed in lysosomes of HepG2 cells. These differences in internal distribution are likely to result in different toxicities for the enantiomers.

Enantiomers of some NMs can also possess different toxicities because of the differential interaction with cells, and not only differences in the rates of uptake or intracellular distribution. For example, L-GSH cadmium telluride (CdTe) QDs were shown to elicit a greater dose-and chirality-dependent autophagy and cytotoxicity than D-GSH-QDs (Li et al. 2011). These toxic effects were attributed to differences in their interaction with the autophagy machinery rather than differences in cellular uptake. D-GSH coated 4-nm Au NCs also caused more prominent effects in human gastric cancer (MGC-803) cells and human gastric epithelial (GES-1) cells than L-GSH coated Au NCs because of differences in the ability to cause ROS-induced i mitochondrial membrane depolarization, cell cycle arrest and apoptosis (Zhang et al. 2015). Furthermore, Xu et al. (2022) reported different *in vitro* and *in vivo* immune responses for achiral, left-handed and right-handed gold biomimetic NPs due to differential effects on the production of immune signaling complexes and the maturation of mouse bone-marrow-derived dendritic cells. The immune responses depended on the *g*-factors of the NMs. Chirality-dependent toxicities have also been reported by other authors including *inter alia* Xin et al. (2017), Sun et al. (2018) and Yuan et al. (2018).

Chiral NMs have numerous applications in biochemical sciences especially in the diagnosis and treatment of diseases, as well as in the area of biochemical catalysis. In disease diagnosis, Erbilen et al. (2019), for example, developed an electrochemical chiral sensor based on reduced graphene oxide, gold NPs modified with poly-L-cysteine, and poly-L-phenylalanine methyl ester. The electrode could discriminate between L-tryptophan (L-Trp) from D-Trp, which is associated with a number of psychiatric and neurological diseases. In addition, L-cysteine-functionalized ZnO NPs could be used to detect urinary dopamine, a neurotransmitter that plays major functions in the cardiovascular and central nervous systems (Balram et al. 2018). Furthermore, Kumar et al. (2018) used chiral surface plasmons to detect amyloids, which are closely associated with various neurodegenerative diseases, such as Parkinson's disease.

D-GSH stabilized Au NPs were shown to possess a higher brain biodistribution and larger binding affinity to Aβ42 that resulted in better treatment of behavioral impairments compared with its L-enantiomer (Hou et al. 2020). Similarly, penicillamine-capped D-Pen@Se NPs were shown to possess less toxicity and is therefore a better treatment for Alzheimer's disease than L-Pen@Se NPs (Sun et al. 2017). Moreover, a number of chiral NMs have been shown to possess anti-tumor properties. For example, Yuan et al. (2018) could selectively induce autophagy and safely suppress tumor growth in breast cancer cells both *in vitro* and *in vivo* using D-PAV-AuNPs. Similarly, Liu et al. (2021) used D-Cu_{2-x}Se NPs in cancer chemodynamic therapy (CDT) and photothermal therapy (PTT) to completely eliminate tumors.

It is important to note that chirality affects the toxicity of some chiral NMs and not of others. Similar to chirality in organic medicines, one enantiomer can be biologically active while the other is less active, inactive or can even exert a different effect. For example, chirality did not affect the toxicity of 10–20 nm CdS-penicillamine tetrapods in the NG108-15 neuroblastoma cells (Govan et al. 2010). On the other hand, D-GSH coated 4-nm Au NCs caused more prominent effects than L-GSH coated Au NCs, even though both enantiomers affected MGC-803 and GES-1 cells (Zhang et al. 2015). Therefore, investigations of the effects of chirality of NMs are needed on a case-by-case basis.

Although most NMs are achiral, it is very important to take into account the effects of chirality whenever it occurs. Indeed, NMs are utilized in the pharmaceutical, pesticide and chemical industries, where some products are chiral. Nanozymes, NMs with enzyme-like characteristics, are continuously being developed to address the limitations of natural enzymes and conventional synthetic enzymes. Some of the nanozymes are chiral. For example, a chiral enzyme based on graphene oxide could discriminate between the chiral Parkinson's disease drug, Levodopa (L-dopa), from its enantiomer D-dopa which is not only inactive but can also cause adverse effects (Xu et al. 2014).

This chapter has shown the occurrence and importance of chirality in many biochemical systems and applications. Therefore, chirality should be considered in the risk assessment of potentially chiral NMs (Utembe and Gulumian 2019). In that regard, since two enantiomers of the same NMs may interact differently with biochemical systems, risk assessment studies need to be conducted on each enantiomer, the racemate and any relevant enantiomeric ratios of the NMs.

Conclusion

Chirality can be bestowed onto NMs by adsorption of chiral molecules and by careful cutting of crystals to expose chiral kinked and stepped surface structures. Since pairs of enantiomers possess exactly the same physiochemical properties, spectroscopic studies of these chiral NMs have been designed based on circular dichroism, ORD and polarimetry. The chirality can result in enantioselectivity of the NMs, resulting in different biological activities for the pairs of enantiomers of the NMs. Chiral NMs have numerous applications in biochemical sciences especially in the diagnosis and treatment of disease. Since chirality affects the biological properties of some NMs and not of others, investigations of the effects of chirality of NMs are needed on a case-by-case basis. It is very important to take into account the effects of chirality whenever it occurs, so that biochemical studies may have to be conducted on each enantiomer, the racemate and any relevant enantiomeric ratios of the NMs.

Disclaimer

This chapter was prepared by the author in his personal capacity. The opinions, findings and conclusions in this article are the author's own and do not necessarily reflect or represent the views of the National Institute for Occupational Health (NIOH), the University of Johannesburg or the University of Cape Town.

References

Al-Khedhairy, A.A. and R. Wahab. 2022. Size-dependent cytotoxic and molecular study of the use of gold nanoparticles against liver cancer cells. Applied Sciences 12: 1–18.

Baddeley, C.J. and N.V. Richardson. 2010. Chirality at metal surfaces. *In*: M. Bowker and P.R. Davies (eds.). Scanning Tunneling Microscopy, in Surface Science, Nanoscience and Catalysis. Wiley-VCH Verlag GmbH & Co. KGaA, Weinheim, Germany

Balram, D., K.-Y. Lian and N. Sebastian. 2018. A novel electrochemical sensor based on flower shaped zinc oxide nanoparticles for the efficient detection of dopamine. Int. J. Electrochem. Sci. 13: 1542–1555.

Buzea, C., I.I. Pacheco and K. Robbie. 2007. Nanomaterials and nanoparticles: Sources and toxicity. Biointerphases 2: MR17–MR71.

Cathcart, N., P. Mistry, C. Makra, B. Pietrobon, N. Coombs, M. Jelokhani-Niaraki et al. 2009. Chiral thiol-stabilized silver nanoclusters with well-resolved optical transitions synthesized by a facile etching procedure in aqueous solutions. Langmuir 25: 5840–5846.

Cho, E.C., J. Xie, P.A. Wurm and Y. Xia. 2009. Understanding the role of surface charges in cellular adsorption versus internalization by selectively removing gold nanoparticles on the cell surface with a I2/KI etchant. Nano Lett. 9: 1080–1084.

Cleary, O., F. Purcell-Milton, A. Vandekerckhove and Y.K. Gun'ko. 2017. Chiral and luminescent TiO$_2$ nanoparticles. Advanced Optical Materials 5: 1601000.

Collins, J.T., C. Kuppe, D.C. Hooper, C. Sibilia, M. Centini and V.K. Valev. 2017. Chirality and chiroptical effects in metal nanostructures: Fundamentals and current trends. Advanced Optical Materials 5: 1700182.

Davankov, V.A. 1997. Analytical chiral separation methods IUPAC Recommendations 1997. Pure and Applied Chemistry 69: 1469–1474.

Deng, J., S. Wu, M. Yao and C. Gao. 2016. Surface-anchored poly (acryloyl-L (D)-valine) with enhanced chirality-selective effect on cellular uptake of gold nanoparticles. Scientific Reports 6: 1–12.

Dolamic, I., B. Varnholt and T. Bürgi. 2015. Chirality transfer from gold nanocluster to adsorbate evidenced by vibrational circular dichroism. Nature Communications 6: 1–6.

Erbilen, N., E. Zor, A.O. Saf, E.G. Akgemci and H. Bingol. 2019. An electrochemical chiral sensor based on electrochemically modified electrode for the enantioselective discrimination of D-/L-tryptophan. Journal of Solid State Electrochemistry 23: 2695–2705.

Fan, Y., R. Luo, H. Han, Y. Weng, H. Wang, J.A. Li, P. Yang, Y. Wang and N. Huang. 2017. Platelet adhesion and activation on chiral surfaces: the influence of protein adsorption. Langmuir 33: 10402–10410.

Gautier, C. and T. Bürgi. 2005. Vibrational circular dichroism of N-acetyl-L-cysteine protected gold nanoparticles. Chemical Communications 43: 5393–5395.

Gautier, C. and T. Bürgi. 2006. Chiral N-isobutyryl-cysteine protected gold nanoparticles: Preparation, size selection, and optical activity in the UV-vis and infrared. Journal of the American Chemical Society 128: 11079–11087.

Gautier, C. and T. Bürgi. 2008. Chiral inversion of gold nanoparticles. Journal of the American Chemical Society 130: 7077–7084.

Gautier, C. and T. Bürgi. 2009. Chiral gold nanoparticles. ChemPhysChem. 10: 483–492.

Gautier, C. and T. Bürgi. 2010. Vibrational circular dichroism of adsorbed molecules: BINAS on gold nanoparticles. The Journal of Physical Chemistry C 114: 15897–15902.

Gautier, C., R. Taras, S. Gladiali and T. Bürgi. 2008. Chiral 1, 1′-binaphthyl-2, 2′-dithiol-stabilized gold clusters: Size separation and optical activity in the UV–vis. Chirality 20: 486–493.

Gellman, A.J. 2010. Chiral surfaces: Accomplishments and challenges. ACS Nano. 4: 5–10.

Gellman, A.J. 2021. An account of chiral metal surfaces and their enantiospecific chemistry. Accounts of Materials Research 2: 1024–1032.

Gellman, A.J., J.D. Horvath and M.T. Buelow. 2001. Chiral single crystal surface chemistry. Journal of Molecular Catalysis A: Chemical 167: 3–11.

Gogoi, A., S. Konwer and G.-Y. Zhuo. 2021. Polarimetric measurements of surface chirality based on linear and nonlinear light scattering. Frontiers in Chemistry 8: 1–24.

Govan, J.E., E. Jan, A. Querejeta, N.A. Kotov and Y.K. Gun'ko. 2010. Chiral luminescent CdS nanotetrapods. Chem. Commun. 46: 6072–6074.

Govorov, A.O., Y.K. Gun'ko, J.M. Slocik, V.A. Gérard, Z. Fan and R.R. Naik. 2011. Chiral nanoparticle assemblies: Circular dichroism, plasmonic interactions, and exciton effects. Journal of Materials Chemistry 21: 16806–16818.

Guerrero-Martínez, A., J.L. Alonso-Gómez, B. Auguié, M.M. Cid and L.M. Liz-Marzán. 2011. From individual to collective chirality in metal nanoparticles. Nano Today 6: 381–400.

Hansjosten, I., M. Takamiya, J. Rapp, L. Reiner, S. Fritsch-Decker, D. Mattern et al. 2022. Surface functionalisation-dependent adverse effects of metal nanoparticles and nanoplastics in zebrafish embryos. Environmental Science: Nano. 9: 375–392.

Hendler, N., L. Fadeev, E.D. Mentovich, B. Belgorodsky, M. Gozin and S. Richter. 2011. Bio-inspired synthesis of chiral silver nanoparticles in mucin glycoprotein—the natural choice. Chemical Communications 47: 7419–7421.

Hidalgo, F. and C. Noguez. 2016. How to control optical activity in organic–silver hybrid nanoparticles. Nanoscale 8: 14457–14466.

Hou, K., J. Zhao, H. Wang, B. Li, K. Li, X. Shi et al. 2020. Chiral gold nanoparticles enantioselectively rescue memory deficits in a mouse model of Alzheimer's disease. Nature Communications 11: 4790.

Huang, P., O. Pandoli, X. Wang, Z. Wang, Z. Li, C. Zhang et al. 2012. Chiral guanosine 5′-monophosphate-capped gold nanoflowers: Controllable synthesis, characterization, surface-enhanced Raman scattering activity, cellular imaging and photothermal therapy. Nano Research 5: 630–639.

Imtani, A.N. and V. Jindal. 2009. Structure of chiral single-walled carbon nanotubes under hydrostatic pressure. Computational Materials Science 46: 297–302.

Jozwiak, K. 2012. Stereochemistry-basic terms and concepts. *In*: J. Krzysztof, W.J. Lough and I.W. Wainer (eds.). Drug Stereochemistry: Analytical Methods and Pharmacology, Informa Healthcare New York, USA.

Karimova, N.V. and C.M. Aikens. 2019. Chiral noble metal nanoparticles and nanostructures. Particle & Particle Systems Characterization 36: 1900043.

King, R. 2001. Nonhanded chirality in octahedral metal complexes. Chirality 13: 465–473.

King, R.B. 2003. Chirality and handedness. Annals of the New York Academy of Sciences 988: 158–170.

Knoppe, S. and T. Bürgi. 2014. Chirality in thiolate-protected gold clusters. Accounts of Chemical Research 47: 1318–1326.

Knoppe, S., L. Lehtovaara and H. Häkkinen. 2014. Electronic structure and optical properties of the intrinsically chiral 16-electron superatom complex [Au20 (PP3)$_4$]$^{4+}$. The Journal of Physical Chemistry A 118: 4214–4221.

Kovrižnych, J.A., R. Sotníková, D. Zeljenková, E. Rollerová, E. Szabová and S. Wimmerová. 2013. Acute toxicity of 31 different nanoparticles to zebrafish (*Danio rerio*) tested in adulthood and in early life stages–comparative study. Interdisciplinary Toxicology 6: 67–73.

Kumar, J., H. Eraña, E. López-Martínez, N. Claes, V.F. Martín, D.M. Solís et al. 2018. Detection of amyloid fibrils in Parkinson's disease using plasmonic chirality. Proceedings of the National Academy of Sciences 115: 3225–3230.

Kumar, J., K.G. Thomas and L.M. Liz-Marzán. 2016. Nanoscale chirality in metal and semiconductor nanoparticles. Chemical Communications 52: 12555–12569.

Kühnle, A., T.R. Linderoth and F. Besenbacher. 2006. Enantiospecific adsorption of cysteine at chiral kink sites on Au (110)-(1× 2). Journal of the American Chemical Society 128: 1076–1077.

Li, J., X. Du, N. Zheng, L. Xu, J. Xu and S. Li. 2016. Contribution of carboxyl modified chiral mesoporous silica nanoparticles in delivering doxorubicin hydrochloride in vitro: pH-response controlled release, enhanced drug cellular uptake and cytotoxicity. Colloids and Surfaces B: Biointerfaces 141: 374–381.

Li, T., B. Albee, M. Alemayehu, R. Diaz, L. Ingham, S. Kamal et al. 2010. Comparative toxicity study of Ag, Au, and Ag–Au bimetallic nanoparticles on Daphnia magna. Analytical and Bioanalytical Chemistry 398: 689–700.

Li, T., H.G. Park, H.-S. Lee and S.-H. Choi. 2004. Circular dichroism study of chiral biomolecules conjugated with silver nanoparticles. Nanotechnology 15: S660.

Li, Y., M. Xu, Z. Zhang, G. Halimu, Y. Li, Y. Li et al. 2022. In vitro study on the toxicity of nanoplastics with different charges to murine splenic lymphocytes. Journal of Hazardous Materials 424: 127508.

Li, Y., Y. Zhou, H.Y. Wang, S. Perrett, Y. Zhao, Z. Tang et al. 2011. Chirality of glutathione surface coating affects the cytotoxicity of quantum dots. Angewandte Chemie International Edition 50: 5860–5864.

Li, Z., L. Shi and Z. Tang. 2017. An introduction to chiral nanomaterials: Origin, construction, and optical application. In: Z.T. Tang (ed.). Chiral Nanomaterials: Preparation, Properties and Applications. Wiley-VCH Verlag GmbH & Co, Germany.

Liu, H., D. Nishide, T. Tanaka and H. Kataura. 2011. Large-scale single-chirality separation of single-wall carbon nanotubes by simple gel chromatography. Nature Communications 2: 309.

Liu, Y., H. Li, S. Li, X. Zhang, J. Xiong, F. Jiang et al. 2021. Chiral Cu2–x Se nanoparticles for enhanced synergistic cancer chemodynamic/photothermal therapy in the second near-infrared biowindow. ACS Applied Materials & Interfaces 13: 60933–60944.

Lovern, S.B. and R. Klaper. 2006. Daphnia magna mortality when exposed to titanium dioxide and fullerene (C60) nanoparticles. Environmental Toxicology and Chemistry: An International Journal 25: 1132–1137.

Maroto, E.E., M. Izquierdo, S. Reboredo, J. Marco-Martínez, S. Filippone and N. Martin. 2014. Chiral fullerenes from asymmetric catalysis. Accounts of Chemical Research 47: 2660–2670.

Martynenko, I., V. Kuznetsova, I. Litvinov, A. Orlova, V. Maslov, A. Fedorov et al. 2016. Enantioselective cellular uptake of chiral semiconductor nanocrystals. Nanotechnology 27: 075102.

McFadden, C.F., P.S. Cremer and A.J. Gellman. 1996. Adsorption of chiral alcohols on "chiral" metal surfaces. Langmuir 12: 2483–2487.

Molotsky, T., T. Tamarin, A.B. Moshe, G. Markovich and A.B. Kotlyar. 2010. Synthesis of chiral silver clusters on a DNA template. The Journal of Physical Chemistry C 114: 15951–15954.

Murty, B.S., P. Shankar, B. Raj, B.B. Rath and J. Murday. 2013. Unique Properties of Nanomaterials. In: Textbook of Nanoscience and Nanotechnology. Springer, Berlin, Heidelberg. Germany.

Nishida, N., H. Yao and K. Kimura. 2008. Chiral functionalization of optically inactive monolayer-protected silver nanoclusters by chiral ligand-exchange reactions. Langmuir 24: 2759–2766.

Noguez, C. and I.L. Garzón. 2009. Optically active metal nanoparticles. Chemical Society Reviews 38: 757–771.

Oila, M.J. and A.M. Koskinen. 2006. Chirally modified gold nanoparticles: Nanostructured chiral ligands for catalysis. ARKIVOC 15: 76–83.

Pérez, E.M. and N. Martín. 2012. Chiral recognition of carbon nanoforms. Organic & Biomolecular Chemistry 10: 3577–3583.

Purdie, N. and K.A. Swallows. 1989. Analytical applications of polarimetry, optical rotatory dispersion, and circular dichroism. Analytical Chemistry 61: 77A–89A.

Rao, C.R. and A. Govindaraj. 2011. Nanotubes and Nanowires. Royal Society of Chemistry, United Kingdom.

Reimers, J.R., M.J. Ford, S.M. Marcuccio, J. Ulstrup and N.S. Hush. 2017. Competition of van der Waals and chemical forces on gold–sulfur surfaces and nanoparticles. Nature Reviews Chemistry 1: 1–13.

Rocco, A., Z. Aturki and S. Fanali. 2013. Chiral separations in food analysis. TrAC Trends in Analytical Chemistry 52: 206–225.

Sánchez-Castillo, A., C. Noguez and I.L. Garzón. 2010. On the origin of the optical activity displayed by chiral-ligand-protected metallic nanoclusters. Journal of the American Chemical Society 132: 1504–1505.

Sharma, V.P., U. Sharma, M. Chattopadhyay and V.N. Shukla. 2018. Advance applications of nanomaterials: A review. Materials Today: Proceedings 5: 6376–6380.

Shemer, G., O. Krichevski, G. Markovich, T. Molotsky, I. Lubitz and A.B. Kotlyar. 2006. Chirality of silver nanoparticles synthesized on DNA. Journal of the American Chemical Society 128: 11006–11007.

Sholl, D.S. and A.J. Gellman. 2009. Developing chiral surfaces for enantioselective chemical processing. AIChE Journal 55: 2484–2490.

Sholl, D.S., A. Asthagiri and T.D. Power. 2001. Naturally chiral metal surfaces as enantiospecific adsorbents. The Journal of Physical Chemistry B 105: 4771–4782.

Shukla, N., M.A. Bartel and A.J. Gellman. 2010. Enantioselective separation on chiral Au nanoparticles. Journal of the American Chemical Society 132: 8575–8580.

Shukla, N., N. Ondeck, N. Khosla, S. Klara, A. Petti and A. Gellman. 2015. Polarimetric detection of enantioselective adsorption by chiral Au nanoparticles–effects of temperature, wavelength and size. Nanomaterials and Nanotechnology 5: 5–1.

Skandani, A.A., R. Zeineldin and M. Al-Haik. 2012. Effect of chirality and length on the penetrability of single-walled carbon nanotubes into lipid bilayer cell membranes. Langmuir 28: 7872–7879.

Smith, K.W., S. Link and W.-S. Chang. 2017. Optical characterization of chiral plasmonic nanostructures. Journal of Photochemistry and Photobiology C: Photochemistry Reviews 32: 40–57.

Smith, S.W. 2009. Chiral toxicology: It's the same thing… only different. Toxicological Sciences 110: 4–30.

Subashkumar, S. and M. Selvanayagam. 2014. First report on: Acute toxicity and gill histopathology of fresh water fish Cyprinus carpio exposed to Zinc oxide (ZnO) nanoparticles. International Journal of Scientific and Research Publications 4: 1–4.

Sudha, P.N., K. Sangeetha, K. Vijayalakshmi and A. Barhoum. 2018. Nanomaterials history, classification, unique properties, production and market. In: A. Barhoum and A.S.H. Makhlouf (eds.). Emerging Applications of Nanoparticles and Architecture Nanostructures, Elsevier.

Sun, D., W. Zhang, Q. Yu, X. Chen, M. Xu, Y. Zhou et al. 2017. Chiral penicillamine-modified selenium nanoparticles enantioselectively inhibit metal-induced amyloid β aggregation for treating Alzheimer's disease. Journal of Colloid and Interface Science 505: 1001–1010.

Sun, M., T. Hao, X. Li, A. Qu, L. Xu, C. Hao et al. 2018. Direct observation of selective autophagy induction in cells and tissues by self-assembled chiral nanodevice. Nature Communications 9: 4494.

Tohgha, U., K.K. Deol, A.G. Porter, S.G. Bartko, J.K. Choi, B.M. Leonard et al. 2013a. Ligand induced circular dichroism and circularly polarized luminescence in CdSe quantum dots. ACS Nano 7: 11094–11102.

Tohgha, U., K. Varga and M. Balaz. 2013b. Achiral CdSe quantum dots exhibit optical activity in the visible region upon post-synthetic ligand exchange with D-or L-cysteine. Chemical Communications 49: 1844–1846.

Utembe, W. and M. Gulumian. 2019. Chirality, a neglected physico-chemical property of nanomaterials? A mini-review on the occurrence and importance of chirality on their toxicity. Toxicology Letters 311: 58–65.

Valev, V.K. 2016. Chiral nanomaterials and chiral light. Optics and Photonics News 27: 34–41.

von Zelewsky, A. 1996. Chiral complexes of platinum metals. Platinum Metals Rev. 40: 102–109.

Wang, T., J. Lu, H. Zhu, J. Liu, X. Lin, Y. Liu et al. 2017. The electronic properties of chiral silicon nanotubes. Superlattices and Microstructures 109: 457–462.

Wang, X., M. Wang, R. Lei, S.F. Zhu, Y. Zhao and C. Chen. 2017. Chiral surface of nanoparticles determines the orientation of adsorbed transferrin and its interaction with receptors. ACS Nano. 11: 4606–4616.

Wilhelm, C., C. Billotey, J. Roger, J. Pons, J.-C. Bacri and F. Gazeau. 2003. Intracellular uptake of anionic superparamagnetic nanoparticles as a function of their surface coating. Biomaterials 24: 1001–1011.

Wong, C.S. 2006. Environmental fate processes and biochemical transformations of chiral emerging organic pollutants. Analytical and Bioanalytical Chemistry 386: 544–558.

Xin, Q., Q. Liu, L. Geng, Q. Fang and J.R. Gong. 2017. Chiral nanoparticle as a new efficient antimicrobial nanoagent. Advanced Healthcare Materials 6: 1601011.

Xu, C., C. Zhao, M. Li, L. Wu, J. Ren and X. Qu. 2014. Artificial evolution of graphene oxide chemzyme with enantioselectivity and near-infrared photothermal effect for cascade biocatalysis reactions. Small 10: 1841–1847.

Xu, L., X. Wang, W. Wang, M. Sun, W.J. Choi, J.-Y. Kim et al. 2022. Enantiomer-dependent immunological response to chiral nanoparticles. Nature 601: 366–373.

Yanez, J.A., P.K. Andrews and N.M. Davies. 2007. Methods of analysis and separation of chiral flavonoids. Journal of Chromatography B 848: 159–181.

Yang, H., J. Yan, Y. Wang, G. Deng, H. Su, X. Zhao et al. 2017. From racemic metal nanoparticles to optically pure enantiomers in one pot. Journal of the American Chemical Society 139: 16113–16116.

Yao, H., K. Miki, N. Nishida, A. Sasaki and K. Kimura. 2005. Large optical activity of gold nanocluster enantiomers induced by a pair of optically active penicillamines. Journal of the American Chemical Society 127: 15536–15543.

Ye, J., M. Zhao, J. Liu and W. Liu. 2010. Enantioselectivity in environmental risk assessment of modern chiral pesticides. Environmental Pollution 158: 2371–2383.

Ye, J., M. Zhao, L. Niu and W. Liu. 2015. Enantioselective environmental toxicology of chiral pesticides. Chemical Research in Toxicology 28: 325–338.

Yuan, L., F. Zhang, X. Qi, Y. Yang, C. Yan, J. Jiang et al. 2018. Chiral polymer modified nanoparticles selectively induce autophagy of cancer cells for tumor ablation. Journal of Nanobiotechnology 16: 55.

Zeng, C. and R. Jin. 2017. Chiral gold nanoclusters: Atomic level origins of chirality. Chemistry–An Asian Journal 12: 1839–1850.

Zhang, C., Z. Zhou, X. Zhi, Y. Ma, K. Wang, Y. Wang et al. 2015. Insights into the distinguishing stress-induced cytotoxicity of chiral gold nanoclusters and the relationship with GSTP1. Theranostics 5: 134.

Zhang, N.-N., H.-R. Sun, S. Liu, Y.-C. Xing, J. Lu, F. Peng et al. 2022. Gold nanoparticle enantiomers and their chiral-morphology dependence of cellular uptake. CCS Chemistry 4: 660–670.

Zhu, Y., J. Guo, X. Qiu, S. Zhao and Z. Tang. 2020. Optical activity of chiral metal nanoclusters. Accounts of Materials Research 2: 21–35.

Chapter 9

Surface Chemistry and Immunochemistry at a Crossroad called COVID-19

Munishwar Nath Gupta

Introduction

When the Covid-19 assumed pandemic status; it caught us unprepared in several respects. That included the scientific establishments and scientists. It looked in those early days of Covid-19 that they were always scrambling to meet the unfolding catastrophe. They reacted rather than acted most of the times. The "kits" malfunctioned; common men were saddled with trying to understand what false positive or a false negative result means in a diagnostic test. Most bizarre was the confusion about droplet transmission vs aerosol transmission. Many "instant experts" nurtured by 24 x 7 TV and social media created more confusion.

However just like good money can drive the bad money out of the market; dissemination of good science can let the light shine through cracks of our civilization. Why that did not happen? One reason is that our scientific training has changed over the last few decades. For a biologist, it is deemed more necessary to understand bioinformatics rather than surface chemistry! We have powerful tools to measure sizes of particles such as light scattering based methods but their output is used mostly to garnish a paper. The biologists are seldom exposed to the basic science behind these tools and neither do most of them learn about the challenges in using such methods to obtain valid data. We are increasingly training biologists to use tools and techniques such as black boxes; with the opaque power of computers shrouding the deep knowledge with algorithms. We have forgotten what the wise people once snickered about GIGO (garbage in, garbage out) as the danger of computer based technology. Unlike the scientists trained in the earlier era, the simple protein

Former Emeritus Professor, Department of Biochemical Engineering and Biotechnology Indian Institute of Technology Delhi, Hauz Khas, New Delhi 110016, India.
Email: mn7gupta@gmail.com

chemistry/immunochemistry behind the diagnostic kits is a mystery to those who use them. Everything is outsourced, with that is the skill to figure out whether the kit malfunctioned or why it did not work. To those who were taught about colloids as part of simple surface chemistry; the confusion about droplets was comically tragic. To those who learned how soaps work, how to sanitise our hands and surfaces would have come easy. Similarly, too much inaccurate information about simple immunochemistry went viral; herd mentality took over and even the people who should have known better dropped the ball about herd immunity.

This chapter is based upon a rationale that for biotechnologists (and medical persons), some simple (re)-education about surface chemistry and immunochemistry may be desirable in the post-covid era. The need has arisen as the current crop of biotechnologists often miss out on these fundamentals. Covid-19 has shown the tragic consequences of that.

Chemistry & biochemistry of colloids

Our first exposure to general science teaches us that there are three states of matter- solid, liquid and gas. In the next stage, we learn how these three states of matter behave ideally or in real life. These theoretical frameworks mostly focus on their behavior when these states of matter exist alone. Also, we are taught when these phases dissolve' in each other. The chemistry of solutions is dealt with quite adequately in school chemistry. There is also a chapter tucked in on colloids. Biochemists tend to not worry about colloids; they generally talk of enzymes, carbohydrates, nucleic acids 'in solutions' (overlooking that many macromolecules actually exist in colloidal state). The binary of solutions and suspensions is further emphasised by the fact that lipids are uniquely defined in terms of their being insoluble in aqueous buffers. Most biochemists tend to ignore that protein 'solutions' are not 'true solutions'; these are actually colloidal in nature. This section of present chapter is aimed at discussing the importance of colloids in biochemistry and biotechnology and explain their relevance to the pandemic Covid-19. Much of the confusion (as pointed out in the introduction) about aerosol vs droplet transmission was avoidable by recalling fundamentals of colloids.

Colloids also represent two phases-dispersion medium and the dispersed phase (Moore 1986, Atkins 1978, Gupta et al. 2021). In a true solution, the sizes of the particles of the dispersed phase (solute) are less than about 1 nm. If the sizes of these particles exceed 100 nm, this leads to suspensions. The sizes of colloidal particles are between 1–100 nm. Such colloidal preparations (also called sols which is not to be taken as a short form of solutions!) have properties different from a true solution. In recent times, there has been much excitement about the nanoparticles having novel properties because of their small sizes and large surface to surface volume ratios (Gupta et al. 2021). In fact, ancient physical chemists realised that the dramatic changes in the size (while moving from true solutions of a solute to its colloidal form) also endows colloids with very different properties. How the reduction in size have diverse consequences in science and technology is illustrated beautifully in the book by Frankel and Whitesides (2009).

Some coloured glasses exemplify solid dispersed in solid; many paints are solid colloidal particles dispersed in a liquid. Aerosols like smoke are solids dispersed in gas whereas gels (e.g., butter) are actually liquids dispersed in solids. Emulsions (milk being an important example) are liquids dispersed in other liquids. Liquids dispersed in gas are also called aerosols, e.g., fog, mist and cloud. Solid foams (e.g., foam rubber) are gases dispersed in solids. Gases dispersed in liquids are also well known; froth on the top of the coffee cup or whipped cream are examples of such foams (Atkins 1978, Moore 1986).

However, to gain further insight into the colloids relevant to our discussion, let us first briefly recall the fundamental concept of surface tension (Moore 1986). Molecules in the (bulk) interior of a system behave differently from those at the surface. And smaller particles have large interfaces with the dispersion medium in colloids. At a liquid-air surface, the potential energy of the molecules is in between those in the air and interior of the liquid. In the interior, molecules are surrounded by other molecules with which they interact. Surface tension is a force directed on the surface molecules towards the interior and represents the gain in potential energy if these molecules were to be in the interior.

Droplets tend to be spherical as it minimises the surface area and allows the maximum number of molecules to be in the bulk rather than surface. As pointed out by Atkins (1978), gravitational forces 'flatten spheres into puddles or oceans'!

Cavities in a liquid are small amounts of gas/vapour molecules trapped in the liquid. Bubbles, on the other hand are thin films of liquids enclosing liquid or vapour. Cavities have a single surface; bubbles have two. Soap bubbles are bit more complicated in structure. Soaps molecules are salts of long chain fatty acids having a polar end and a longer non-polar tail. Such surfactant molecules give rise to many interesting structures in chemistry and biology. Biological membranes, it may be recalled are bilayers of phospholipid (which are good examples of biosurfactants) molecules. Liposomes are spherical micro-vesicles made of the phospholipid bilayers enclosing the liquid. Not all biological membranes in a cell are spherical in nature. This is due to the presence of phospholipids of diverse structures and presence of peripheral proteins. The physical chemistry of this is discussed by Kozlov et al. (2014). Soap bubbles have a thin bilayer of the soap molecules containing water molecules in between the bilayer which encloses air in the centre of the spheres. Micelles are colloids in which surfactant molecules come together so that the polar parts of these amphipathic molecules face the bulk water, thus minimising the contact of non-polar parts with water. On the other hand, reverse micelles are formed when the surfactant molecules are in an organic solvent, the contacts of polar ends with the organic medium is minimised.

Coming back to the interesting consequences of small sizes, as early as 1871, Kelvin showed that because of surface tension, vapour pressure of a liquid in a small droplet is far more than the liquid in the bulk or on the plane surface (Atkins 1978). That is why the volatile liquids in aerosol sprays evaporate very fast (Moore 1986).

Droplet vs aerosol transmission of viral particles

There has been less clarity about some basic concepts related to the transmission of infective pathogens. Respiratory droplets which are > 5–10 microns in size fall (because of the gravity) to a nearby surface within 1 m (Stetzenbach et al. 2004). The fluid in the exhaled droplets evaporates; with increase in temperature the evaporation is faster and the droplet size would rapidly decrease as it travels. Also, the particles are never restricted to this size range to start with. Hence, it is surprising that early message from regulatory institutions and experts was that CoV-2 does not spread by airborne transmission. Airborne transmission is due to smaller particles which can travel further than many meters. In practice, the air borne particles spread in the three dimensional space; there is rapid dilution of the infective particles. Thus, both for droplet transmission and airborne transmission, the highest probability of catching infection is at a short distance from the source. Indoors, crowded spaces especially with poor ventilation were found to be the sites of 'super-spreading' (Chaudhuri et al. 2020, Qu et al. 2020].

One of the best sources of information on this were blogs by Morgenstern (2020, 2020a). These were far more critical, exhaustive and accurate than many reviews/research papers which flooded most of the respectable journals. Strictly speaking, aerosols are colloids which are liquids or solids suspended in air. Some like fog may be visible, dust/pollens are not. In the context of COVID-19 transmission, aerosol transmission has generally meant transmission via airborne particles. Contact spread refers to spread by touching a "fomite" (contaminated surface) or by being in the spray (breathing, talking, sneezing, coughing) zone around a person who is shedding infective pathogens. Infective water sources can also generate infective aerosols-like flushing a toilet. In fact, detecting the pathogen in a water body around a community has been used as a marker of virus spread. Presymptomatic or asymptotic spread indicates airborne transmission whereas symptomatic patients exhale mostly droplets. It should be added that that there is no universal agreement about the size limit beyond which a droplet would be considered large (to qualify as the cause of droplet transmission); different workers have stated the cutoff between 2–100 microns. Ambient conditions (humidity, temperature, level of ventilation, etc.) influence the transmission quite a lot. The size of the infective particle matters a lot. As pointed out by Morgenstern (2020), a 100-micron particle will take about 3 seconds to fall at about 1 m; 1 micron aerosol will take 30,000 seconds. Such estimates assume that the 100-micron droplet will not reduce in size. Also, it is likely to be accompanied by smaller aerosols particles.

It is a common mistake to correlate Ro value with airborne transmission. Ro relates to infectivity. As has been pointed out earlier, even airborne (or less accurately called aerosol transmission) transmission is not necessarily effective over a large distance. At 1 m, exposure to aerosol is 2000 x more effective than exposure to the droplets (Morgenstern 2020a).

Another level of complexity is added by the presence of super spreaders. Some estimates are that 10–19% of the people were responsible for 80% of the transmission. Many workers think that the dispersion factor k is more important than Ro factor (Kupferschimidt 2020). The dispersion factor measures the infectivity within a population. Smaller k indicates the high clusters-that is fewer individuals are responsible for larger number of transmissions. Some estimates of k for COVID-19 put it around 0.1. Another observation has been that most COVID patients are highly infectious for a very short time.

Which mask is good during covid-19?

A mask has two functions in a social setting. If you are infected and releasing virus particles while exhaling; it traps those and protects people around you. That could be your familiy and friends. Thus you have blocked the transmission of the virus and spread of the infection. The second protection is for you if an infected person happens to be around you or even passing by you (in case of more infectious variants of the virus). A huge variety of masks are now available in the market. Some look very good and expensive but are not necessarily effective. Also many well known personalities appeared on TV during the early days and their message was -just get hold of any piece of cloth, create loops for your ears and you are good to go. Yes, something is better than nothing but that depends upon many factors. If the infection is highly prevalent around you and if you are in a crowded and non-ventilated space, contrary to what the celebrities and TV studio experts spoke, just two loops and a good old rag does not have a high probability of preventing infection.

Filtration, especially membrane filtration is a well-developed area in biotechnology. For many decades, it has been known that the pore size of the membranes for particles has to be selected according to the size range of the particles which one wishes to retain. Microfiltration for cell debris and ultrafiltration for even smaller particles (biopolymers) have been known for quite some time. The process of retaining the virus away from the non-infected through a mask is essentially a filtration process.

Even a surgical mask can filter droplets but maybe quite ineffective against smaller particles. CDC (USA) recommends well-fitting respirators (masks) approved by National Institute of Occupational Safety & Health (NIOSH). N-95 masks (approved by NIOSH) can filter upto 95% of the airborne particles. A cloth face covering, on the other hand, has as high as 75% leakage from inhaled or exhaled airborne particles. N-95 masks were originally designed and manufactured by 3M company and are made up of non-woven (polypropylene) material with electrostatic charges (both positive and negative). The fit in the design is as important as the filtration efficiency of the material. These are also claimed to be conducive to easy breathing.

KN95 are masks tested by Chinese manufactures whereas KF94 are tested as per their standards by Korean ministry of Food and Drug Safety. Similarly FFP2 masks follow EU standards and have at least 94% filtration efficiency for airborne particles.

Brief introductions to infectious pathogens and viruses

The "germ theory of disease" has a long and interesting history (Pelczar et al. 2010). Robert Koch and Pasteur are generally mentioned for advancing this theory. Long before them were Giralamo Fracastoro (1483–1533) and von Plenciz (who stated in 1762 that different germs caused different diseases). Oliver Wendell Holmes; Ignaz Philipp Semmelweis and Joseph Lister also made notable and early contribution to identification of microbes as the causative agents of many diseases. It is Koch, who in 1876 identified Bacillus anthracis as the cause of anthrax-the first microbe to be identified for a specific infectious disease. Koch also went on to identify in 1883 Vibrio cholerae as the infectious agent for Cholera. Koch, of course, is well known for Koch's 4 postulates which were criteria for establishing a microbe as the pathogen responsible for a particular disease (Pelczar et al. 2010).

Diverse bacteria are always present in humans as what has become well known in recent years as 'gut flora' (Guo et al. 2022). A much smaller number of fungi (mostly yeasts) and protozoa may also inhabit our bodies. Some viruses are also present (called orphan or enteric cytopathogenic human orphans or echoviruses); some have been known to be associated with diseases like nonspecific febrile illness; acute respiratory disease; enteritis, etc. Chronic adenovirus infections are known to occur without causing any disease but their presence in the lymphoid tissues have been detected after culturing in the laboratory (Pelczar et al. 2010). It is interesting to note that recently a large number of cases of hepatitis in children are suspected to be due to the adenovirus 41 F which became severely pathogenic. It is not clear what triggered this but many hypothesis suspect the role of Covid-19 in this (Brodin 2022).

The normal flora is not limited to gut. For example, skin discourages microbial colonization by dryness, low pH and secretory substances like lysozyme, long chain unsaturated fatty acids, etc. Yet, skin does have a normal flora with sweat and sebaceous glands secreting nutrients like amino acids, urea, salts and fatty acids for their growth. Species of *Staphylococcus* and aerobic corynebacteria are prominent in skin flora. The normal flora is also found in nose, eyes, mouth cavity, teeth and gastrointestinal and urogenital tracts (Pelczar et al. 2010, Janeway et al. 2001).

Most of the microbes in normal flora are harmless. In fact, healthy gut flora is critical for good health (Guo et al. 2022).

Infections occur at external surfaces (like skin, conjunctiva) or internal surfaces (like respiratory, intestinal and urogenital tracts) (Hood et al. 1984, Tizard 1988, Janeway et al. 2001). After successful colonization, some may remain local while others may spread to result in generalised infections. Some may enter cells while in other cases, infection is limited to extracellular space. In some cases, microbial secretions called toxins cause disease. Broadly, infections may be classified as acute; chronic; fulminating (sudden and severe); localised; generalised; mixed; primary and secondary (for example, pneumonia preceded by flu). Pathogens use virulence factors (diverse in nature) to overcome the defence systems of the invaded host (Janeway et al. 2001).

In many cases, most microbes when cultured repeatedly in media or used for infecting other animals become a virulent. Such strains are useful to be used as live microorganism vaccines. Sabin's oral polio vaccine is the most well-known example of this (Male 2014).

Let us now discuss viruses briefly.

Viruses can enter animal bodies through skin or mucous membranes of respiratory, gastrointestinal and urogenital tracts. The infection can remain local or it may spread to different organs-that depends upon the virus as well as general health and immune status of the host (Pelzcar et al. 2010).

Airborne pathogens (including viruses) enter respiratory tract (Janeway et al. 2001). These originate in the air from nose and throat secretions of the already infected persons. Respiratory infections also can spread from fomites (any surface which was in contact with the earlier infected person). The coughing, sneezing and even talking can generate aerosols which remain in the air for some time. Disturbing fomites can also send back the viral particles back in the air. Influenza is a well-known example of an airborne disease (Janeway et al. 2001).

Waterborne viruses will initially infect intestines (Pelczar et al. 2010). Typhoid and cholera exemplify such diseases. Viruses present in the fecal matter of the earlier patients contaminate drinking water sources and that is how transmission of such diseases takes place. Foodborne diseases like botulism or food poisoning are caused by ingestion of food containing the pathogen. In this case, the food from the animal source may be infected; food may be infected from the spores because of contamination through soil or a carrier (handling food preparation may be infected). Some pathogens are transmitted by direct contact. Sexually transmitted diseases are the prime examples but contact with infected animal tissues by hunters, butchers, slaughterhouse workers or veterinarians may occur. Open wounds are vulnerable to infections in diverse ways. Finally, there are vector (an organism carrying the pathogen) borne diseases. Even a common housefly can transmit pathogens which cause salmonellosis, polio, hepatitis. Yellow fever, dengue fever are viral diseases involving mosquito as the vector. Another arthropod vector is wood tick which cause tick fever by transmission of tick fever virus (Peczar et al. 2010).

Viruses are considered obligate intracellular parasites with diverse range of sizes (10–400 nm) and shapes (Gupta et al. 2021). Their extracellular form is called virion or (more frequently) simply virus particle. However, viruses need hosts to survive and multiply. Thus, it is only inside hosts that they can mutate and produce variants with different infectivity and pathogenicity. Viruses with bacteria, fungi, plants and animals as hosts are known. In some cases and under some circumstances they can overcome specie barrier -hence the debates about CoV-2 having originated in bats, etc. Viruses are essentially nucleoproteins but lipids are also present in some cases. Depending upon nature of the genome; a virus can be a DNA virus or a RNA virus. The nucleic acid may be single stranded or double stranded; linear or circular (in case of DNA, which can be a close circle on one strand or both; in case of RNA viruses, only some plant ones are circular) and may have positive sense, negative sense or ambi-sense (Watson et al. 1987, Alberts et al. 1994).

Further discussion is limited to animal viruses. The protein portion called capsid can be either a hollow helix or icosahedral. Some animal viruses have an envelope (made up of lipid bilayer) surrounding the capsid or nucleocapsid (as it is sometime called) in case of enveloped viruses. Tagovirus, retrovirus, coronavirus are examples of enveloped +strand RNA viruses. Herpesvirus and poxvirus are examples of dsDNA viruses. Examples of nonenveloped viruses include parvovirus (ssDNA), papovavirus and adenovirus (both dsDNA), picornavirus (ssRNA), reovirus (dsRNA). RNA viruses have genomes in the narrower range (7–20 kb); genomes of DNA viruses fall in the range of 3–200 kb. Virions have single copy of the genome (are thus haploids); though retroviruses have 2 copies of ssRNA (Alberts et al. 1994).

In animal viruses, capsid proteins (in case of nonenveloped viruses) and virally encoded glycoproteins called peplomers or spike proteins (projecting from the envelope in case of enveloped viruses) bind to the receptors on the host animal cells (Watson et al. 1987, Baker et al. 2011). These receptors are actually host cell surface proteins with defined cellular functions but are exploited as receptors. There is a specificity involved-different viruses use different host cell surface proteins as receptors. The capsid protein VP1 of polio virus exploits CD 155, a member of Ig superfamily found on many cells but only those of humans and primates. In case of HIV, its envelope protein gp120 binds to CD4 of human T-cells. In fact, CCR5 and CXCR4 act as coreceptors. Flu virus can bind to many cells as its protein hemagglutinin just needs sialic acid on the host cell membrane proteins. Human coronavirus 229E uses aminopeptidase N as receptor whereas human coronavirus OC43 just needs N-acetyl-9-O-acetylneuraminic acid. So, these interactions cover a wide range in terms of their nature and specificities. They also define the range of hosts for a virus as without this step, there is no infection (Alberts et al. 1994).

The next step is viral entry into the host cell. For nonenveloped viruses, virions are engulfed by phagocytosis and the capsid is removed in the phagocytic vacuole by lysosomal proteases. For enveloped viruses, the fusion of the envelope with host cell membrane takes place, nucleocapsid enters after which the viral genome is uncoated within the cytoplasm (Alberts et al. 1994).

Viruses require host cell machinary to replicate and that includes protein synthesis (Alberts et al. 1994). Thus, the host cell damage starts with this hijacking of cellular resources for its replication by the virus. The viral proteins are synthesized in two phases- early and late. Early proteins are synthesized before viral gemome replication as that recquires these proteins. The first set of early proteins (called immediate early) just need cellular RNA polymerases. The late proteins are generally synthesized along with viral genome replication. Many viral proteins require post-translational modifications. Phosphorylation and glycosylation are more common but palmitoylation,acylation and myristoylation are also known. There are several ways by which mRNA is formed from the viral genome. These strategies constitute the basis of Baltimore classification of viruses. In case of some viruses (e.g., influenza, ebola and mumps), RNA polymerase is the one which is part of nucleocapsid which is inside the cell after entry. So, there are different schemes which operate in different cases. The protein synthesis can be controlled at both transcription (of viral genome) and translation level. The host cell ribosomes are used in all cases. Proteins which form nucleocapsid, matrix (the space between capsid and envelope in case of enveloped viruses) and envelope are all late proteins and constitute what are known as structural proteins of viruses. Apart from these, virus also have several enzymatic and regulatory proteins-polymerase, protease, kinase and those which are involved in gene regulation. These also include some which diminish original protein synthesis by the host cells; favoring the protein synthesis required for viral replication. The virus genome also must be replicated for the various new viral particles to form (Alberts et al. 1994).

The next stage is the assembly of the new viral particles. The cellular location for this is different in case of different viruses. Capsid part is assembled either in the nucleous or cytoplasm. In general (with few exceptions); assembly of DNA viruses takes place in the nucleous whereas RNA viruses are assembled in the host cell cytoplasm. Next stages of the assembly take place in different regions of the cell. The release (egress) of the newly replicated viruses from the cell also follows different schemes depending upon the nature of the virus. In case of some animal viruses, host cells burst to release the viral particles. In other cases, release is sort of reverse of entry (Alberts et al. 1994).

The envelope of the animal viruses is actually derived from the host membrane. This process called budding may involve nuclear or plasma or endoplasmic reticulum membrane. The necessary (viral) glycoproteins are in place in specific regions of the membrane which are involved in budding. In some animal viruses, this does not involve host cell membrane damage as the viruses leave through some channels (Alberts et al. 1994).

Corona viruses and CoV-2

CoV-2 is not the first corona virus which has caused wide spread alarm. Earlier SARS-CoV and MERS-CoV have caused endemics in 2002 and 2012, respectively. Many human corona viruses hCoV-229 E, hCoV-NL63, hCoV-OL43 and hCoV-HKU1 have been known for decades but they all cause only common cold symptoms (Vu and Menachery 2021).

Less pathogenic human corona viruses restrict their infection and replication to the upper respiratory tract only. CoV-2 spreads and replicates throughout the respiratory tract (Fig. 1). Also, while the high transmission is generally associated with less pathogenesis, CoV-2, at least some of its variants, have show high transmission and high degree of pathogenesis. Vu and Menachery (2021) have discussed how binding affinity of the virus to human receptors and some other parameters affect the pathogenesis throughout the various organs in the human host.

Essential immunology

The above discussion shows that there are diverse kinds of infectious agents against which the immune system has to protect us. In the order of their sizes, these are worms (e.g., tapeworm, filaria); protozoa (e.g., amoeba, malaria); fungi (e.g.,

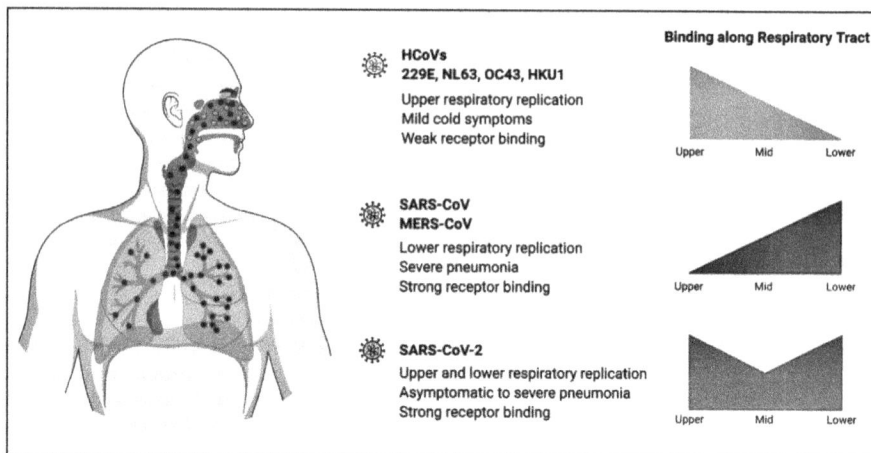

Figure 1. hCoV replications in the respiratory tract. (Figure reproduced from Vu and Menachery, 2021 published under CC BY license).

candida); bacteria (e.g., mycobacterium, staphylococcus) and viruses (e.g., pox, polio, influenza, corona viruses). Infections occur through exposed surfaces (Roitt 1988, Roitt et al. 1993). Our body surfaces have both non-immunological as well as immunological mechanisms operating right at the surfaces (Roitt 1988, Tizard 1988). The microbial flora present on the skin and intestines not only competes with the infectious agent, there are secondary mechanisms like pH (lactic acid in the skin), enzyme lysozyme (in tears), etc., which are inhibitory to the growth of the invaders. Urine and milk not only have flushing action, low pH of the former and enzymes and other proteins in the latter have protective effects locally. Glycogen in vagina encourages the growth of lactobacilli which produce lactic acid as a protective agent. In the respiratory tract, the size of the openings of upper respiratory tract, bronchi, bronchioles and alveoli progressively decrease to filter out the invading agents/particles. The particles are also trapped by the mucous in the upper respiratory tract; mucous has lysozyme as well (Tizard 1988). Surfaces also have immunological molecules/mechanisms in place.

Innate immunity

One type of immunity is innate immunity (Roitt, 1988). Innate immunity is not antigen (or pathogen) specific and is the first line of defence which quickly kicks in once we are exposed to a pathogen. Innate immunity has no memory; so previous encounters with the same/similar pathogen/antigen are not of any consequence. Innate immunity involves both soluble molecules as well as cells. It is present right at the birth and protects the infant before it develops the adoptive immunity. The cells which are part of innate immunity are phagocytes, natural killer (NK) cells, mast cells and dendritic cells (Roitt 1988, Tizzard 1988).

Phagocytes can take up antigens/pathogens and degrade them. These are of three kinds: monocytes, macrophages and neutrophils. NK cells kill the infected host cells.

For a further description of the cells responsible for innate immunity, any standard text of immunity may be consulted (Roitt 1988, Tizard 1988).

Adoptive immunity and antibodies

Adaptive immunity (or acquired immunity as it is sometime called) is antigen specific and takes time to develop. It also has memory; i.e., a repeat encounter produces swifter and stronger response (Roitt 1988, Leydard et al. 2011).

One component of adoptive immunity is antibody mediated immunity (Roitt 1988). Antibodies are produced in response to the presence of antigens in the system. A foreign substance has to be antigenic for triggering the immune system. Larger size and complex (non-repetitive) structures tend to enhance antigenicity. Thus proteins (with the different side chains of the constituent amino acids) are much better antigens as compared to polysaccharides or nucleic acids. Immune systems can distinguish between self and non-self molecules. Hence the requirement of foreign nature. A pathogen triggers immune system as it carries antigens as part of its system. Thus, an antigen can have a high molecular weight complex molecule, a cell, an organism or even these entities immobilised on some particles (e.g., latex particles) (Roitt 1988).

For a beginner, it is helpful to always keep in mind that immunologists tend to switch over seamlessly between experimental immunology and functioning of the immune system. Much of the early immunological work was based upon obtaining crude preparation of antibodies in the form of sera from immunised animals (such as cats, goats, chickens). Many of the serological techniques (which are still routinely used) were developed with such sera. These include agglutination; precipitin reaction (in solution or in agar gel); Ouchterlony double diffusion and immunoelectrophoretic techniques (Roitt 1988). Radio-immunoassay and enzyme-linked immunosorbent assay (ELISA) have been two other well known techniques which have been around for a long time. Later many fluorescence based techniques such as immunofluorescence; flow cytometry and fluorescence-activated cell sorter have emerged as powerful tools based upon antigen-antibody interactions (Roitt 1988).

Given that the antigen-antibody interactions are fairly specific (however see elsewhere in this chapter for the cross-reactivity of antibodies and protein promiscuity); some of the early affinity adsorbents were developed with antigens and reasonably high purity antibody preparations were soon available (Roitt 1988). Generally, some substances termed adjuvants are mixed with antigens before injecting these in the animals; the incorporation of adjuvants results in better response. Water-in-oil emulsions were described by Jules Freund (and are called Freund's incomplete adjuvant) and were improved upon by incorporating heat killed bacteria such as *M. tuberculosis* (constituting what are known as complete Freund's adjuvant). There are diverse kinds of adjuvants which are used, the mechanisms of their function depends upon their nature. The common basis is longer half-life of the antigen preparation in the animal because of the use of the adjuvant.

Right in the early days of immunology, it puzzled scientists that the immune system is always able to produce a specific antibody against an antigen which it had never encountered before (Rich and Davidson 1968). This led to two schools of thought with their respective theories. One was led by Linus Pauling and believed that the origin was somatic in nature, the antigen acted as a 'template' and 'instructed' the antibody molecules to fold to produce the appropriate binding site. These theories are called template theories or instruction theories. Biologists (Jerne, Talmage, Burnet, Lederberg) on the other hand believed in the genetic basis and suggested that the B-lymphocytes (white blood cells which secrete antibodies) pre-existed with potential to produce antibodies of diverse specificities and any antigen 'selected' the particular lymphocytes to start producing the corresponding antibody molecules (Hood et al. 1984). These were called selection theories. To cut a rather long story (which lasted for a number of years and involved heated controversies) short, at present "clonal selection theory" is widely accepted (Hood et al. 1984, Male 2014).

The clonal selection theory is briefly described here. B-lymphocytes are continuously produced in bone marrow and have immunoglobulin M (with different binding sites for antigens) on their surfaces. Hence a library of B lymphocytes pre-exists and a set of specific clones gets triggered by a pathogen (which carries many antigens!). Every antigen has some different regions which participate in this selection. These regions were called "epitopes" and are currently more often called 'antigenic determinants'. A specific antigenic determinant selects a limited number of sets out of this population of B-cells by binding to the immunoglobulin M on the surface (Hood et al. 1984).

This leads these subset of clones to expand in number and also become active in terms of synthesising immunoglobulins. As these sets (clones) are selected, the theory is called clonal selection theory. After this selection, these clones multiply; cells become bigger in size and start secreting out immunoglobulin M molecules. The first kind of immunoglobulins to be secreted are immunoglobulin M. As the infection persists, the B lymphocytes switch over to synthesise and secrete immunoglobulin G. This phenomenon is called "class switching" and can also produce A, E and D immunoglobulins in different organs and under various situations. These immunoglobulins are soluble proteins and this part of immune response is called humoral response (Hood et al. 1984, Male 2014).

Experimental immunology consists of raising antibodies by injecting an antigen into the animal (Hood et al. 1984, Janeway et al. 2001). Initially, immunoglobulin M are produced at a low level. This is followed by production of immunoglobulin G. A 'booster dose' of the same antigen given to the same primed animal after few weeks results in faster production of immunoglobulin G and at a much higher level in the next few weeks. The blood from the animal after this can be collected and is called polyclonal sera. So much of immunological work can be done with this crude preparation of antibodies. Individual antibodies for an antigenic determinant in more pure form can be isolated from this crude sera by the usual protein purification methods (Hood et al. 1984).

As mammalian tissue culture technology matured, hybridoma cells were prepared which were fusions of a cancer cell (which proliferates rapidly) and a B-lymphocyte (Male 2014). One feature of clonal selection theory is that each clone of identical lymphocytes produces identical antibodies. Hence, antibody molecules produced by hybridoma cell culture were identical, more pure and highly specific. These are what are known as monoclonal antibodies.

Antibodies produced in other animals are antigenic for humans to a varying degree (depending upon how far is the animal from humans on the evolutionary tree). Use of recombinant technology facilitates producing these antibodies in which some portions (not part of the binding site) of the molecule is replaced by the corresponding portions of any antibody produced in humans. These *humanised* antibody molecules retain the specificity but are less antigenic when used for therapeutic purposes (Male 2014).

Cell mediated immunity is another kind of adoptive immunity.

So far, we have followed sort of chronological development of immunology. What we have discussed has been limited to one part of adoptive immunity. Adoptive immunity itself has two main components. One is humoral immunity which is production of antibodies by B-lymphocytes. At some point in time; immunologists discovered another kind of lymphocytes called T-cells (Hood et al. 1984, Male 2014). T cell responses are called cell mediated immunity. It turned out that there are actually three different kinds of T-cells. One kind, T-helper cell helps in immune response of B-cells to some antigens. Another kind is killer T-cell which kills the infected cells and hence limits the spread of infection in the body. Different pathogens have different capabilities for triggering T cell responses (Male 2014).

The third kind is regulatory T-cell which controls the immune response. When this falters, we have many kinds of hypersensitivity responses. Two of these are commonly known: anaphylactic shock upon administration of insulin to some people and allergies like hay-fever. Autoimmune diseases also exemplify hypersensitivity (Hood et al. 1984, Janeway et al. 2001). T-cells secrete various cytokines (lymphokines). These cytokines are part of defence mechanisms; unfortunately the term "cytokine storm" used frequently in the context of Covid-19 has been restricted to endow these molecules only with negative connotations. However, these soluble lymphokines (cytokines secreted by T-lymphocytes) are critical for the role of cell-mediated immunity by T-cells and in the role of helper T-cells in effective functioning of what is known as B cell-T cell cooperation (Male 2014).

As the infection fades away, some B lymphocytes and T lymphocytes get transformed into the corresponding memory cells. How large is the population of these memory cells and how long these lasts varies from one pathogen to another. Memory cells enable us to respond to exposure to the same pathogen in future more swiftly (Janeway et al. 2001).

Acquired immunity & vaccines

Immunity against pathogens can be acquired by humans in various ways (Lydyard et al. 2011). One of course is to become infected. The immunity resulting from such prior infections can last for a varying period of time which depends upon various parameters including the nature of the pathogen (or its variant).

There is passive immunisation in which the immune system of the patient is not involved (Tizzard 1988, Benjamin et al. 1996). In this, antibodies produced by another experimental animal/other humans are isolated and injected. Antibodies against tetanus raised in horses is an important example of vaccination by passive immunisation of humans. The antibodies present in the sera of the horses are purified. The immunity is rather short-lived and that is why the doctors invariably ask about the last time anti-tetanus was injected when you are likely to be vulnerable to tetanus. The horse antibodies are viewed by our immune systems as foreign proteins, i.e., antigens and can produce "serum sickness" or even anaphylaxis (hypersensitivity reaction). In recent times, polyclonal antibody preparations can be substituted by monoclonals which because of higher specificity can be more effective.

Passive immunisations has also been used in humans for measles, hepatitis A and B, rubella, botulism, diphtheria and rabies. In almost all the cases, passive immunisation is used after the infection has occurred. It is especially useful for immunocompromised persons (Hood et al. 1984, Janeway et al. 2001).

The "antibody therapy" talked about in the context of Covid-19 refers to this approach. Apart from the above described negative features, passive immunisation has to be evaluated carefully in case of each pathogen. With more data available, it would be possible to have a more sound judgement about this approach to be used in case of Covid-19.

Active immunisation (to trigger the immune system against the pathogen to prime the immune response for preventive purposes) or vaccination is more common and these vaccines are administered as subcutaneous or intradermal injections. Few vaccines like rotavirus or Sabin polio vaccine are given orally. Even fewer (for example flu vaccine) use nasal route (Male 2014).

Some of the vaccines being used for prevention of Covid-19 are quite novel in design. These have been introduced for the first time to meet the sudden requirement due to this pandemic.

Before we briefly discuss them, it is worthwhile to describe prior designs of vaccines which have been in use for other diseases.

A major class of vaccines are subunit vaccines which are based upon use of experimental animals for obtaining antibodies against microbial antigens (Tizzard 1988, Janeway et al. 2001).

Toxoids are also examples of subunit vaccines. Few diseases like tetanus and diphtheria are caused due to toxins produced by the pathogen. Toxoids are chemically modified toxins. Generally formaldehyde is used to ensure that toxicity is abolished while retaining the antigenicity.

In other cases, capsular polysaccharides (e.g., meningococci), capsid proteins (e.g., rabies or foot-and-mouth disease) or their (synthetic) epitopes (e.g., influenza or hapatitis B or foot-and-mouth disease) are used to design subunit vaccines (Janeway et al. 2001).

The well known history of "vaccination" (mostly concerned with prevention of smallpox which used to be a dreaded infectious disease) tells us that the early "vaccinations" were live organisms (Tizzard 1988). Chinese used scabs from the patients which were very mild cases of smallpox. Jenner used cow pox patients as the antibodies in the sera of these patients cross-reacted with smallpox virus. A very important class of vaccines have been inactivated or attenuated forms of the pathogen (Tizzard 1988).

Challenges in designing vaccines against CoV-2

A vaccine is expected to prevent any serious consequence in the infected but vaccinated patient. The first challenge in designing an effective vaccine has been the high mutation in the CoV-2 which has given rise to many variants of the virus. As the virus has no life outside a host, these variants are arising inside infected hosts. Ideally, if we can stop the spread of the infection by social distancing, wearing proper masks and keep the already infected persons in a proper containment; not only the spread of infection will stop; the variants will not be produced! The ideal situation has been difficult to reach globally and given our means of fast travel; achieving it locally or even parts of the world is not good enough. That is where mass vaccination is desirable , not just in the rich nations but in the relatively less empowered nations as well.

The variants of the virus documented are alpha, beta, gamma, delta, lambda and mu (Chen and Wei 2022). One of the current worries has been the 3 subvariants of omicron (BA.1, BA.2 , BA.3). As BA.2 is known to infect persons which were earlier infected by BA.1; it illustrates the worrisome aspect of the virus mutating faster than many viruses. Just to make it clear-broadly infection and vaccination both trigger the immune system. Of course, vaccination is expected to trigger it in such a way that there are no ill effects due to the immune responses. So, in this case; a variant which infects the person previously infected/vaccinated (with an earlier viral mutant) essentially means that any immune response generated due to the earlier infection/ vaccination has not protected completely in these cases. This can happen broadly due to two reasons. One is that the immune system recognises the new variant as having very different antigenic determinants. Second possibility is that the memory cells generated during the earlier infection are not there in enough numbers. Both possibilities are not mutually exclusive.

It has been reported that omicron BA.2 is not only more contagious but also about 17 times more likely to evade the benefit of current vaccines! In case of this variant, the origin of its higher infectivity is in its stronger binding to ACE 2 (Chen and Wei 2022). Most of the current vaccines target the spike protein on the virus (so do antibodies). There are 60 mutations in BA.1 and 32 are on the spike protein. 15 of these affect binding with the host angiotensin-converting enzyme 2 (ACE-2) via influencing receptor binding domain (RBD) on the virus. RBD-ACE-2 interaction initiates the virus entry. Chen and Wei (2022) also refer to the earlier studies which confirm that natural selection favours mutations which favour production of variants which are more infectious. That is the only way, virus can multiply and survive. The mutations also enable the variant to escape the neutralization of antibodies in the host due to any prior infection by the earlier variants.

The m-RNA based vaccines introduced for the first time and given emergency authorisation for mass vaccination are based upon a novel concept. The concept is to introduce synthetic mRNA in the animal which would express the desirable antigens in the host animal. So, animal now would mount an attack via its immune system. In case of these vaccines for Covid-19, mRNA corresponding to the viral spike protein fragments (which initiates the infection by binding to the cell surface of the animal) have been the most frequently used for the vaccine designs.

An excellent review written on the early development of mRNA vaccines was written before Covid-19 happened (Pardi et al. 2018). It is necessary to point out that mRNA based vaccine design was not suddenly developed after Covid-19! Even before this DNA based vaccines have been described in the literature. Unlike those, mRNA based vaccines offer two clear advantages. Firstly, mRNA does not have to cross the nuclear membrane. Secondly, RNA is degraded much rapidly to act as an antigen itself in the animal. High potency and a more convenient platform (in terms of changing the sequence if the particular virus evolves to create new variants) was already appreciated (Pardi et al. 2018).

Late 1980s and early 1990s saw exploration of injecting mRNA to produce proteins in the model animals (Pardi et al. 2018). It was realised that there was a clear need of a carrier to facilitate uptake and expression of the RNA. A large number of options-polymers, nano-emulsions, nanomaterials and dendrimers were tried. Encapsulating mRNA in lipid nanoparticles have been determined to be best choice. By early 2020, first sequence of CoV-2 RNA was available. By end of that year, Moderna and BioNTech-Pfizer had developed their mRNA based vaccines and obtained regulatory approval! The two vaccines essentially differ in composition of the lipid nanoparticle. That and the formulation affects the storage stability and dosages (and their schedules). A large number of vaccines have been under development, some are slowly getting regulatory approvals. Not all are mRNA based. The key parameter is to increase the storage stability. Another direction being pursued is the delivery mode-vaccines in the form of nasal spray is one such exciting possibility as respiratory tract is the major (in fact the only one known so far) route for infection by CoV-2. Enhancing immune memory; eliminating vaccine escape and increasing effectiveness against evolving viral variants are other important areas of research in development of future vaccines for the future.

Let us now shift to another challenging aspect of the vaccine design to prevent the consequences of infection by this virus. For that, we need to briefly discuss perhaps the most frequently observed feature of pathogenesis after the infection which is hypoxia (Danta 2021). Hypoxia is low oxygen saturation level in the tissues. This is easily measured by sensing oxygen level in the blood. Low saturation level of oxygen in the blood is called hypoxemia. That is why pulse oximeters became one of the early and critical tool to determine whether a covid patient needed hospitalization. Hypoxia (and some other independent consequences of the viral infection) affect functioning of the pathways such as inflammation, cytokine signaling (leading to so called cytokine storm) and calcium signaling (Danta 2021). Intracellular calcium level increases in many cells. This in turn affects many other pathways. Hypoxia and inflammation are interdependent processes, influencing each other. Hypoxia is associated with loss of lung perfusion control; low oxygen vasoconstriction, etc. The viral infection increases glycolysis rate. Combined with low oxygen level (just like in exercised muscles!); the end product is lactic acid. This in turn reduces blood pH, produces respiratory acidosis. Higher intracellular calcium levels result in increase of reactive oxygen species in myocytes which damage tissues. Many viruses are known to replicate faster under conditions of hypoxia. This is due to a hypoxia-associated factor (HIF-1) which affects several host metabolic pathway. HIF-1 under hypoxia exerts transcriptional control on several genes associated with glycolysis, iron metabolism and angiogenesis (Danta 2021).

Now what transpires is that the immune system triggered by the severe infection due to CoV-2 itself causes pathogenesis. So, any vaccine design should aim at safer immune response (Thames et al. 2020). The immune mediated pathogenesis includes acute respiratory syndrome, pulmonary edema and fibris and acute lung injury. A safe and effective vaccine would trigger specific and effector functions of the antibodies, T-cell responses (cytotoxic T cells being most critical) but prevent/minimise damage to the lung and other organs (as time goes by, we are coming across the pathogenesis associated with other organs; some associated with long term effects of the infection by the virus). Looking back at what was learnt during the 2003 outbreak of SARS CoV has been useful. Unfortunately, the follow-ups of those studies were put on the back burner till Covid-19 happened and caught the scientific world by surprise. Most of these studies were made with animals especially primates. Even in the case of SARS CoV, lung damage was observed in case of prior infection/vaccinated primates. It was also realised that IgG level in the blood was not a good indicator of protection available from that virus.

With CoV-2, activation of proinflammatory macrophages which leads to the cytokine storm seems to be a major cause of pathogenesis; the injury to the lung has been identified as a major outcome of this (Thames et al. 2020). Both infection and the vaccination produces antibodies, their Fc component attracts the macrophages and triggers them to produce cytokines. IL-1, IL-6 and IL-8 have been identified as the important cytokines secreted by the macrophages as being responsible for causing injury to the lungs. The virus specific antibodies also facilitate uptake of the virus by macrophages which again creates cytokine storm.

One may ask that if antibodies cause these problems, why the convalescent plasma therapy (via passive immunization route) has given better results in some cases. Again, the animal studies in case of SARS CoV throw some light. It had been reported that while the antibodies raised against nucleocapsid proteins leads to the lung damage, those raised against RBD were neutralising antibodies. That is in line with just IgG level not being a good indicator of the protection against the virus as not all antibody molecules present would be neutralising antibodies. So, one hypothesis has been that the early antibody population (before any affinity maturation sets in) are the cause of the lung injury (Thames et al. 2020). The antibodies against some parts of RBD domain have been identified as being better in neutralisation/clearance of the virus. It is also shown that IgA, which mediates immune responses in the respiratory tract appears much later after the infection. Low levels of CD 3+ and CD 8+ (subpopulations of lymphocytes so classified because of the presence/absence of the surface markers CD 3 or CD 8) have been found to be good predictors of mortality due to this viral infection. One useful observation has been that "dendritic cells loaded with immunodominant epitopes" result in selective production of CD 8+ T cell subset without any innate immunity response (Thames et al. 2020).

Which test is better and other conundra

Most of the tests for the virus are done with a sample of nasopharyngeal fluid. These include multiplex rRT-PCR, CRISPR/Cas 12, CRISPR/cas 3, lateral flow immunoassay, paper-based biomolecular sensors, specific high-sensitivity enzymatic reporter unblocking (SHERLOCK), loop-mediated isothermal amplification (LAMP), etc. (McLennan et al. 2013, Miller et al. 2018, Joung et al. 2020, Rhoads et al. 2021, Tombuloglu et al. 2021).

The two most commonly talked about (and widely used) are rapid antigen test (RAT) and rtPCR. RAT detects the presence of viral (antigens) as it is; rtPCR amplifies the viral antigens. It works with the viral genome extracted and selected antigens therefrom are amplified. Hence it can detect viral presence with much greater sensitivity than RAT. On the other hand, RAT does not require any machine and can be done by the patient itself. It is akin in principle to pregnancy detection tests which now have been available for decades.

The infection has broadly 3 phases. First is the incubation period when virus is multiplying inside the host. Till sufficient concentration is there, no discernible symptoms appear and hence it is called pre-asymptotic period. This can last for 2–14 days after the infection but has been around 5 days with early variants. This has been shorter in case of some latter variants like omicron. Some people never develop any symptoms ever and such cases are called asymptotic. The incubation period is followed by infectious phase when the person is spreading the virus by transmitting it to the surroundings. Asymptotic persons also shed the virus even though they may not have any symptoms of the infection. Finally, there is a post-infectious phase when the patient is no longer has enough viral presence in the body to spread. rtPCR is sensitive enough to detect the virus in all the 3 phases, i.e., for a while even during the post-infectious stage. RAT only detects the viral presence during the infectious phase or later part of the incubation period. False positives are thus common in case of RAT during the early periods. False negatives are rare if the test kit is of a good quality, has not expired and the test is carried out exactly as the vendor specifies. rtPCR is more reliable in this respect.

RAT

RAT is a lateral flow test (LFT), the full name being lateral flow immunochromatography assay (Guglielmi 2020, Liu et al. 2021). As it can be carried out at point of care and even at home; many countries used it for initial testing during the peak periods of the pandemic and advised rtPCR if the patient is symptomatic but shows a negative outcome in RAT. The sample is applied on a pad (paper or any other suitable material), the adsorbed material flows to an adjoining pad which has an ELISA (antibody-enzyme conjugate). There if the sample has the antigen being looked out for, a coloured response indicates a positive test.

rRT-PCR

This technique essentially refers to real time reverse transcription polymerase chain reaction (Tombuloglu et al. 2021). Let us decode this. Polymerase chain reaction amplifies a small amount of DNA and crime fiction and TV related to crime genre has made it universally known. The amplification is carried out in cycles, with each cycle gradually increasing the amount of DNA at the end. This became possible due to the discovery of a DNA polymerase which is thermally stable and is called Taq polymerase. This is because PCR involves high temperature necessary to unfold DNA in order to amplify. As in many cases, one needs to detect a small amount of RNA , one needs an enzyme which is able to make a DNA copy of the RNA before amplification. The process is called reverse transcription and the enzyme is called reverse transcriptase. As CoV-2 is an RNA virus, we need the reverse transcription in its detection as well. RT-PCR involves post-amplification detection/identification. In -rtPCR, that is in-built by use of fluorescent labels. The multiplex versions allow handling of multiple samples together.

The droplet-based digital PCR (ddPCR) combines the power of microfluidics with PCR technology so that one can work with femtoliters-nanoliters sample volumes and is an extremely powerful technology.

In all these formats, the important parameter to understand is cycle threshold value (CT value). This is essentially refers to how many cycles of amplification are required for a given sample in order to be able to detect the viral RNA. Higher the viral presence in the sample, less number of cycles are required; lower is the CT value. So, lower CT value is a bad news. Any value higher than 37–40 is considered a negative test. The CT values of 25 or less have been considered severe cases and were found to require serious attention in a hospital.

Therapeutic intervention for infectious diseases

With above background in place, let us discuss what all options we have in terms of therapeutic interventions beyond the use of vaccines (as already discussed).

To start with, there are non-infectious diseases for which pathogens are not responsible. Identifying the causative factor[s] tends to be bit more complicated and often it is a mix of several factors like heredity, life style or nutritional deficiencies (Pelczar et al. 2010, Janeway et al. 2001). Infectious disease is caused due to a particular pathogen. The major infectious diseases are hepatitis B, TB, malaria, pertussis, AIDS, measles, diarrhoea, flu. The examples of diseases caused by viruses include flu,measles and AIDS. Bacterial diseases include Strep throat and septicemia. Thrush exemplifies fungal diseases (Lydyard et al. 2011).

It was Robert Koch who identified criteria (which are generally known as Koch's postulates) for confirming that a pathogen is responsible for a disease. These postulates are not applicable to the viruses and were meant for animal diseases and extendable to human diseases (Pelczar et al. 2010).

All pathogens occur in a reservoir before they infect a human. This reservoir may be another (carrier/infected) human or soil, water or even some surfaces. There are two kinds of transmission from a reservoir to a person. Direct transmission routes are: contact with an infected person; droplet transmission; contact with the soil; animal bite and transmission from the mother to the fetus. Indirect transmission occurs via another arthropod, etc.; through water, food and blood; air; surfaces (fomite-born) and hands (which have come in the contact with the pathogen recently) (Pelczar et al. 2010).

Conclusions and future perspectives

It is increasingly realised that the main lesson from the pandemic has been that all stake holders have to come out of their silos. This requires understanding of the basic concepts in different areas like chemistry, statistics and data analysis, biochemistry and nanotechnology (Gupta 2012, Gupta and Mukherjee 2015, Gupta and Roy 2020, Gupta et al. 2020, Gupta et al. 2021). Understanding how the tool boxes in these areas operate and what are their limitations are two critical components before an effective communication can take place. This will ensure that the decision making committees operate on the basis of science and not over-run by the cognitive biases of the individuals (Gupta and Thelma 2019).

It is under-appreciated that specificity of the biological entities is at the heart of so many issues concerning management of this and similar endemics and pandemics (Gupta et al. 2020, Blundell et al. 2020). Drugs bind specifically to the receptors. The virus is able to infect as it fools the cellular receptors by faking the specificity involved in the normal ligand-receptor infections. The adoptive immunity works through the specific recognitions by the components of the immune system to various 'foreign invaders'. At the heart of biological specificity is the old structure-function paradigm biology.

Even among biologists, it is not uncommon to ignore that this paradigm has undergone drastic revisions (Gupta et al. 2011, Gupta and Mukherjee 2013). Biology seems to exploit both structure and intrinsic disorder in healthy and diseased organisms (Goh et al. 2012, Gupta and Uversky 2022). The biological specificity is seldom absolute and often misunderstood-it is often a simple selection from what is available in real time and space for binding (Gupta and Uversky 2022). At another time and place, the "binding specificity" can change. Both moonlighting and promiscuity by proteins result from this (Petsko and Ringe 2009, Gupta et al. 2019, Gupta et al. 2020). Any drug binding to a receptor is a promiscuous binding; it is not a normal metabolic process. Drug repurposing is simply further exploitation/ re-examination of the promiscuous binding! (Li et al. 2019, Li and Clercq 2020, Gupta and Roy 2021).

There is a mistaken belief that just having some big data and its analysis will solve many problems. There is a need to understand the value of the correct designs of experiments which will generate valid data (Gupta and Thelma 2019). Otherwise, as they say in the legal parlance, the data and analysis will be the 'fruits of a poisoned tree'.

Our struggles in management of Covid-19 was not only due to lack of understanding about the behavior of the virus. These also resulted from poor application of what we knew. This chapter is an attempt to revisit what we know and should have applied that knowledge instead of ceding space often to poor epidemiology, poor science and more often to social media. Perhaps, even a quick look at it will enhance better communication among experts from different disciplines.

References

Alberts, B., D. Bray, J. Lewis, M. Raff, K. Roberts and J.D. Watson. 1994. Molecular Biology of the Cell. Garland Publishing Inc., New York, pp. 274–290.

Atkins, P.W. 1978. Physical Chemistry. W.H. Freeman, Oxford, pp. 171–198.

Baker, S., C. Griffiths and Nicklin, J. 2011. Microbiology. Garland Science, New York, pp. 248–324.

Benjamin, E., G. Sunshine and S. Leskowitz. 1996. Immunology, A Short Course, Wiley-Liss. New York.

Blundell, T.L., M.N. Gupta and S.E. Hasnain. 2020. Intrinsic disorder in proteins: Relevance to protein assemblies, drug design and host-pathogen interactions. Prog. Biophy. Mol. Biol. 156: 34–42.

Brodin, P. 2022. SARS-CoV-2 infections in children: Understanding diverse outcomes. Immunity 55: 201–209.

Chaudhuri, S., S. Basu, P. Kabi, V.R. Unni and A. Saha. 2020. Modeling the role of respiratory droplets in COVID-19 type high pandemics. Physics of Fluids 32: 063309.

Chen, J. and G.W. Wei. 2022. Omicron BA.2 [B.1.1.529.2]: High potential for becoming the next dominant variant. J. Phys. Chem. Lett. 13(17): 3840–3849.

Danta, C.C. (2021). SARS-CoV-2, hypoxia, and calcium signaling: The consequences and therapeutic options. ACS Pharmacol. Transl. Sci. 4(1): 400–402.

Frankel, F.C. and G.M. Whitesides. 2009. No Small Matter, Science on the Nanoscale, The Belknap Press of Harvard University Press, Cambridge.

Goh, G.K., A.K. Dunker and V.N. Uversky. 2012. Understanding viral transmission behavior via protein intrinsic disorder prediction: Coronaviruses. J. Pathogens 2012: 738590.

Guglielmi, G. 2020. Fast corona virus tests: What they can & can't do. Nature 585(7826): 496–498.

Guo, Q., H. Qin, X. Liu, X. Zhang, Z. Chen, T. Qin, L. Chang, W. Zhang, J. Joung, A. Ladha, M. Saito, N.-G. Kim, A.E. Woolley, M. Segel, R.P.J. Barretto, A. Ranu, R.K. Macrae, G. Faure, E.I. Ioannidi, R.N. Krajeski, R. Bruneau, M.-L.W. Huang, X.G. Yu, J.Z. Li, B.D. Walker, D.T. Hung, A.L. Greninger, K.R. Jerome, J.S. Gootenberg, O.O. Abudayyeh and F. Zhang. 2022. The emerging role of human gut microbiota in gastrointestinal cancer. Front. Immunol. 13: 915047.

Gupta, M.N. 2012. Some deleterious consequences of birth of new disciplines in science: The case of biology. Curr. Sci. 103: 126–127.

Gupta, M.N. and B.K. Thelma. 2019. Ethics in measurement practices. pp. 45–63. In: K. Muralidhar, A. Ghosh and A.K. Singhvi (eds.). Ethics in Science Education, Research and Governance. Ind. Natn. Sci. Acad., New Delhi.

Gupta, M.N. and I. Roy. 2020. How corona formation impacts nanomaterials as drug carriers. Mol. Pharmaceutics 17: 725–737.

Gupta, M.N. and I. Roy. 2021. Drugs, host proteins and viral proteins: How their promiscuities shape antiviral design. Biol. Rev. 96: 205–222.

Gupta, M.N. and J. Mukherjee. 2013. Enzymology: Some paradigm shifts over the years. Curr. Sci. 104: 1178–1186.

Gupta, M.N. and J. Mukherjee. 2015. Skills in biocatalysis are essential for establishing a sound biotechnological base in India. Proc. Ind. Natn. Sci. Acad. 81: 1113–1132.

Gupta, M.N. and V.N. Uversky (eds.). 2023. Structure and Intrinsic Disorder in Enzymology. Academic Press, Cambridge (USA).

Gupta, M.N., A. Alam and S.E. Hasnain. 2020. Protein promiscuity in drug discovery, drug-repurposing and antibiotic resistance. Biochimie 175: 50–57.

Gupta, M.N., M. Kapoor, A.B. Majumder and V. Singh. 2011. Isozymes, moonlighting proteins and promiscuous enzymes. Curr. Sci. 100: 1152–1162.

Gupta, M.N., M. Perwez and M. Sardar. 2020. Protein crosslinking: uses in chemistry, biology and biotechnology. Biocat. Biotransform. 38: 178–201.

Gupta, M.N., S. Pandey, N.Z. Ehtesham and S.E. Hasnain. 2019. Medical implications of protein moonlighting. Ind. J. Med. Res. 149: 322–325.

Gupta, M.N., S.K. Khare and R. Sinha (eds.). 2021. Interfaces Between Nanomaterials and Microbes. CRC Press. Boca Raton.

Hood, L., I.L. Weissman, W.B. Wood and J.H. Wilson. 1984. The Benjamin/Cummings Publishing, Menlo Park.

Janeway, C.A., P. Travers, M. Walport and M. Shlomchik. 2001. Immunobiology 5, The Immune System in Health and Disease. Garland Publishing. New York.

Joung, J., A. Ladha, M. Saito, N.-G. Kim, A.E. Woolley, M. Segel, R.P.J. Barretto, A. Ranu, R.K. Macrae, G. Faure, E.I. Ioannidi, R.N. Krajeski, R. Bruneau, M.-L.W. Huang, X.G. Yu, J.Z. Li, B.D. Walker, D.T. Hung, A.L. Greninger, K.R. Jerome, J.S. Gootenberg, O.O. Abudayyeh and F. Zhang. 2020. Detection of SARS-CoV-2 with SHERLOCK one-pot testing. New Eng. J. Med. 383: 1492–1494.

Kozlov, M.M., F. Campelo, N. Liska, L.V. Chernomordic, S.J. Marrink and H.T. McMahon. 2014. Mechanisms shaping cell membranes. Curr. Opin. Cell Biol. 29: 53–60.

Kupferschmidt, K. 2020. Why do some COVID-19 patients infect many others, whereas most don't spread the virus at all? Doi: 10.1126/ Science abc 8931.

Li, C.C., X.J. Wang and H.R. Wang. 2019. Repurposing host-based therapeutics to control coronavirus and influenza virus. Drug Discovery Today 24: 726–736.

Li, G. and E. De Clercq. 2020. Therapeutic options for the 2019 novel coronavirus (2019-nCoV). Nature Reviews Discovery 19: 149–150.

Liu, Y., L. Zhan, Z. Qin, J. Sackrison and J.C. Bischof. 2021. Ultrasensitive and highly specific lateral flow assays for point-of-care diagnosis. ACS Nano 15(3): 3593–3611.

Lydyard, P., A. Whelan and M. Fanger. 2011. Immunology. Garland Science. New York.

Male, D. 2014. Immunology, An Illustrated Outline. Garland Science. New York.

McLennan, A., A. Bates, P. Turner and M. White. 2013. Molecular Biology. Garland Science. New York, pp. 303–308.

Miller, J.M., M.J. Binnicker, S. Campbell, K.C. Carroll, K.C. Chapin, P.H. Gilligan, M.D. Gonzalez, R.C. Jerris, S.C. Kehl, R. Patel, B.S. Pritt, S.S. Richter, B. Robinson-Dunn, J.D. Schwartzman, J.W. Snyder, S. Telford, E.S. Telford 3rd, Theel, E.S., R.B. Jr, Thomson, M.P. Weinstein and J.D. Yao. 2018. A guide to utilization of the microbiology laboratory for infectious diseases: 2018 update by the infectious diseases society of America and the American society for microbiology. Clinical Infectious Diseases 67(6): e1–e94.

Moore, W.J. 1986. Basic Physical Chemistry, Prentice-Hall of India, New Delhi, pp. 404–421.

Morgenstern, J. 2020. Aersols, droplets, and airborne spread: Everything you could possibly want to know. First 10M, April 6, 2020. http://doi.org/10.51684/FIRS 17317.

Morgenstern, J. 2020a. Covid-19 is spread by aerosols (airborne): An evidence review. First 10 EM, Nov. 30, 2020. http://doi.org/10.51684/FIRS.52248.

Pardi, N., M.J. Hogan, F.W. Porter and D. Weissman. 2018. mRNA vaccines—A new era in vaccinology. Nature Rev. Drug Discov. 17: 261–279.

Pelczar, M.J., E.C.S. Chan and N.R. Krieg. 2010. Microbiology. Tata McGraw-Hill, New Delhi.

Petsko, G. and D. Ringe. 2009. Protein Structure and Function. Oxford University Press. Oxford.

Qu, G., X. Li, L. Hu and G. Jiang. 2020. An imperative need for research on the role of environmental factors in transmission of novel coronaviruses (COVID-19). Environ. Sci. Technol. 54: 3730–3732.

Rhoads, D., D.R. Peaper, R.C. She, F.S. Nolte, C.M. Wojewoda, N.W. Anderson and B.S. Pritt. 2021. College of american pathologists (CAP) microbiology perspective: Caution must be used in interpreting the cycle threshold (CT) value. Clinical Infectious Diseases 72(10): e685–e686.

Rich, A. and N. Davidson. (eds.). 1968. Structural Chemistry and Molecular Biology. W.H. Freeman, San Francisco, pp. 153–186.

Roitt, I.M. 1988. Essential Immunology. Blackwell Scientific Publications. Oxford.

Roitt, I.M., J. Brostoff and D. Male. 1993. Immunology. Garland Publishing. New York.

Stetzenbach, L.D., M.P. Buttner and P. Cruz. 2004. Detection and enumeration of airborne biocontaminants. Curr. Opin. Biotechnol. 15(3): 170–174.

Thames, A.H., K.L. Wolnaik, S.I. Stupp and M.C. Jewett. 2020. Principles learned from the international race to develop a safe and effective COVID-19 vaccine. ACS Cent. Sci. 6(8): 1341–1347.

Tizard, I.R. 1988. Immunology, An Introduction. Saunders College Publishing. Orlando.

Tombuloglu, H., H. Sabit, E. Al-Suhaimi, R.A. Jindan and K.R. Alkharsah. 2021. Development of multiplex reactive RT-PCR assay for the detection of SARS-CoV-2. PLoS ONE 16(4): e0250942.

Vu, M.N. and V.D. Menachery. 2021. Binding and entering: Covid finds a new home. PLoS Pathog. 17(8): e1009857.

Watson, J.D., N.H. Hopkins, J.W. Roberts, J.A. Steitz and A.M. Weiner. 1987. Molecular Biology of the Gene. The Benjamin/Cummings Publishing Co., Menlo Park, USA.

Index